CAMBRIDGE MONOGRAPHS ON APPLIED AND COMPUTATIONAL MATHEMATICS

45 Minimal Cubature Rules

The *Cambridge Monographs on Applied and Computational Mathematics* series reflects the crucial role of mathematical and computational techniques in contemporary science. The series publishes expositions on all aspects of applicable and numerical mathematics, with an emphasis on new developments in this fast-moving area of research.

State-of-the-art methods and algorithms as well as modern mathematical descriptions of physical and mechanical ideas are presented in a manner suited to graduate research students and professionals alike. Sound pedagogical presentation is a prerequisite. It is intended that books in the series will serve to inform a new generation of researchers.

A complete list of books in the series can be found at www.cambridge.org/mathematics.

Recent titles include the following:

29. Partial differential equation methods for image inpainting, *Carola-Bibiane Schönlieb*
30. Volterra integral equations, *Hermann Brunner*
31. Symmetry, phase modulation and nonlinear waves, *Thomas J. Bridges*
32. Multivariate approximation, *V. Temlyakov*
33. Mathematical modelling of the human cardiovascular system, *Alfio Quarteroni, Luca Dede', Andrea Manzoni & Christian Vergara*
34. Numerical bifurcation analysis of maps, *Yuri A. Kuznetsov & Hil G.E. Meijer*
35. Operator-adapted wavelets, fast solvers, and numerical homogenization, *Houman Owhadi & Clint Scovel*
36. Spaces of measures and their applications to structured population models, *Christian Düll, Piotr Gwiazda, Anna Marciniak-Czochra & Jakub Skrzeczkowski*
37. Quasi-interpolation, *Martin Buhmann & Janin Jäger*
38. The Christoffel–Darboux kernel for data analysis, *Jean Bernard Lasserre, Edouard Pauwels & Mihai Putinar*
39. Geometry of the phase retrieval problem, *Alexander H. Barnett, Charles L. Epstein, Leslie Greengard & Jeremy Magland*
40. Discrete variational problems with interfaces, *Roberto Alicandro, Andrea Braides, Marco Cicalese & Margherita Solci*
41. Spectral and spectral element methods for fractional ordinary and partial differential equations, *Mohsen Zayernouri, Li-Lian Wang, Jie Shen & George Em Karniadakis*
42. High-accuracy finite difference methods, *Bengt Fornberg*
43. An introduction to immersed boundary methods, *Roberto Verzicco, Marco D. de Tullio & Francesco Viola*
44. Differential equations and variational methods on graphs, *Yves van Gennip & Jeremy Budd*

Minimal Cubature Rules

Theory and Practice

YUAN XU
University of Oregon

CAMBRIDGE
UNIVERSITY PRESS

CAMBRIDGE
UNIVERSITY PRESS

Shaftesbury Road, Cambridge CB2 8EA, United Kingdom

One Liberty Plaza, 20th Floor, New York, NY 10006, USA

477 Williamstown Road, Port Melbourne, VIC 3207, Australia

314–321, 3rd Floor, Plot 3, Splendor Forum, Jasola District Centre,
New Delhi – 110025, India

103 Penang Road, #05–06/07, Visioncrest Commercial, Singapore 238467

Cambridge University Press is part of Cambridge University Press & Assessment,
a department of the University of Cambridge.

We share the University's mission to contribute to society through the pursuit of
education, learning and research at the highest international levels of excellence.

www.cambridge.org
Information on this title: www.cambridge.org/9781009663922

DOI: 10.1017/9781009663939

When citing this work, please include a reference to the DOI 10.1017/9781009663939

First published 2026

A catalogue record for this publication is available from the British Library

A Cataloging-in-Publication data record for this book is available from the Library of Congress

ISBN 978-1-009-66392-2 Hardback

To Litian
with deep appreciation

Contents

Preface

A numerical integration rule is a finite linear combination of point evaluations approximating an integral. It is called a *quadrature* rule if the integral is of one variable and a *cubature* rule if the integral is of several variables. For a nonnegative weight function w defined on a domain $\Omega \subset \mathbb{R}^d$, a cubature rule of degree m is a finite sum so that

$$\int_{\Omega} f(\mathbf{x}) w(\mathbf{x}) d\mathbf{x} = \sum_{k=1}^{N} \lambda_k f(\mathbf{x}_k)$$

for all polynomials f of total degree at most m if $d > 1$, and it is a quadrature rule if $d = 1$, where the points $\mathbf{x}_k$, called nodes, are points in $\mathbb{R}^d$ and λ_k, called weights, are scalars. On the computational side, these rules are indispensable tools in scientific computing and applied sciences whenever evaluating, or discretizing, integrals is called for. On the theoretical side, their study intersects many topics, including approximation theory, computational algebraic geometry, discrete Fourier analysis, numerical analysis, orthogonal polynomials, and sampling theory, to name a few. The quadrature rules have been studied extensively and are fairly well understood. Most studies of cubature rules have been on the computational side, and many theoretically fundamental problems remain open. This monograph aims at cubature rules of a fixed degree m that use a minimal or near-minimal number N of nodes. In other words, we are interested in those cubature rules that use fewer nodes among all rules of a fixed degree.

1. Cubature rules and their constructions. Unlike quadratures of one variable, cubature rules are generally much more involved because of the complexity of domains in geometry and the intricacy between polynomial ideals and varieties in algebra. For integrals with a constant weight function on a product domain, such as a cube, one has the product cubature rules derived as the product of quadrature rules of one variable. Product-type cubature rules can also be defined by appropriate parametrization of integrals on several other domains, such as the simplex and the unit ball. However, such constructions are not ideal. First of all, a cubature rule of degree $2n-1$ of this type will need at least n^d number of nodes, much more than necessary. Second, the distribution of nodes often does not reflect the geometry of the domain; for example, for the unit ball, the nodes will be heavily concentrated at the origin if the integral is parametrized by polar coordinates and not well distributed. Moreover, integrals over most domains cannot be parametrized as a product of integrals of one variable to permit product-type cubature rules.

For applications such as in the finite element method (FEM), cubature rules of lower-to-moderate degrees, on a triangle or tetrahedra, are often constructed numerically by solving a system of moment equations that has both nodes and weights of the cubature rules as variables, with judicious choices on the formation or symmetry of the nodes to reduce the complexity. The most straightforward setup comes from choosing the monomial basis $\{\mathbf{x}^\alpha : |\alpha| \le m,\, \alpha \in \mathbb{N}_0^d\}$, where $|\alpha| = \alpha_1 + \cdots + \alpha_d$, of the space of polynomials of degree at most m in d variables and considering the system

$$\sum_{k=1}^{N} \lambda_k \mathbf{x}_k^\alpha = m_\alpha = \int_\Omega \mathbf{x}^\alpha w(\mathbf{x}) \mathrm{d}\mathbf{x}, \quad |\alpha| \le m, \quad \alpha \in \mathbb{N}_0^d,$$

where m_α are the moments of the integral. With $\lambda_k \in \mathbb{R}$ and $\mathbf{x}_k \in \mathbb{R}^d$ as variables, the solution of this system of equations gives a cubature rule of degree m. In theory, every cubature rule corresponds to a solution of such a nonlinear system of equations. However, solving nonlinear systems of equations is a challenging problem. Little can be said about the existence and uniqueness of the solution, and less if the solution is required to be real. Most search for such cubature rules is done by brute force numerical computation. So, while useful for practical applications, this approach has limitations such as solvability, stability, and accuracy in solving nonlinear systems, and it is applicable only for rules of lower-to-moderate degrees.

From a theoretical point of view, the numerical approach has another potential deficiency. The number of equations in the system of moment equations is equal to $\binom{m+d}{d}$, which is the cardinality of $\{\alpha \in \mathbb{N}_0^d : |\alpha| = m\}$ or the dimension of the space Π_m^d of polynomials of degree at most m, and the number of variables is $(d+1)N$, with dN from N nodes and N from the weights. To solve the nonlinear system of equations numerically, it is often necessary to assume that the number of variables is no less than the number of equations so that $N \ge \frac{1}{d+1}\binom{m+d}{d}$. However, as discussed in the next section, this lower bound for the number of nodes could be severely overestimated since there are cubature rules that use far fewer nodes and retain the same degree of precision for some integrals.

In the ideal world, we would like to construct cubature rules with as few nodes as possible, which are called *minimal cubature rules*, and understand the structures of such rules. This is the fundamental problem in the theory of cubature rules and the main topic of this monograph. We describe the problem in the next section.

2. Minimal cubature rules. For integrals on the real line, a minimal quadrature rule is the Gauss quadrature rule, which uses n nodes if the rule is of degree $m = 2n - 1$ or $2n - 2$. Such a rule always exists under mild assumptions on the integral, and its nodes are the zeros of the orthogonal polynomial of degree n associated with w if $m = 2n - 1$. For integrals in more than one variable, however, it is a challenging and highly nontrivial problem to find a minimal cubature rule. Let us consider, for example, the cubature rules of total degree $2n - 1$. A moment of reflection shows that the number of nodes, N, of such a rule needs to be at least $\binom{n+d-1}{d}$, which is the dimension of the space Π_{n-1}^d of polynomials of degree at most $n - 1$, and any polynomial of degree n that vanishes on all nodes of the cubature rule must be an orthogonal polynomial, just as in one variable. We call a cubature rule of degree $2n - 1$ Gauss if it has $N = \binom{n+d-1}{d}$ number of nodes. A Gauss cubature rule of degree $2n - 1$ exists if, and only if, its nodes are zeros of all orthogonal polynomials of degree n, which consist of a polynomial subspace of dimension $\binom{n+d-1}{d-1}$. For $d = 2$, for example, this means that a basis of $n + 1$ orthogonal polynomials of degree n must all be zero on $\frac{1}{2}(n + 1)n$ many real distinct points on $\mathbb{R}^2$. As a result, it is not surprising that Gauss cubature rules exist rarely, and what is surprising is that they do exist in some particular cases. Notice that such a rule will not fit the presumption of more variables than equations for solving the system of moment equations since it corresponds to a system of far more equations than variables.

When the Gauss cubature rule does not exist for an integral, any cubature rule of degree $2n - 1$ or $2n - 2$ for the integral will have to require more than $\dim \Pi_{n-1}^d$ nodes. We are interested in having as few nodes as possible and call a cubature rule *minimal* if it has a minimal number of nodes among all cubature rules of the same degree. The existence of a minimal cubature rule is a tautology of its definition. Just about all other questions on such rules are open. First, in most cases, the number of nodes of minimal rules is unknown. Second, finding minimal cubature rules depends heavily on the geometry of the domain Ω and the weight function w. To determine minimal cubature rules, we need to carry out two essential quests:

1. Find a lower bound on the number of nodes for cubature rules.
2. Construct a cubature rule that attains the lower bound.

While an answer for the first quest will provide a benchmark for identifying minimal cubature rules, its proof and characterization of when the lower bound is attainable could also give clues to the second quest. The Gauss cubature rules, when they exist, are minimal since they attain the classical lower bound $N \ge \dim \Pi_{n-1}^d$. Moreover, the existence of such cubature rules is characterized by the common zeros of orthogonal polynomials, and the latter is used to construct the rules when they exist. Some stronger lower bounds than the classical ones are known, most notably Möller's lower bounds and the lower bound derived from the linear programming technique. A stronger lower bound also implies the nonexistence of certain cubature rules; for example, Möller's lower bound shows that the Gauss cubature rules do not exist when both Ω and w are centrally symmetric. Like most problems that deal with discrete data in higher dimensions, constructing cubature rules of a high degree of precision is much more challenging than in one dimension, more so still for minimal or near-minimal cubature rules.

Another angle of looking at minimal cubature rules uses the language of ideal and variety in algebraic geometry. In essence, a cubature rule of degree m exists if and only if there is a polynomial ideal I, generated by m-orthogonal polynomials, and its variety V consists of a finite set of real points, and the codimension of the ideal, $\dim \Pi_m / I$, is equal to $|V|$, the size of the variety. This abstract point of view pinpoints the root of the difficulty in understanding the minimal cubature rules and indicates a theoretical roadmap. In a nutshell, constructing a cubature rule of degree m amounts to finding a set of m-orthogonal polynomials (a basis for an ideal) with many real common zeros. In particular, it is necessary to study the structure of m-orthogonal polynomials of several variables.

3. Aim and scope. The monograph aims to provide an account of what is known about minimal or near-minimal cubature rules. These cubature rules are fascinating objects when they exist, and there have been continuing searches for such rules, both theoretically and numerically. The results on the subject are scattered among a wide range in the literature. The standard reference on cubature rules has been, up to now, Stroud's book [146] of 1971. It contains little on minimal cubature rules, as most of the results on such cubature rules are developed more recently. We will concentrate on the structures of minimal cubature rules, give an account from a theoretical perspective, and provide explicit examples whenever they fit our narrative. Most of the content in this monography has not appeared in any books.

Per our consideration, we will not discuss nor give a list of known cubature rules of lower-to-moderate degrees on regular domains. There are hundreds of papers on constructing cubature rules of low-to-moderate degrees for integrals on squares, disks, triangles, and the Gauss integrals, as well as their extensions, mostly in two or three dimensions. Such rules are useful for computations and provide tools for numerical methods such as, say, *hp* finite element methods. We refer to [26] for an account of how cubature rules of lower degrees are found and for references and to [30, 27] for a list of known cubature rules, and there is also an online database as described in [28]. We mention in particular a large online repository of cubature rules in Python that includes formulas on several regular domains; its address is in [136]. For our study of cubature rules with few nodes, these known cubature rules provide empirical examples for the existence and nonexistence of cubature rules attaining a given bound for the number of nodes. However, they often provide little insight into theoretical understanding.

Several families of cubature rules with few nodes will be given in the monograph. Most useful for numerical computations are given for weighted integrals on the square. The main focus, however, is not numerical. For readers interested in the numerical evaluation of integrals in higher dimensions on regular or irregular domains, a common method is the quasi-Monte Carlo method, which we refer to [21, 41, 42] and references therein.

Another topic that we will not go into detail is cubature rules on the unit sphere, especially those having equal cubature weights, which are called spherical designs because of their applications in combinatorics. The most relevant to our purpose is the lower bound for the number of nodes of the cubature rules, which will be treated fully in Chapter 4. There are hundreds of papers on these cubature rules, but most are only tangibly related to our aims. We refer to [25] for the background and to [5] for a fairly recent survey.

The subject of this monograph intersects several areas in mathematics, from numerical analysis, orthogonal polynomials, and discrete Fourier analysis to discrete geometry, symmetry, and algebraic geometry. Our selection of topics is intended to provide a narrative that focuses on the structure of cubature rules with few nodes. Many open questions are waiting to be explored. We hope the monograph will help to bring attention to this research area.

4. Organization. The monograph is organized as follows: Chapter 1 reviews essential results for quadrature rules and orthogonal polynomials in one variable. General properties and results of cubature rules are given in Chapter 2. After a summary of the necessary results on orthogonal polynomials of several variables in Chapter 3, the Gauss cubature rules are studied in Chapter 4, which contains a characterization of common zeros of orthogonal polynomials, examples, and ends with the nonexistence of these rules. Various lower bounds for the number of nodes of the cubature rules are discussed in Chapter 5, the most important one being Möller's lower bound. Chapter 6 contains a characterization of a subset of cubature rules for which the lower bound is attainable, which includes the case of minimal cubature rules that attain Möller's lower bound and the first set of examples on the square. Further examples of minimal and near-minimal rules on the square are presented in Chapter 7. A close relation between the discrete Fourier analysis and cubature rules is discussed in Chapter 8, which leads to the second example of the Gauss cubature rule. Chapter 9 discusses a characterization of cubature rules in the language of ideal and variety. As an epilogue, we give several open questions and discuss a family of cubature rules on the cube in Chapter 10. Finally, as suggested by two reviewers of the book proposal, an addendum chapter on cubature rules on the triangle and simplex, singled out as examples due to their applications in the FEM, or hp-FEM method, is included as Chapter 11.

Each chapter starts with a brief introduction of its contents and ends with a section entitled Notes and Further Results, which contains references and further discussions. This last section allows us to keep the references to a bare minimum in the text.

1

Quadrature Rules and Orthogonal Polynomials

We start with quadrature rules on the real line. The theory of quadrature rules has been extensively developed and is now well understood. We select those results and poofs that could be extended to several variables as a prelude to the study of cubature rules. Section 1.1 is introductory; it ends with a definition of Gauss quadrature rules. Section 1.2 gives a concise account of orthogonal polynomials on the real line, which is needed for studying the Gauss and minimal quadrature rules in Section 1.3. Section 1.4 consists of one characterization of positive quadrature rules of a higher degree of precision via a special class of m-orthogonal polynomials.

1.1 Quadrature Rules

Let w be a nonnegative weight function such that $\int_{\mathbb{R}} |t|^n w(t)\mathrm{d}t$ is finite for all $n \in \mathbb{N}_0$. We call such a weight function regular. A quadrature, or a quadrature rule, $\mathrm{Q}(f)$ associated with w is a finite sum

$$\mathrm{Q}(f) = \sum_{k=1}^{N} \lambda_k f(t_k), \qquad t_k \in \mathbb{R}, \quad \lambda_k \in \mathbb{R},$$

that approximates the integral $\int_{\mathbb{R}} f(t)w(t)\mathrm{d}t$. The points t_k, called nodes, are assumed to be pairwise distinct, $t_j \neq t_k$ for $j \neq k$, and λ_k, called weights, are assumed to be nonzero, so that N is the number of nodes of the quadrature rule. If all weights are positive, the quadrature rule is called positive.

Let Π_m denote the space of polynomials of degree at most m in one variable.

Definition 1.1.1. The quadrature $\mathrm{Q}(f)$ is said to have the *degree of precision* m, or degree m in short, if

$$\int_{\mathbb{R}} f(t)w(t)\mathrm{d}t = \mathrm{Q}(f) \quad \text{for all } f \in \Pi_m, \tag{1.1}$$

and the identity fails to hold for at least one polynomial of degree $m+1$.

Let $t_1,\ldots,t_n$ be a set of pairwise distinct points in $\mathbb{R}$, the Lagrange interpolation problem asks if there is a unique polynomial $L_n f$ of degree $n-1$ that satisfies

$$L_n f(t_j) = f(t_j), \qquad 1 \le j \le n,$$

for a generic function on $\mathbb{R}$. The uniqueness is a consequence of the fundamental theorem of algebra, and an explicit solution to the problem is given by the Lagrange form of interpolation polynomial

$$L_n f(t) = \sum_{k=1}^{n} f(t_k)\ell_k(t), \quad \text{where} \quad \ell_k(t) = \prod_{j=1, j\neq k}^{n} \frac{t-t_j}{t_k - t_j}. \tag{1.2}$$

The polynomials ℓ_k, called fundamental interpolating polynomials, satisfy the relation $\ell_k(t_j) = \delta_{k,j}$, where $\delta_{k,j}$ is the Kronecker delta. Since $L_n f = f$ for $f \in \Pi_{n-1}$ by the uniqueness, integrating $L_n f$ leads to a quadrature rule of degree at least $n-1$,

$$\int_{\mathbb{R}} f(t)w(t)\mathrm{d}t = \sum_{k=1}^{n} \lambda_k f(t_k), \quad \text{where} \quad \lambda_k = \int_{\mathbb{R}} \ell_k(t)w(t)\mathrm{d}t, \tag{1.3}$$

which is often called the *interpolatory* quadrature. Such a quadrature, however, may not be positive for a generic set of nodes. Positivity is important for numerical stability, and it implies, for example, that all quadrature weights are bounded by $\int_{\mathbb{R}} w(t)\mathrm{d}t$. In particular, if w is normalized to have a unit integral, then all weights of a positive quadrate satisfy $0 < \lambda_k < 1$. We are interested in the interpolatory quadrature rules that are positive and have a degree of precision greater than $n-1$.

A natural question asks to determine the highest degree of precision of a quadrature rule with a fixed number of nodes. Equivalently, the question asks to determine the minimal number of nodes among all quadrature rules of a fixed degree.

Proposition 1.1.2. *The number of nodes N of a quadrature rule of degree m satisfies $N \ge n$, where $m = 2n-1$ or $m = 2n-2$.*

Proof Let Q_m be a quadrature rule of degree m for the integral $\int_{\mathbb{R}} f(t)w(t)\mathrm{d}t$. Let $q(t) = (t-t_1)\ldots(t-t_N)$. Then, $\mathrm{Q}_m(q^2) = 0$ since q vanishes on the nodes of $\mathrm{Q}_m(f)$. Since $\int_{\mathbb{R}} [q(t)]^2 w(t)\mathrm{d}t > 0$, it follows that the precision of $\mathrm{Q}_m(f)$ must be less than the degree of q^2 or $m < 2N$. Thus, $N \ge \lfloor \frac{m}{2} \rfloor + 1 = n$. □

Among quadrature rules with n nodes, the highest possible degree of precision is $m = 2n-1$. We call a quadrature rule of degree $2n-1$ with n nodes *Gauss quadrature*.

If a Gauss quadrature Q_{2n-1} of degree $2n-1$ exists, then the polynomial $q_n(t) = (t-t_1)\cdots(t-t_n)$ of degree n that vanishes on all its nodes satisfies

$$\int_{\mathbb{R}} q_n(t)p(t)w(t)\mathrm{d}t = \mathrm{Q}_{2n-1}(q_n p) = 0 \qquad \forall p \in \Pi_{n-1}.$$

This means that q_n is orthogonal to all polynomials of lower degrees in terms of the inner product of $L^2(w)$, which we call an orthogonal polynomial of degree n with respect to w. The Gauss quadrature rule exists for all n and all regular weight functions, and they are always positive. To show this is the case, we first consider orthogonal polynomials.

1.2 Orthogonal Polynomials on Real Line

Let w be a regular weight function on the real line supported on the interval $[a,b]$, where $-\infty \le a < b \le \infty$. We consider the inner product

$$\langle f,g\rangle_w = \int_a^b f(t)g(t)w(t)\mathrm{d}t$$

of $L^2(w)$. A polynomial p_n of degree n is called orthogonal with respect to w if $\langle p_n,q\rangle_w = 0$ for all q of degree at most $n-1$.

For each nonnegative integer n, we assume that the nth moment matrix

$$M_n = (\mu_{i+j})_{0\le i,j\le n} \quad \text{with} \quad \mu_k = \int_a^b t^k w(t)\mathrm{d}t$$

is positive definite, which ensures the existence of orthogonal polynomials for the inner product $\langle f,g\rangle_w$. Indeed, the polynomial defined by

$$\frac{1}{\det M_{n-1}} \det \begin{bmatrix} \mu_0 & \mu_1 & \cdots & \mu_n \\ \mu_1, & \mu_2 & \cdots & \mu_{n+1} \\ \vdots & \vdots & \ddots & \vdots \\ \mu_{n-1}, & \mu_n & \cdots & \mu_{2n-1} \\ 1 & t & \cdots & t^n \end{bmatrix} = t^n + \ldots$$

is an orthogonal polynomial of degree n with the leading coefficient 1. In general, we denote by $p_n(w)$ an orthogonal polynomial of degree n given by

$$p_n(w;t) = \gamma_n t^n + \gamma_{n,n-1}t^{n-1} + \cdots + \gamma_{n,0}, \quad \gamma_n = \gamma_n(w) > 0, \tag{1.4}$$

which has the leading coefficient γ_n, and it satisfies the orthogonality

$$\int_a^b p_n(w;t)p_m(w;t)w(t)\mathrm{d}t = h_n(w)\delta_{n,m}, \qquad n \neq m,$$

where $h_n(w)$ is the square of the $L^2(w)$ norm of $p_n(w)$. The polynomial is called *orthonormal* if $h_n(w) = 1$. If $p_n(w)$ is an orthogonal polynomial, then $\hat{p}_n(w) = p_n(w)/\sqrt{h_n(w)}$ is orthonormal.

Orthogonal polynomials enjoy many properties, and they are well documented. We state some of their properties that will be needed in our study without giving full proof.

1.2.1 Properties of Orthogonal Polynomials

The first property is the three-term recursive relation, which follows from writing $tp_n(w;t)$ as a linear expansion in $p_k(w)$ for $0 \le k \le n+1$ and determines coefficients of the expansion by applying orthogonality.

I. Recursive relation. There are real numbers b_n and c_n such that

$$p_{n+1}(w;t) = (t - b_n)p_n(w;t) - c_n p_{n-1}(w;t), \quad n \ge 0, \tag{1.5}$$

where we assume $p_{-1}(w;t) = 0$, and b_n and c_n are determined by

$$b_n = \frac{1}{h_n(w)} \langle tp_n(w), p_n(w) \rangle_w \quad \text{and} \quad c_n = \frac{1}{h_{n-1}(w)} \langle tp_n(w), p_{n-1}(w) \rangle_w.$$

The recursive relation characterizes the orthogonality by Favard's theorem, which states that if a sequence of polynomials $\{p_n\}$ is defined by (1.5) for some $b_n > 0$ and c_n, then it must be orthogonal with respect to a positive linear functional $\mathcal{L}$ in the sense that $\mathcal{L}(p_n p_m) = 0$, $n \ne m$.

When the polynomials are orthonormal, the recursive relation is often written in a more symmetric form as given below.

II. Three-term relation. If $\{p_n(w) : n \ge 0\}$ is a sequence of orthonormal polynomials, then there are sequences of $\{a_n\}$ and $\{b_n\}$ such that

$$tp_n(w;t) = a_n p_{n+1}(w;t) + b_n p_n(w;t) + a_{n-1} p_{n-1}(w;t), \quad n \ge 0. \tag{1.6}$$

Moreover, $a_n = \langle tp_n(w), p_{n+1}(w) \rangle_w = \gamma_n / \gamma_{n+1} > 0$ and $b_n(w) = \langle tp_n(w), p_n(w) \rangle_w$.

The three-term relation gives a representation of the product operator $t : f \mapsto tf$ in terms of the Jacobi matrix defined via the coefficients of (1.6), and we are also interested in the final Jacobi matrix.

III. Jacobi matrix. Let a_n and b_n be given in (1.6). Define

$$J := \begin{bmatrix} b_0 & a_0 & & & \bigcirc \\ a_0 & b_1 & a_1 & & \\ & a_1 & b_2 & a_1 & \\ \bigcirc & & \ddots & \ddots & \ddots \end{bmatrix}$$

and, for $n \ge 1$,

$$J_n := \begin{bmatrix} b_0 & a_0 & & & \bigcirc \\ a_0 & b_1 & a_1 & & \\ & \ddots & \ddots & \ddots & \ddots \\ & & a_{n-3} & b_{n-2} & a_{n-2} \\ \bigcirc & & & a_{n-2} & b_{n-1} \end{bmatrix}. \tag{1.7}$$

Then, J is the matrix representation of the product operator $Y \mapsto tY$ for $Y \in \ell^2$, and $tY = JY$, where $Y = [p_0(w), p_1(w), \ldots]^{\mathsf{T}}$ is a column vector. Moreover, the polynomial $p_n(w)$ is a multiple of the characteristic polynomial of the finite Jacobi matrix J_n,

$$p_n(w;t) = \gamma_n(w)\det(tI_n - J_n), \tag{1.8}$$

where I_n denotes the $n \times n$ identity matrix. It is easy to verify the identity

$$\det(tI_{n+1} - J_{n+1}) = (t - b_n)\det(tI_n - J_n) - a_{n-1}^2 \det(tI_{n-1} - J_{n-1}).$$

Comparing it with the three-term relation (1.6) proves (1.8).

A consequence of the three-term relation is a closed-form formula for the kernel

$$K_n(w;s,t) := \sum_{k=0}^{n} p_k(w;s) p_k(w;t),$$

where $p_k(w)$ are orthonormal polynomials, which is the reproducing kernel for Π_n in $L^2(w)$ as

$$\int_a^b K_n(w;x,y) p(y) w(y) \mathrm{d}y = p(x), \quad \forall p \in \Pi_n.$$

IV. Christoffel–Darboux formula:

$$K_n(w;s,t) = a_n \frac{p_{n+1}(w;s)p_n(w;t) - p_{n+1}(w;t)p_n(w;s)}{s-t}, \quad y \neq x, \tag{1.9}$$

and, taking the limit,

$$K_n(w;t,t) = a_n \left[p_{n+1}'(w;t)p_n(w;t) - p_{n+1}(w;t)p_n'(w;t)\right]. \tag{1.10}$$

The identity (1.9) follows from applying the three-term relation (1.6) on the right-hand side and using induction.

By (1.8), zeros of $p_n(w)$ are eigenvalues of the Jacobi matrix J_n. Since J_n is a symmetric matrix, all zeros of $p_n(w)$ are real. By (1.10),

$$a_n p_n'(w;t_k) p_{n-1}(w;t_k) = K_{n-1}(w;t_k,t_k) > 0$$

if t_k is a zero of $p_n(w)$, which shows in particular that all zeros of $p_n(w)$ are simple. The identity further implies that the zeros of $p_n(w)$ and $p_{n-1}(w)$ are interlacing and all zeros of $p_n(w)$ are inside (a,b). We sum up the properties as follows.

V. Zeros of orthogonal polynomials. The polynomial $p_n(w)$ has n simple real zeros in (a,b), and the zeros of $p_n(w)$ and $p_{n-1}(w)$ interlace.

Orthogonal polynomials $p_n(w)$ inherit the parity of the weight function w.

VI. Orthogonal polynomials for symmetric weight. If $b = -a$ and w is even on $[-a,a]$, then $p_n(w)$ is an even polynomial if n is even and an odd polynomial if n is odd. In particular, $b_n = 0$ in the three-term relation (1.6).

The Fourier orthogonal series is defined and converges in $L^2(w)$ by the standard theory of Hilbert space. The reproducing kernel $K_n(\cdot,\cdot)$ is also the kernel for the nth partial sum of the Fourier orthogonal series.

VII. Fourier orthogonal series. *For* $f \in L^2(w,[a,b])$,

$$f = \sum_{n=0}^{\infty} \hat{f}_n p_n(w), \quad \text{where} \quad \hat{f}_n = \frac{1}{h_n(w)} \int_a^b f(s) p_n(w;s) w(s) \mathrm{d}s.$$

Moreover, the nth partial sum of the series satisfies

$$s_n f(t) := \sum_{k=0}^{n} \hat{f}_k p_k(w;t) = \int_a^b f(s) K_n(w;t,s) w(s) \mathrm{d}s.$$

1.2.2 Classical Orthogonal Polynomials

We recall classical orthogonal polynomials: Jacobi polynomials and several other related families, Laguerre polynomials, and Hermite polynomials. They will be used extensively.

Jacobi polynomials $P_n^{(\alpha,\beta)}$. For $\alpha,\beta > -1$, the Jacobi weight is defined by

$$w_{\alpha,\beta}(t) := (1-t)^{\alpha}(1+t)^{\beta}, \qquad -1 < t < 1. \tag{1.11}$$

Its normalization constant $c'_{\alpha,\beta}$, defined by $c'_{\alpha,\beta}\int_{-1}^{1} w_{\alpha,\beta}(t)\mathrm{d}t = 1$, is given by

$$c'_{\alpha,\beta} = \frac{1}{2^{\alpha+\beta+1}} c_{\alpha,\beta} \quad \text{with} \quad c_{\alpha,\beta} := \frac{\Gamma(\alpha+\beta+2)}{\Gamma(\alpha+1)\Gamma(\beta+1)}. \tag{1.12}$$

In terms of the hypergeometric function ${}_2F_1$, the Jacobi polynomial is given by

$$P_n^{(\alpha,\beta)}(t) = \frac{(\alpha+1)_n}{n!} {}_2F_1\left(\begin{matrix} -n, n+\alpha+\beta+1 \\ \alpha+1 \end{matrix} ; \frac{1-t}{2} \right).$$

These polynomials satisfy the orthogonal relations

$$c'_{\alpha,\beta} \int_{-1}^{1} P_n^{(\alpha,\beta)}(t) P_m^{(\alpha,\beta)}(t) w_{\alpha,\beta}(t) \mathrm{d}t = h_n^{(\alpha,\beta)} \delta_{m,n}, \tag{1.13}$$

where $h_n^{(\alpha,\beta)}$ is the square of the L^2 norm given by

$$h_n^{(\alpha,\beta)} = \frac{(\alpha+1)_n(\beta+1)_n(\alpha+\beta+n+1)}{n!(\alpha+\beta+2)_n(\alpha+\beta+2n+1)}. \tag{1.14}$$

The Jacobi polynomials are eigenfunctions of a second-order differential operator:

$$\left[(1-t)\partial^2 - (\alpha-\beta+(\alpha+\beta+2)t)\partial\right]u = -n(n+\alpha+\beta+1)u,$$

where $\partial f = f'$ and $u = P_n^{(\alpha,\beta)}$ for $n = 0,1,2,\ldots$.

Gegenbauer polynomials C_n^λ. These polynomials are orthogonal with respect to the weight function w_λ defined by

$$w_\lambda(t) = (1-t^2)^{\lambda-\frac{1}{2}}, \qquad \lambda > -\tfrac{1}{2}, \tag{1.15}$$

which is the Jacobi weight $w_{\lambda-\frac{1}{2},\lambda-\frac{1}{2}}$. The polynomials, denoted by C_n^λ, are a constant multiple of $P_n^{(\lambda-\frac{1}{2},\lambda-\frac{1}{2})}$, and they also satisfy

$$C_n^\lambda(t) = \frac{(\lambda)_n 2^n}{n!} t^n {}_2F_1\left(\begin{matrix} -\frac{n}{2}, \frac{1-n}{2} \\ 1-n-\lambda \end{matrix}; \frac{1}{t^2}\right).$$

The orthogonality of these polynomials is given explicitly by

$$c_\lambda \int_{-1}^1 C_n^\lambda(t) C_m^\lambda(t) w_\lambda(t)\mathrm{d}t = h_n^\lambda \delta_{n,m}, \tag{1.16}$$

where the normalization constant c_λ and the norm square h_n^λ are given by

$$c_\lambda = \frac{\Gamma(\lambda+1)}{\Gamma(\frac{1}{2})\Gamma(\lambda+\frac{1}{2})} \quad \text{and} \quad h_n^\lambda = \frac{\lambda}{n+\lambda} C_n^\lambda(1) = \frac{\lambda}{n+\lambda}\frac{(2\lambda)_n}{n!}. \tag{1.17}$$

Chebyshev polynomials T_n and U_n. The case $\lambda = 0$ corresponds to the Chebyshev polynomial of the first kind, $\lim_{\lambda\to 0} \frac{1}{\lambda} C_n^\lambda(t) = T_n(t)$, and the case $\lambda = 1$ is the Chebyshev polynomial of the second kind, $C_n^1(t) = U_n(t)$. These polynomials are given explicitly by

$$T_n(t) = \cos n\theta \quad \text{and} \quad U_n(t) = \frac{\sin(n+1)\theta}{\sin\theta}, \qquad t = \cos\theta,$$

and they satisfy the orthogonality

$$\frac{1}{\pi}\int_{-1}^1 T_n(t)T_m(t)\frac{\mathrm{d}t}{\sqrt{1-t^2}}\mathrm{d}t = \frac{1}{\pi}\int_0^\pi \cos n\theta \cos m\theta \,\mathrm{d}\theta = \delta_{n,m}\begin{cases} \frac{1}{2} & n>0 \\ 1 & n=0 \end{cases}$$

and

$$\frac{2}{\pi}\int_{-1}^1 U_n(t)U_m(t)\sqrt{1-t^2}\mathrm{d}t = \frac{2}{\pi}\int_0^\pi \sin(n+1)\theta \sin(n+1)\theta \,\mathrm{d}\theta = \delta_{n,m}.$$

In particular, U_n is orthonormal, and the orthonormal Chebyshev polynomials of the first kind, denoted by $\hat{T}_n$, are given by $\hat{T}_0 = T_0$ and $\hat{T}_n = \sqrt{2}\,T_n$ for $n \ge 1$.

Generalized Gegenbauer polynomials $C_n^{(\lambda,\mu)}$. These polynomials are orthogonal with respect to the weight function $\omega_{\lambda,\mu}$ defined by

$$\omega_{\lambda,\mu}(t) = |t|^{2\mu}(1-t^2)^{\lambda-\frac{1}{2}}, \qquad \lambda > -\tfrac{1}{2}, \quad \mu \ge 0. \tag{1.18}$$

Denoted by $C_n^{(\lambda,\mu)}$, they can be given in terms of the Jacobi polynomials

$$\begin{aligned} C_{2n}^{(\lambda,\mu)}(t) &= \frac{(\lambda+\mu)_n}{(\mu+\frac{1}{2})_n} P_n^{(\lambda-\frac{1}{2},\mu-\frac{1}{2})}(2t^2-1), \\ C_{2n+1}^{(\lambda,\mu)}(t) &= \frac{(\lambda+\mu)_{n+1}}{(\mu+\frac{1}{2})_{n+1}} t P_n^{(\lambda-\frac{1}{2},\mu+\frac{1}{2})}(2t^2-1). \end{aligned} \tag{1.19}$$

The orthogonality of these polynomials is given explicitly by

$$c_{\lambda-\frac{1}{2},\mu-\frac{1}{2}} \int_{-1}^{1} C_n^{(\lambda,\mu)}(t) C_m^{(\lambda,\mu)}(t)\,\omega_{\lambda,\mu}(t)\mathrm{d}t = h_n^{\lambda,\mu}\delta_{n,m},$$

where the norm square $h_n^{\lambda,\mu}$ is given by

$$\begin{aligned} h_{2n}^{(\lambda,\mu)} &= \frac{\lambda+\mu}{2n+\lambda+\mu} C_{2n}^{(\lambda,\mu)}(1) = \frac{(\lambda+\frac{1}{2})_n(\lambda+\mu)_n(\lambda+\mu)}{n!(\mu+\frac{1}{2})_n(\lambda+\mu+2n)}, \\ h_{2n+1}^{(\lambda,\mu)} &= \frac{\lambda+\mu}{2n+1+\lambda+\mu} C_{2n+1}^{(\lambda,\mu)}(1) = \frac{(\lambda+\frac{1}{2})_n(\lambda+\mu)_{n+1}(\lambda+\mu)}{n!(\mu+\frac{1}{2})_{n+1}(\lambda+\mu+2n+1)}. \end{aligned} \tag{1.20}$$

For $\mu = 0$, the polynomial $C_n^{(\lambda,0)} = C_n^\lambda$ is the ordinary Gegenbauer polynomials and, in particular, the (1.19) with $\mu = 0$ gives a relation, called quadratic transform, between the Gegenbauer and the Jacobi polynomials.

Laguerre polynomials L_n^α. For $\alpha > -1$, the Laguerre polynomials are orthogonal with respect to the weight function w_α defined by

$$w_\alpha(t) = t^\alpha \mathrm{e}^{-t}, \qquad t \in \mathbb{R}_+ := \{t : t \ge 0\}.$$

The normalization constant of w_α is $1/\Gamma(\alpha+1)$. The Laguerre polynomials are given by

$$L_n^\alpha(t) = \frac{(\alpha+1)_n}{n!} \sum_{j=0}^{n} \frac{(-n)_j}{(\alpha+1)_j j!} t^j = \frac{(\alpha+1)_n}{n!} {}_1F_1\left(\begin{matrix} -n \\ \alpha+1 \end{matrix}; t\right)$$

and their orthogonality is given explicitly by

$$\frac{1}{\Gamma(\alpha+1)} \int_0^\infty L_n^\alpha(t) L_m^\alpha(t) w_\alpha(t)\mathrm{d}t = \frac{(\alpha+1)_n}{n!}\delta_{m,n}.$$

Hermite polynomials H_n. The Hermite polynomials are orthogonal with respect to the weight function e^{-t^2} on $\mathbb{R}$, and they are given by

$$H_n(t) = \sum_{0\le j\le n/2} \frac{n!}{(n-2j)!j!}(-1)^j(2t)^{n-2j}.$$

The orthogonality of these polynomials is given explicitly by

$$\frac{1}{\sqrt{\pi}}\int_{-\infty}^{\infty} H_n(t)H_m(t)\mathrm{e}^{-t^2}\,\mathrm{d}t = 2^n n!\delta_{m,n}.$$

1.3 Gauss Quadrature Rules

We start with the existence of the Gauss quadrature rule that has n nodes and is of degree $2n-1$. Without loss of generality, we assume that the weight function is normalized by $\int_a^b w(t)\mathrm{d}t = 1$, so that $p_0(w;t) = 1$.

Theorem 1.3.1. *Let w be a regular weight function supported on $[a,b]$. Let $t_1,\dots,t_n$ be the zeros of orthogonal polynomial $p_n(w)$. Then, $\mathrm{Q}_n(f)$ defined by*

$$\int_a^b f(t)w(t)\mathrm{d}t = \sum_{k=1}^{n}\lambda_k f(t_k) =: \mathrm{Q}_n(f) \tag{1.21}$$

is the Gauss quadrature of degree $2n-1$, where λ_k is given by

$$\lambda_k = \frac{1}{K_{n-1}(w;t_k,t_k)} = \frac{\gamma_{n+1}}{\gamma_n}\frac{1}{p_n'(w;t_k)p_{n-1}(w;t_k)}.$$

In particular, it is a positive quadrature.

Proof Since $t_1,\dots,t_n$ are zeros of the $p_n(w)$, the Lagrange interpolation polynomial can be given by

$$L_n f(x) = \sum_{k=1}^{n} f(t_k)\ell_k(t), \qquad \ell_k(t) = \frac{K_{n-1}(w;t,t_k)}{K_{n-1}(w;t_k,t_k)}, \tag{1.22}$$

since ℓ_k given in (1.22) is a polynomial of degree at most $n-1$ and satisfies $\ell_k(t_j) = \delta_{k,j}$ by the Christoffel–Darboux formula. Integrating $L_n f$ gives the interpolatory quadrature $\mathrm{Q}_n(f)$, which is a quadrature rule of degree $n-1$, and the first identity of λ_k shows that it is a positive quadrature by $K_{n-1}(x,x) \ge p_0(w;t)^2 = 1$. Now, if f is a polynomial of degree at most $2n-1$, we can write $f = qp_n(w)+r$, where q and r are polynomials of degree at most $n-1$. It follows that

$$\int_a^b f(t)w(t)\mathrm{d}t = \int_a^b r(t)w(t)\mathrm{d}t = \mathrm{Q}_n(r) = \mathrm{Q}_n(qp_n(w)+r) = \mathrm{Q}_n(f),$$

where the first equality follows from the orthogonality of $p_n(w)$, the second one from applying Q_n on r, and the third one holds since $p_n(w)$ vanishes on all nodes of Q_n. □

We call a quadrature rule *minimal* if it uses the smallest number of nodes among all quadrature rules of the same degree. A Gauss quadrature rule of n nodes is a minimal quadrature rule of degree $2n-1$ and is unique for each given integral and fixed n. By Proposition 1.1.2, a minimal quadrature of an even integer $2n-2$ has at least n nodes. If $\mathrm{Q}_n(f)$ is such a quadrature rule with n nodes $t_1,\ldots,t_n$, then the polynomial $q(t) = \gamma_n(t-t_1)\cdots(t-t_n)$, where γ_n is the leading coefficient of $p_n(w)$, satisfies $\int_a^b q(t)r(t)w(t)\mathrm{d}t = \mathrm{Q}_n(pq) = 0$ for all polynomials r of degree at most $n-2$. Writing q as the expansion of $p_k(w)$, the orthogonality shows $q = p_n(w) + \rho p_{n-1}(w)$ for some scalar ρ. The polynomial

$$q_{n,1}(t) := p_n(w;t) + \rho p_{n-1}(w;t), \qquad \rho \in \mathbb{R}, \quad \rho \neq 0,$$

is called a *quasi-orthogonal* polynomial of degree n.

Theorem 1.3.2. *The quasi-orthogonal polynomial $q_{n,1}$ has n distinct real zeros, and they are nodes of a minimal quadrature rule of degree $2n-2$,*

$$\int_a^b f(t)w(t)\mathrm{d}t = \sum_{k=1}^{n} \lambda_k f(t_k), \qquad \lambda_k = \frac{1}{K_{n-1}(w;t_k,t_k)}.$$

In particular, it is a positive quadrature.

Proof Using the identity (1.8), it is easy to verify that

$$q_{n,1}(t) = \gamma_n \det(tI - J_{n,1}), \qquad J_{n,1} = J_n - \begin{bmatrix} 0 & 0 \\ 0 & a_{n-1}\rho \end{bmatrix},$$

which shows, in particular, that the zeros of $q_{n,1}$ are eigenvalues of the Jacobi matrix $J_{n,1}$. Since the matrix is symmetric, all zeros are real. Moreover, by the Christoffel–Darboux formula, we deduce readily that

$$K_{n-1}(w;s,t) = \frac{q_{n,1}(s)p_{n-1}(w;t) - q_{n,1}(t)p_{n-1}(w;s)}{s-t}, \tag{1.23}$$

which leads to $0 < K_{n-1}(w;t_k,t_k) = q_{n,1}'(t_k)p_{n-1}(w;t_k)$, so that $q_{n,1}'(t_k) \neq 0$ and, consequently, all zeros of $q_{n,1}$ are simple. Moreover, the Lagrange interpolation polynomial can be given as in (1.22), which allows us to follow the proof of Theorem 1.3.1 almost verbatim to show that the interpolatory quadrature rule on the zeros of $p_{n,1}$ is a quadrature rule of degree $2n-2$. □

The identity (1.23) also shows that the zeros of $q_{n,1}$ lie in the interior of $[a,b]$ for $\rho > 0$ ($\rho < 0$), except for the greatest or least zero that lies in $[a,b]$ only for

$\rho \geq -p_n(w;b)/p_{n-1}(w;b)$ $[\rho \leq -p_n(w;a)/p_{n-1}(w;a)]$ (cf. [148, p. 46]). If $\rho = -p_n(w;b)/p_{n-1}(w;b)$, then the quasi-orthogonal polynomial $q_{n,1}$ has n simple real zeros $t_1,\ldots,t_{n-1},b$ inside $(a,b]$, and if $\rho = -p_n(w;a)/p_{n-1}(w;a)$, then the quasi-orthogonal polynomial $q_{n,1}$ has n simple real zeros $a,t_1,\ldots,t_{n-1}$. In both cases, the zeros $t_1,\ldots,t_{n-1}$ can be stated as zeros of the orthogonal polynomial of degree $n-1$ for a modified nonnegative weight function on $[a,b]$,

$$w_a(t) := (t-a)w(t) \quad \text{or} \quad w_b(t) := (b-t)w(t), \quad a \leq t \leq b.$$

The corresponding quadrature rule of degree $2n-2$ that has one node on a boundary of the interval is called the *Gauss–Radau quadrature.*

Let w be the weight function on $[a,b]$ that has at least one of the endpoints is finite. We denote by $\lambda_{k,n-1}(w)$ the weights of the Gauss quadrature rules Q_{n-1} of degree $2n-3$ in (1.21) in the following theorem.

Theorem 1.3.3. *Let $t_1,\ldots,t_{n-1}$ be the zeros of $p_{n-1}(w_a)$. Then the Gauss–Radau quadrature*

$$\int_a^b f(t)w(t)\mathrm{d}t = \lambda_{0,n}f(a) + \sum_{k=1}^{n-1} \lambda_{k,n}f(t_k) \tag{1.24}$$

of degree $2n-2$ is positive with

$$\lambda_{0,n} = \frac{\int_a^b [p_{n-1}(w_a,t)]^2 w(t)\mathrm{d}t}{p_{n-1}^2(w_a,a)} \quad \textit{and} \quad \lambda_{k,n} = \frac{\lambda_{k,n-1}(w_a)}{t_k - a}, \quad 1 \leq k \leq n-1.$$

Moreover, another Gauss–Radau quadrature

$$\int_a^b f(t)w(t)\mathrm{d}t = \sum_{k=1}^{n-1} \lambda_{k,n}f(t_k) + \lambda_{n,n}f(b) \tag{1.25}$$

of degree $2n-2$ is positive with

$$\lambda_{k,n} = \frac{\lambda_{k,n-1}(w_a)}{b-t_k}, \quad 1 \leq k \leq n-1, \quad \textit{and} \quad \lambda_{n,n} = \frac{\int_a^b [p_{n-1}(w_a,t)]^2 w(t)\mathrm{d}t}{p_{n-1}^2(w_b,b)}.$$

Proof The existence of the quadrature (1.24) has already been established. To determine the nodes and weights, we observe that (1.24) with $f(t) = (t-a)g(t)$, where $g \in \Pi_{n-3}$, becomes

$$\int_a^b g(t)w_a(t)\mathrm{d}t = \int_a^b g(t)(t-a)w(t)\mathrm{d}t = \sum_{k=1}^{n-1} \lambda_{k,n}(t_k - a)g(t_k), \forall g \in \Pi_{2n-3}.$$

Comparing with the Gauss quadrature for w_a shows $\lambda_k(w_a) = \lambda_{k,n}(t_k - a)$, which determines $\lambda_{k,n}$ for $1 \le k \le n-1$, whereas $\lambda_{0,n}$ is determined by setting $f(t) = p_{n-1}^2(w_a,t)$ in (1.24) so that

$$\lambda_{0,n} p_{n-1}^2(w_a,a) = \int_a^b [p_{n-1}(w_a,t)]^2 w(t)\mathrm{d}t > 0.$$

The nodes and weights of (1.25) are determined similarly. □

1.4 Positive Quadrature Rule of Higher Degree

An interpolatory quadrature with n nodes has a degree of precision of at least $n-1$. We are interested in those with higher degrees. Let r and n be integers and $0 \le r \le n-1$. The goal of this subsection is to give a characterization of positive quadrature rules with n nodes and of degree $2n-1-r$.

Definition 1.4.1. Let $0 \le r \le n-1$. A quadrature rule $\mathrm{Q}(f)$ with n nodes and of degree $2n-1-r$ is called a $(2n-1-r,n)$-quadrature. If $q_{n,r}$ is the polynomial of degree n that vanishes on the nodes of $\mathrm{Q}(f)$, we say that the $(2n-1-r,n)$-quadrature is generated by $q_{n,r}$.

The Gauss quadrature rule of degree $2n-1$ corresponds to $r=0$ and is generated by the orthogonal polynomial of degree n. The minimal quadrature of degree $2n-2$ corresponds to $r=1$ and is generated by a quasi-orthogonal polynomial of degree n. If $\mathrm{Q}(f)$ is a $(2n-r-1,n)$-quadrature generated by the polynomial $q_{n,r}$ of degree n, then

$$\int_a^b q_{n,r}(t)h(t)w(t)\mathrm{d}t = \mathrm{Q}(q_{n,r}h) = 0, \qquad \deg h \le n-r-1,$$

whenever $q_{n,r}h$ is a polynomial of degree at most $2n-r-1$, which shows that $q_{n,r}$ is orthogonal to all polynomials of degree $n-r-1$. By orthogonality, the polynomial $q_{n,r}$ can be written as, up to a multiple constant,

$$q_{n,r}(t) = p_n(w;t) + \rho_{n,1}p_{n-1}(w;t) + \cdots + \rho_{n,r}p_{n-r}(w;t), \tag{1.26}$$

where $\rho_{n,1},\dots,\rho_{n,r}$ are real scalars, we assume the first coefficient in front of $p_n(w)$ on the right-hand side is 1. For convenience, we give the following definition.

Definition 1.4.2. Let m be a fixed positive integer. A polynomial p is called m-orthogonal in $L^2(w)$ if

$$\int_a^b p(t)q(t)w(t)\mathrm{d}t = 0, \qquad \deg p + \deg q \le m,$$

for any polynomial q of degree at most $m - \deg p$.

Per this definition, a $(2n-1)$-orthogonal polynomial is of degree at least n, and it is the ordinary orthogonal polynomial $p_n(w)$ when its degree is n. A quasi-orthogonal polynomial $p_n(w)+\rho p_{n-1}(w)$ of degree n is a $(2n-2)$-orthogonal polynomial. More generally, a $(2n-r-1)$-orthogonal polynomial of degree n is a constant multiple of $q_{n,r}$ in (1.26) if $\rho_{n,r}\neq 0$. In particular, a $(2n-r-1,n)$ quadrature rule is generated by a $(2n-r-1)$-orthogonal polynomial of degree n.

Our goal is to give one characterization of positive $(2n-1-r,n)$-quadrature rules in terms of the structure of their generating polynomials. First, we consider a class of $(2n-1)$-orthogonal polynomials defined in terms of a modified Jacobi matrix given in (1.7).

Throughout the rest of this section, we assume that $p_n(w)$ is the orthonormal polynomial and J_n is the Jacobi matrix in (1.7) with a_n and b_n being the coefficients in the three-term relation (1.6). Let s be a positive integer and $s\le n$, and let $\sigma_1,\ldots,\sigma_s$ and $\tau_1,\ldots,\tau_{s-1}$ be real numbers. We define a symmetric and tri-diagonal $n\times n$ matrix $J_{n,r}$ as follows: for $r=2s-1$,

$$J_{n,2s-1}=J_n-\begin{bmatrix} \bigcirc & & & & \bigcirc \\ & a_{n-s}\sigma_1 & a_{n-s}\tau_1 & & \\ & a_{n-s}\tau_1 & a_{n-s+1}\sigma_2 & a_{n-s+1}\tau_2 & \\ & \ddots & \ddots & \ddots & \\ & & a_{n-3}\tau_{s-2} & a_{n-2}\sigma_{s-1} & a_{n-2}\tau_{s-1} \\ \bigcirc & & & a_{n-2}\tau_{s-1} & a_{n-1}\sigma_s \end{bmatrix} \tag{1.27}$$

and, for $r=2s-2$, $J_{n,2s-2}$ is the same as $J_{n,2s-1}$ but with $\sigma_1=0$.

Lemma 1.4.3. *The characteristic polynomial of $J_{n,r}$ is $(2n-r-1)$-orthogonal,*

$$\gamma_n\det(tI_n-J_{n,r})=p_n(w;t)+\rho_{r,1}^{(n)}p_{n-1}(w;t)+\cdots+\rho_{r,r}^{(n)}p_{n-r}(w), \tag{1.28}$$

where $\rho_{1,r}^{(n)},\ldots,\rho_{r,r}^{(n)}$ are functions of $\tau_1,\ldots,\tau_{s-1}$ and $\sigma_1,\ldots,\sigma_s$.

Proof Since $J_{n,2s-1}$ becomes $J_{n,2s-2}$ when $\sigma_1=0$, we apply induction on $J_{n,2s-1}$. For $s=1$, $J_{n,1}$ is the matrix in the proof of Theorem 1.3.2. Assuming (1.26) has been established for $r=2j-1$ for $1\le j\le s$, we show that it also holds for $r=2s+1$. Let $q_{n,r}(t)=\gamma_n\det(tI_n-J_{n,r})$. Expanding the determinant by its last row, we obtain

$$\begin{aligned} q_{n,2s+1}(t)&=\gamma_n(t-b_{n-1}+a_{n-1}\sigma_{s+1})\det(tI_{n-1}-J_{n-1,2s-1})\\ &\qquad+\gamma_n a_{n-2}^2(1-\tau_s)^2\det(tI_{n-2}-J_{n-2,2s-3})\\ &=a_{n-1}^{-1}(t-b_n+a_{n-1}\sigma_{s+1})q_{n-1,2s-1}(t)+a_{n-1}^{-1}a_{n-2}(1-\tau_s)^2q_{n-2,2s-3}(t), \end{aligned}$$

where we have used $a_{n-1} = \frac{\gamma_{n-1}}{\gamma_n}$. Define $\rho_{0,r}^{(n)} = 1$. By the induction hypothesis,

$$q_{n-1,2s-1}(t) = \sum_{k=0}^{2s-1} \rho_{k,2s-1}^{(n-1)} p_{n-1-k}(w;t), \quad q_{n-2,2s-3}(t) = \sum_{k=0}^{2s-3} \rho_{k,2s-3}^{(n-2)} p_{n-2-k}(w;t).$$

Moreover, applying the three-term relation satisfied by $p_m(w)$, we can further write

$$\begin{aligned} tq_{n-1,2s-1}(t) &= a_{n-1}\, p_n(w;t) + \left(\rho_{1,2s-1}^{(n-1)} a_{n-2} + \rho_{0,2s-1}^{(n-1)} b_{n-1}\right) p_{n-1}(w;t) \\ &\quad + \sum_{k=2}^{2s-1} \left(\rho_{k,2s-1}^{(n-1)} a_{n-k-1} + \rho_{k-1,2s-1}^{(n-1)} b_{n-k} + \rho_{k-2,2s-1}^{(n-1)} a_{n-k}\right) p_{n-k}(w;t) \\ &\quad + \left(\rho_{2s-1,2s-1}^{(n-1)} b_{n-2s} + \rho_{2s-2,2s-1}^{(n-1)} a_{n-2s}\right) p_{n-2s}(w;t) \\ &\quad + \rho_{2s-1,2s-1}^{(n-1)} a_{n-2s+1}\, p_{n-2s-1}(w;t). \end{aligned}$$

Putting these identities together, we see that $q_{n,2s+1}$ has the desired representation specified in (1.26) with $\rho_{k,2s+1}^{n}$, $1 \le k \le 2s$, given by

$$\begin{aligned} \rho_{k,2s+1}^{(n)} = a_{n-1}^{-1} \Big[&\rho_{k,2s-1}^{(n-1)} a_{n-k-1} + \rho_{k-1,2s-1}^{(n-1)} (b_{n-k} - b_n + a_{n-1}\sigma_{s+1}) \\ &+ \rho_{k-2,2s-1}^{(n-1)} a_{n-k} + \rho_{k-2,2s-3}^{(n-2)} a_{n-2} (1-\tau_s)^2 \Big], \end{aligned}$$

where we have defined $\rho_{j,k}^{(m)} = 0$ if $j > k$ or $j < 0$, and for $k = 2s+1$,

$$\rho_{2s+1,2s+1}^{(n)} = a_{n-1}^{-1} a_{n-2s-1} \rho_{2s-1,2s-1}^{(n-1)}.$$

This proves (1.28) for r being an odd integer. Furthermore, iterating the last identity gives

$$\begin{aligned} \rho_{2s+1,2s+1}^{(n)} &= a_{n-1}^{-1} \cdots a_{n-s}^{-1} a_{n-2s-1} \cdots a_{n-s-1} \rho_{1,1}^{(n-s)} \\ &= a_{n-1}^{-1} \cdots a_{n-s}^{-1} a_{n-2s-1} \cdots a_{n-s-1} \sigma_1. \end{aligned}$$

Hence, if $\sigma_1 = 0$, then $\rho_{2s+1,2s+1}^{(n)} = 0$ and, as a result, $q_{n,2s+1}$ becomes $q_{n,2s}$, which shows that (1.28) also holds for r being an even integer. □

Definition 1.4.4. A $(2n-r-1)$-orthogonal polynomial is said to have a matrix representation if it is the characteristic polynomial of $J_{n,r}$ defined in (1.27). It is said to have a positive matrix representation if the sub-diagonal elements of $J_{n,r}$ are all positive.

The matrix $J_{n,r}$ in (1.27) has positive sub-diagonal elements if all $\tau_j < 1$. Proposition 1.4.5 justifies the consideration of this class of $(2n-r-1)$-orthogonal polynomials.

Proposition 1.4.5. *If $q_{n,r}$ is a $(2n-r-1)$-orthogonal polynomial satisfying a positive matrix representation, then its zeros are real and simple. Moreover, $q_{n,r}$ is a generating polynomial of a positive $(2n-r-1,n)$-quadrature rule.*

Proof Let $q_{n,r} = \gamma_n \det(tI_n - J_{n,r})$ with $J_{n,r}$ defined in (1.27). The matrix $J_{n,k}$ is of the form (1.7) with a_k and b_k replaced by $\hat{a}_k$ and $\hat{b}_k$ defined by

$$\hat{a}_k = a_k, \quad 1 \le k \le n-s-1 \quad \text{and} \quad \hat{a}_k = a_k(1-\tau_{s+k-n+1}), \quad n-s \le k \le n-2,$$
$$\hat{b}_k = b_k, \quad 1 \le k \le n-s-1 \quad \text{and} \quad \hat{b}_k = b_k - a_k\sigma_{s+k-n+1}, \quad n-s \le k \le n-1,$$

where $r-2s-1$ or $2s-2$. For $0 \le k \le n-1$, we denote by $J_{n-k,r-2k}$ the principle sub-matrix of $J_{n,r}$ of size $n-k$, the one at the upper left corner, and define

$$q_{n-k,r-2k} = \gamma_{n-k}\det(tI_{n-k} - J_{n-k,r-2k}), \qquad 0 \le k \le n-1. \tag{1.29}$$

Then, $q_{n-k,r-2k} = p_{n-k}(w)$ if $n - \lfloor \frac{r}{2} \rfloor \le k \le n$, and it agrees with that of (1.26) for $k < n - \lfloor \frac{r}{2} \rfloor$. In particular, the polynomials $P_k = q_{n-k,r-2k}$, $0 \le k \le n$, satisfy the recursive relations

$$\gamma_{k+1}^{-1}P_{k+1} = (x - \hat{b}_k)\gamma_k^{-1}P_k - \hat{a}_{k-1}^2\gamma_{k-1}^{-1}P_{k-1}, \qquad 0 \le k \le n-1.$$

Since $a_k = \gamma_k/\gamma_{k+1}$ and $\hat{a}_k = a_k(1-\tau_k)$, normalizing P_k as Q_k defined by

$$Q_{n-k} = \begin{cases} P_{n-k}(w), & \lfloor \frac{r}{2} \rfloor + 1 \le k \le n, \\ (1-\tau_1)^{-1}\cdots(1-\tau_{s-k}^{-1})P_k, & 0 \le k \le \lfloor \frac{r}{2} \rfloor, \end{cases}$$

we see that Q_k, $0 \le k \le n$, satisfy the three-term relations

$$tQ_k(t) = \hat{a}_k Q_{k+1}(t) + \hat{b}_k Q_k(t) + \hat{a}_{k-1}Q_{k-1}(t), \qquad 0 \le k \le n-1,$$

which are the same as (1.6). As a result, we obtain analogs of Christoffel–Darboux formulas (1.9) and (1.10), which shows

$$\hat{K}_{n-1}(s,t) = \sum_{k=0}^{n-1} Q_k(s)Q_k(t) = A_{n-1}\frac{q_{n,r}(s)q_{n-1,r-2}(t) - q_{n,r}(t)q_{n-1,r-2}(s)}{s-t},$$

where $A_{n-1} = a_{n-1}(1-\tau_1)^{-2}\dots(1-\tau_{s-1})^{-2} > 0$, and

$$0 < \hat{K}_{n-1}(t,t) = A_{n-1}\left[q_{n,r}'(t)q_{n-1,r-2}(t) - q_{n,r}(t)q_{n-1,r-2}'(t)\right].$$

In particular, if t_k is a zero of $q_{n,r}$, then the last identity shows that $q_{n,r}'(t_k) \ne 0$, which proves that all zeros of $q_{n,r}$ are simple.

Let $t_1,\dots,t_n$ be the zeros of $q_{n,r}$. The Lagrange interpolation polynomial on $t_1,\dots,t_n$ is given by (1.22) with K_{n-1} replaced by $\hat{K}_{n-1}$, which gives a positive interpolatory quadrature rule $\mathrm{Q}(f)$ of degree at least $n-1$. Moreover, if f is a polynomial of degree $2n-r-1$, then we can write $f = hq_{n,r} + r$ with h and r being polynomials of degree at most $n-1$. Using the $(2n-r-1)$-orthogonality of $q_{n,r}$, the last part of the proof of Theorem 1.3.1 can be followed verbatim to show that $\mathsf{Q}(f)$ is of degree $2n-r-1$. □

The matrix representation of a $(2n-r-1)$-orthogonal polynomial is given in terms of the Jacobi matrix associated with the weight function w. These polynomials characterize $(2n-r-1,n)$-quadrature rules for w.

Theorem 1.4.6. *Let w be a regular weight function. A quadrature $\mathsf{Q}_{n,r}$ is a positive $(2n-1-r,n)$-quadrature rule for w if, and only if, its generating polynomial $q_{n,r}$ has a positive matrix representation.*

Proof The sufficient direction is established in Proposition 1.4.5. To prove the other direction, we assume that $\mathsf{Q}_{n,r}$ is a positive $(2n-1-r,n)$-quadrature with nodes $t_1,\ldots,t_n$. Let $q_{n,r}(t)=\gamma_n(t-t_1)\ldots(t-t_n)$. Then $q_{n,r}$ is of degree n and $(2n-1-r)$-orthogonal with respect to w. Every polynomial f can be written uniquely as $f=hq_{n,r}+R_q(f)$, where h is a polynomial and $R_q(f)$ is a polynomial of degree at most $n-1$. Accordingly, we define a linear functional L_q associated with $q_{n,r}$ on the space Π_{2n-1} by

$$\mathsf{L}_q:\Pi_{2n-1}\mapsto\mathbb{R},\qquad \mathsf{L}_q(f)=\int_a^b R_q(f;t)w(t)\mathrm{d}t.$$

If f is a nonnegative polynomial in Π_{2n-1}, then it can be written as a sum of squares, $f=p_1^2+p_2^2$, for some $p_1,p_2\in\Pi_{n-1}$. Such an f, if not identical zero, cannot vanish at all n nodes of the quadrature formula. Hence, using $R_q(f,t_k)=f(t_k)$, we obtain

$$\mathsf{L}_q(f)=\int_a^b R_q(f;t)w(t)\mathrm{d}t=\mathsf{Q}_{n,r}(R_q(f))=\mathsf{Q}_{n,r}(f)>0,$$

which shows that L_q is a positive linear functional. Moreover, for every $p\in\Pi_{n-1}$, we have $pq_{n,r}\in\Pi_{2n-1}$ and $\mathsf{L}_q(pq_{n,r})=0$ by the above displayed identity. Thus, $q_{n,r}$ is orthogonal to all polynomials in Π_{n-1} for the linear functional L_q. Applying the Gram–Schmidt process, the positivity of L_q on Π_{2n-1} leads to polynomials $q_0,q_1,\ldots,q_{n-1}$ that are orthonormal with respect to L_q and $q_k=\gamma_k(q)x^k+\ldots$ with $\gamma_k(q)>0$. Since $\mathsf{Q}_{n,r}(f)$ is of degree $2n-1-r$,

$$\int_a^b h(t)q_{n,r}(t)w(t)\mathrm{d}t=\mathsf{Q}_{n,r}(q_{n,r}h)=0,$$

if $\deg h\le n-r-1$ so that $\mathsf{L}_q(f)=\int_a^b f(t)w(t)\mathrm{d}t$ for all $f\in\Pi_{2n-1-r}$, which implies in particular that $q_k=p_k(w)$, $0\le k\le n-[r/2]$. The orthonormal polynomials $q_0,q_1,\ldots,q_{n-1}$ satisfy a three-term relation, which can be written as

$$tq_k(t)=\hat{a}_kq_{k+1}(t)+\hat{b}_kq_k(t)+\hat{a}_{k-1}q_{k-1}(t),\quad 1\le k\le n-2,$$

where, by the orthogonality and the positivity of L_q,

$$\hat{a}_k = \mathsf{L}_q(tq_kq_{k+1}) = \frac{\gamma_k(q)}{\gamma_{k+1}(q)}\mathsf{L}_q\left(q_{k+1}^2\right) = \frac{\gamma_k(q)}{\gamma_{k+1}(q)} > 0, \quad 0 \le k \le n-2.$$

Moreover, $q_{n,r}$ can be written as $q_{n,r} = a_{n-1}^{-1}tq_{n-1}(t) + \sum_{k=0}^{n-1} c_kq_k(t)$. By the orthogonality of q_k, $0 \le k \le n-1$, and $q_{n,k}$ with respect to L_q, we obtain

$$q_{n,r}(t) = \hat{a}_{n-1}^{-1}(tq_{n-1}(t) - \hat{b}_{n-1}(t)q_{n-1}(t) - \hat{a}_{n-1}q_{n-1}(t)),$$

where $\hat{b}_{n-1} = \mathsf{L}_q(tq_{n-1}^2)$. Together, these three-term relations show that $q_{n,r}$ admits a symmetric matrix representation, which is positive since $\hat{a}_k > 0$ for $0 \le k \le n-2$. More precisely, let $\tau_k = 1 - \hat{a}_k/a_k$ and define σ_k, $1 \le k \le s$ by $\hat{b}_{n-s+k-1} = b_{n-s+k-1} - a_{n-s+k-1}\sigma_k$, we conclude that the matrix representation is of precisely the form of (1.27). □

Several remarks are in order. First, for the sufficient part, the positivity of the matrix can be replaced by the requirement that none of the sub-diagonal elements of $J_{n,r}$ are zero, which amounts to replacing $\tau_k < 1$ with $\tau_k \ne 1$, as can be seen from the proof of Proposition 1.4.5. Second, the proof of Theorem 1.4.6 relies on orthonormal polynomials for the positive linear functional L_q, and it shows, in particular, that the zeros of $q_{n,r}$ are the nodes of the Gauss quadrature rule of degree $2n-1$ for L_q. The linear functional L_q depends on n, but it could have an integral representation independent of n in some cases. For example, for $b < 1$ and $n \ge 2$, the polynomial $q_{n,1}(t) = U_n(t) - bU_{n-1}(t)$ is a $(2n-2,n)$-orthogonal polynomial for the Chebyshev weight $w_1(t) = \sqrt{1-t^2}$, and it is also an orthogonal polynomial for the weight function [24, p. 205],

$$\hat{w}(t) = \frac{\sqrt{1-t^2}}{1+b^2-2bt}, \qquad -1 \le t \le 1.$$

Hence, the zeros of $q_{n,1}$ are nodes of a $(2n-2,n)$-quadrature for w_1 as well as nodes of a Gauss quadrature for $\hat{w}$.

A $(2n-r-1)$-orthogonal polynomial satisfying a positive matrix representation has real simple zeros. The next theorem shows when the zeros may not all be in the support set $[a,b]$ of w.

Theorem 1.4.7. *Let w be a regular weight function supported on $[a,b]$ and let $q_{n,r}$ be a $(2n-r-1)$-orthogonal polynomial of degree n that has a positive matrix representation. Then, the zeros of $q_{n,r}$ are all located inside $[a,b]$ if, and only if,*

$$q_{n-k,r-2k}(b) > 0 \quad \textit{and} \quad (-1)^{n-k}q_{n-k,r-2k}(a) > 0, \quad 0 \le k \le \left\lfloor \frac{r}{2} \right\rfloor,$$

where the polynomials $q_{n-k,r-2k}$ are related to $q_{n,r}$ as defined in (1.29).

Since the location of the nodes is secondary in our study of cubature rules, we shall skip the proof of this theorem; see [162].

1.5 Notes and Further Results

The quadrature rules are discussed in many books on numerical analysis and scientific computing. The books [20, 36, 55] are devoted entirely to the theory of quadrature rules and consist of topics way beyond our narrative. The Gauss quadrature rules have been thoroughly studied (cf. [147]) and widely utilized, and their nodes and weights are extensively calculated numerically; see, for example, [56]. They are also discussed in books on orthogonal polynomials.

The classical book on orthogonal polynomials is [148] by G. Szegő, which remains a standard reference for classical orthogonal polynomials, and more recent books are [4, 24, 55, 73], to name a few. The Digital Library of Mathematical Functions [118] is an online database of mathematical reference data for special functions, including orthogonal polynomials, and it has become a convenient and reliable resource.

The characterization of positive quadrature rules on the real line is not a standard material in books on quadrature rules and numerical analysis. The problem was studied in [124, 125, 126, 138, 162, 163], where the m-orthogonality appears naturally. Our presentation in Section 1.4 follows that of [163] since the approach therein could be adapted, at least partially, for studying cubature rules.

2
Cubature Rules: Basics

Cubature rules are quadrature rules in higher dimensions. Their degrees are defined by the highest degrees of polynomials integrated exactly, where the degree means the total degree throughout this monograph. We cover the basic results in this chapter. After fixing the notations on polynomials and cubature rules in Section 2.1, the interpolatory cubature rules, which are not necessarily positive, are discussed in Section 2.2. Positive cubature rules always exist for a regular integral, known as Tchakaloff's theorem, and proved in Section 2.3. In Section 2.4, we consider several families of integrals on regular domains that can be parametrized as a product of integrals in one variable so that cubature rules can be derived from the product of quadrature rules accordingly. In Section 2.5, we consider invariant cubature rules for integrals on regular domains that are invariant under a finite subgroup of a rotation group. Finally, in Section 2.6, we give a brief overview of constructing cubature rules numerically.

2.1 Polynomials and Cubature Rules

In this preliminary section, we fix notations for polynomials and cubature rules. Let us start with polynomial spaces. We will use the standard multi-index notation. For $\mathbf{x} = (x_1, \ldots, x_d) \in \mathbb{R}^d$, a monomial in $\mathbf{x}$ is defined as

$$\mathbf{x}^\alpha = x_1^{\alpha_1} \cdots x_d^{\alpha_d}, \qquad \alpha = (\alpha_1, \ldots, \alpha_d) \in \mathbb{N}_0^d.$$

The degree of $\mathbf{x}^\alpha$ is defined as the *total degree* $|\alpha| := \alpha_1 + \cdots + \alpha_d$. A polynomial P is a linear combination of monomials, and its degree is the highest degree of its monomials. For $d \geq 2$, we denote by $\mathcal{P}_n^d$ the space of homogeneous polynomials of degree n,

$$\mathcal{P}_n^d := \left\{ P : P(\mathbf{x}) = \sum_{|\alpha|=n} c_\alpha \mathbf{x}^\alpha \right\} = \operatorname{span}\left\{ \mathbf{x}^\alpha : \alpha \in \mathbb{N}_0^d, \quad |\alpha| = n \right\},$$

and by Π_n^d the space of polynomials of degree at most n, which is a direct sum of $\mathcal{P}_k^d$, $0 \le k \le n$,

$$\Pi_n^d := \left\{ P : P(\mathbf{x}) = \sum_{k=0}^{n} P_k(\mathbf{x}),\ P_k \in \mathcal{P}_k^d \right\} = \operatorname{span}\left\{ \mathbf{x}^\alpha : \alpha \in \mathbb{N}_0^d, \quad |\alpha| \le n \right\}.$$

All polynomials are assumed to be real unless indicated otherwise. The dimensions of these spaces are

$$\dim \mathcal{P}_n^d = \binom{n+d-1}{n} \quad \text{and} \quad \dim \Pi_n^d = \binom{n+d}{d}.$$

For $d > 1$, we need a well-defined order among monomials. There are many such orders, but none is quite natural. One such order is

Lexicographic order. $\mathbf{x}^\alpha \succ \mathbf{x}^\beta$ if the first nonzero entry in the difference $\alpha - \beta$ is positive.

This order, however, does not respect the total degree of monomials. Throughout the monograph, we shall use the order defined by

Graded lexicographic order. $\mathbf{x}^\alpha \succ \mathbf{x}^\beta$ if $|\alpha| > |\beta|$ or if $|\alpha| = |\beta|$ and the first nonzero entry in the difference $\alpha - \beta$ is positive.

This graded order preserves the total degree of monomials since monomials of lower total degrees are ordered before monomials of higher degrees, and those of the same degree are ordered lexicographically. For $d = 2$, monomials of the degree m are ordered by

$$x_1^m \succ x_1^{m-1} x_2 \succ \cdots \succ x_1 x_2^{m-1} \succ x_2^m.$$

The order for monomials of more than two variables is more involved.

Let Ω be a domain in $\mathbb{R}^d$ and W be a nonnegative weight function defined on Ω. We study cubature rules for the integral

$$\mathcal{I}(f) = \int_\Omega f(\mathbf{x}) W(\mathbf{x}) \mathrm{d}\mathbf{x},$$

where $W(\mathbf{x})\mathrm{d}\mathbf{x}$ can also be replaced by $\mathrm{d}\mu$ with a positive Borel measure μ. The weight function W, or μ, is assumed to be *regular* if $\int_\Omega |\mathbf{x}^\alpha| W(\mathbf{x}) \mathrm{d}\mathbf{x} < \infty$ for all $\alpha \in \mathbb{N}_0^d$, which ensures that all moments, $m_\alpha = \mathcal{I}(\mathbf{x}^\alpha)$ for $\alpha \in \mathbb{N}_0^d$, are finite.

A cubature rule of degree m for the integral $\mathcal{I}(f)$ is a finite linear combination of function evaluations

$$\mathsf{C}(f) := \sum_{k=1}^{N} \lambda_k f(\mathbf{x}_k), \qquad \mathbf{x}_k \in \mathbb{R}^d, \quad \lambda_k \in \mathbb{R},$$

that approximates the integral. It is said to have degree m if

$$\int_\Omega f(\mathbf{x}) W(\mathbf{x}) \mathrm{d}\mathbf{x} = \mathsf{C}(f), \qquad \text{for all } f \in \Pi_m^d,$$

and the identity fails to hold for at least one polynomial of degree $m+1$, just as for quadrature rules. We also retain the name *nodes* for $\mathbf{x}_k$ and *weights* for λ_k and assume $\mathbf{x}_k$ are distinct and $\lambda_\kappa \neq 0$ so that N is the *number of nodes*.

In terms of the moments, the existence of a cubature rule of degree m means that $\mathbf{x}_k$ and λ_k satisfy the system of *moment equations*

$$\sum_{k=1}^{N} \lambda_k \mathbf{x}_k^\alpha = \int_\Omega \mathbf{x}^\alpha W(\mathbf{x}) \mathrm{d}\mathbf{x} = \mathcal{I}(\mathbf{x}^\alpha), \qquad |\alpha| \le m,\ \alpha \in \mathbb{N}_0^d, \tag{2.1}$$

where the monomials in $\{\mathbf{x}^\alpha : |\alpha| \le m\}$ are ordered by the graded lexicographical order, and the system consists of $\dim \Pi_m^d = \binom{m+d}{d}$ equations. We are interested in cubature rules with fewer points or high precision, for which we need to treat $\mathbf{x}_k$ in (2.1) as variables so that the system (2.1) is nonlinear.

The existence of a cubature rule of degree m requires a large enough number N of points, which means that N is bounded below by a bound depending on m. The first lower bound of N is analogous to the lower bound in Proposition 1.1.2 for quadrature rules.

Theorem 2.1.1. *If $m = 2n-1$ or $m = 2n-2$, then the number of nodes N of a cubature rule of degree m satisfies*

$$N \ge \dim \Pi_{n-1}^d = \binom{n+d-1}{d}. \tag{2.2}$$

Proof Let $\mathsf{C}(f)$ be a cubature rule of degree m with nodes $\mathbf{x}_1, \ldots, \mathbf{x}_N$. If $N < \dim \Pi_{n-1}^d$ with $n = \lfloor \frac{m}{2} \rfloor + 1$, then the system of equations

$$P(\mathbf{x}_k) = \sum_{|\alpha| \le m} c_\alpha \mathbf{x}_k^\alpha = 0, \qquad 1 \le k \le N,$$

consists of N equations and $\dim \Pi_{n-1}^d > N$ variables $\{c_\alpha : |\alpha| \le m\}$. Since there are more variables than equations, the system has at least one nonzero solution, which yields a nonzero polynomial P of degree $n-1$ that vanishes on all nodes of $\mathsf{C}(f)$. Since C is of degree m, this leads to $0 = \mathsf{C}(P^2) = \mathcal{I}(P^2) > 0$. The contradiction shows that N cannot be less than $\dim \Pi_{n-1}^d$. □

A cubature rule is called *Gauss cubature* if its number of nodes attains the lower bound (2.2). Gauss cubature rules will be characterized and studied in Chapter 4. As will be seen then, such rules exist rarely but do exist in some cases. This provides a sharp contrast between the complexity of cubature rules and quadrature rules. Let us entertain a heuristic argument to give an idea of what may be in stock.

For $d = 1$, the Gauss cubature of degree $2n-1$ agrees with the Gauss quadrate of degree, which has n nodes. For $d = 1$, a quadrature rule of $2n-1$ with n nodes exists if t_k and λ_k satisfy

$$\sum_{i=1}^{n} \lambda_i t_k^j = \int_{\mathbb{R}} t^j w(t) \mathrm{d}t, \qquad 0 \le j \le 2n-1,$$

which is the system of moment equations (2.1), and the system has $2n$ equations. Regarding both λ_k and t_k as variables, the system has $2n$ variables so that the number of variables equals the number of equations. Although matching the two numbers spells nothing about solutions for the *nonlinear* system, the argument is enticing.

For $d \ge 2$, the heuristic argument falls flat. Let us consider the case $d = 2$, for which the system (2.1) for a Gauss cubature rule of degree $2n-1$ becomes

$$\sum_{i=1}^{N} \lambda_i x_i^j y_i^k = \int_{\Omega} x^j y^k W(x,y) \mathrm{d}x\mathrm{d}y, \qquad 0 \le j+k \le 2n-1, \tag{2.3}$$

which is a nonlinear system of $\binom{2n+1}{2}$ equations. Treating both (x_j, y_j) and λ_j as variables, the system has $3N$ variables. The heuristic argument suggests that we should have at least as many variables as equations, that is,

$$N \ge \frac{1}{3}\binom{2n+1}{2} = \frac{2n^2+n}{3}. \tag{2.4}$$

The right-hand side, however, is strictly greater than $\dim \Pi_{n-1}^2 = \frac{n^2+n}{2}$ for $n \ge 2$, the number of nodes of the Gauss cubature rule. In other words, the existence of the Gauss cubature rule implies that the system (2.4) has a solution even when N is far less than the right-hand side of (2.4). In particular, the usual presumption that there ought to be more variables than equations no longer holds.

The existence of the Gauss cubature rule shows that the system (2.4) could, in some cases, have solutions if N is far less than the right-hand side of (2.4), despite there being far fewer variables than equations. On the one hand, this is not surprising in view that the system of moment equations is nonlinear and that little is known, in general, about the nonlinear systems of equations. On the other hand, it is intriguing and suggestive of something highly nontrivial in high dimension. The latter is indeed the case, as we shall see in Chapter 4 and the rest of this monograph.

2.2 Interpolatory Cubature Rules

Interpolatory cubature rules arise from integrating interpolation polynomials as in the case of quadrature rules. However, polynomial interpolation itself is more subtle and intricate in higher dimensions. A straightforward extension of the interpolation problem for several variables is as follows.

Problem 2.2.1. *(Lagrange interpolation). Let $\mathcal{X} = \{\mathbf{x}_1, \ldots, \mathbf{x}_N\}$, with $\mathbf{x}_i \in \mathbb{R}^d$ and $N = \binom{m+d}{d}$, be a set of pairwise distinct points in $\mathbb{R}^d$; that is, $\mathbf{x}_i \neq \mathbf{x}_j$ if $i \neq j$. Determine if there is a unique polynomial $P \in \Pi_m^d$ such that*

$$P(\mathbf{x}_i) = f(\mathbf{x}_i), \qquad 1 \le i \le N. \tag{2.5}$$

Unlike the case of one variable, this problem does not always have a unique solution when $d > 1$. To illustrate, we consider an example in the simplest case of $d = 2$ and $n = 3$.

Example 2.2.2. If $\mathcal{X}$ consists of six pairwise distinct points on the unit circle $x_1^2 + x_2^2 = 1$, then the interpolation problem has no unique solution in Π_2^2.

Indeed, if P is an interpolation polynomial of degree 2 that satisfies (2.5), then so is $P(x_1, x_2) + c(x_1^2 + x_2^2 - 1)$ for every $c \in \mathbb{R}$. Thus, the solution is not unique.

Whether the interpolation problem has a unique solution depends on the geometry of the data. If P is a polynomial in Π_m^d, we call the set $\{\mathbf{x} \in \mathbb{R}^d : P(\mathbf{x}) = 0\}$ an *algebraic hypersurface* of degree m in $\mathbb{R}^d$. For $d = 2$, the algebraic hypersurface is the usual algebraic curve. In particular, the unit circle is the set $\{\mathbf{x} : P(\mathbf{x}) = 0\}$ with $P(\mathbf{x}) = 1 - x_1^2 - x_2^2$.

Theorem 2.2.3. *The Lagrange interpolation of Problem 2.2.1 has a unique solution if, and only if, the set $\mathcal{X}$ is not on (i.e. not a subset of) an algebraic hypersurface of degree m.*

Proof Writing a polynomial $P \in \Pi_m^d$ as $P(\mathbf{x}) = \sum_{|\alpha| \le m} c_\alpha \mathbf{x}^\alpha$, the interpolation requirement (2.5) becomes a linear system of equations, $X_m \mathbf{c} = \mathbf{b}$, where the matrix X_m is given by

$$X_m = \left[\mathbf{x}_k^\alpha\right]_{k=1,|\alpha|\le m}^N.$$

The interpolation problem has a unique solution if, and only if, $X_m \mathbf{c} = 0$ has only a trivial solution. Moreover, $X_m \mathbf{c} = 0$ has a nontrivial solution is equivalent to that $P(\mathbf{x}) = \sum_{|\alpha| \le m} c_\alpha \mathbf{x}^\alpha$ vanishes on $\mathcal{X}$ or that $\mathcal{X}$ is a subset of the hypersurface of degree m defined by $P(\mathbf{x}) = 0$. $\square$

Since a hypersurface in $\mathbb{R}^d$ has measure zero, the interpolation problem has a unique solution for almost all $\mathcal{X}$ that consists of $\binom{m+d}{d}$ distinguished points. The integration of the interpolation polynomial gives a cubature rule.

Proposition 2.2.4. *If $\mathcal{X} = \{\mathbf{x}_k : 1 \le k \le \dim \Pi_m^d\}$ is a set of pairwise distinct points in $\mathbb{R}^d$ and $\mathcal{X}$ is not on an algebraic hypersurface of degree m, then $\mathcal{X}$ is the set of nodes of a cubature rule of degree m.*

Proof By Theorem 2.2.3, there is a unique interpolation polynomial of degree m on $\mathcal{X}$. In particular, the fundamental interpolation polynomial ℓ_k that satisfies $\ell_k(\mathbf{x}_j) = \delta_{k,j}$, $1 \le j \le N$, exists and is unique, and the interpolation polynomial on $\mathcal{X}$ can be written as

$$L_{\mathcal{X}} f(\mathbf{x}) = \sum_{k=1}^{N} f(\mathbf{x}_k) \ell_k(\mathbf{x}).$$

By the uniqueness, if $f \in \Pi_m^d$, then $L_{\mathcal{X}} f = f$. Hence, integrating $L_{\mathcal{X}}$ by an integral $\mathcal{I}(f)$ gives

$$\mathcal{I}(f) = \mathcal{I}(L_{\mathcal{X}} f) = \sum_{k=1}^{N} \lambda_k f(\mathbf{x}_k), \qquad \lambda_k = \mathcal{I}(\ell_k),$$

for all polynomials $f \in \Pi_m^d$, which is a cubature rule of degree m. $\square$

The statement applies to all integral $\mathcal{I}(f)$, and it shows that a cubature rule of degree m with $\dim \Pi_{m-1} = \binom{m+d}{d}$ exists for almost all choices of nodes in $\mathbb{R}^d$. Thus, the interpolatory cubature rules almost always exist. However, these rules may not be positive since $\lambda_k = \mathcal{I}(\ell_k)$ are not generally positive. One could ask if a positive cubature rule of degree m exists. The answer is positive, and it is the content of Section 2.3.

We end this section with a characterization of point sets that are not on an algebraic hypersurface. To state the result, we need a definition.

Definition 2.2.5. A point in $\mathbb{R}^d$ is called a *common zero* of a set of polynomials of d variables if it is a zero of every polynomial in the set.

The characterization stated in the next proposition will be useful for establishing the existence of cubature rules later on.

Proposition 2.2.6. *The pairwise distinct points x_k, $1 \le k \le \dim \Pi_{n-1}^d$, on $\mathbb{R}^d$ are not on an algebraic hypersurface of degree $n-1$ if, and only if, there exist* *monic polynomials* $P_\alpha(\mathbf{x}) = \mathbf{x}^\alpha + R_\alpha(\mathbf{x})$, $R_\alpha \in \Pi^d_{n-1}$, *with* $|\alpha| = n$ *and* $\alpha \in \mathbb{N}_0^d$ *so that these points are common zeros of* $\{P_\alpha : |\alpha| = n\}$.

Proof Let $N = \dim \Pi^d_{n-1}$. Assume that $\mathbf{x}_k$, $1 \le k \le N$, are not on a hypersurface of degree $n-1$, which means that the matrix $X_n = \left[\mathbf{x}_k^\beta\right]_{|\beta|=n-1, 1\le k\le N}$ is nonsingular, where the columns are arranged according to the graded lexicographical order in $\{\beta : 0 \le |\beta| \le n-1\}$. Hence, for each $\alpha \in \mathbb{N}_0^d$, the linear system of equations

$$\sum_{|\beta|\le n-1} c_{\alpha,\beta}\mathbf{x}_k^\beta = \mathbf{x}_k^\alpha, \qquad k = 1,\ldots,N, \tag{2.6}$$

has a unique solution $\{c_{\alpha,\beta} : |\beta| \le n-1\}$, which gives a monic polynomial P_α of degree n by $P_\alpha(\mathbf{x}) = \mathbf{x}^\alpha - \sum_{|\beta|\le n-1} c_{\alpha,\beta}\mathbf{x}^\beta$. Then, P_α vanishes on all $\{\mathbf{x}_k : 1 \le k \le N\}$ for each $|\alpha| = n$. This proves the necessary direction.

In the other direction, suppose that there are monic polynomials $P_\alpha^n = \mathbf{x}^\alpha + R_\alpha$, $|\alpha| = n$, of degree n that have $\mathbf{x}_k$ as common zeros. If the points $\mathbf{x}_k$ are on an algebraic hypersurface of degree $n-1$, then the matrix $X_n = \left[\mathbf{x}_k^\beta\right]_{|\beta|\le n-1, 1\le k\le N}$ is singular. Let $\operatorname{rank} X_n = M-1$ with $M \le N$. Without loss of generality, we assume the first $M-1$ rows of this matrix are linearly independent. For each k, define a vector $\xi_k = \left[\mathbf{x}_k^\beta\right]_{|\beta|\le n-1}$ in $\mathbb{R}^N$; then, $\xi_1,\ldots,\xi_{M-1}$ are linearly independent, and there exist scalar $c_1,\ldots,c_M$, not all zero, such that $\sum_{k=1}^M c_k\xi_k = 0$. Moreover, considering the first component of this vector equation, it follows that at least two c_j are nonzero. Assume $c_1 \ne 0$ and $c_M \ne 0$. Then, the equations $P_\alpha^n(\mathbf{x}_k) = 0$ can be written as in (2.6), and its left-hand side can be written as $C_\alpha^{\mathsf{T}}\xi_k$ with a vector C_α in $\mathbb{R}^N$. It follows from $\sum_{k=1}^M c_k\xi_k = 0$ that $\sum_{k=1}^M c_k\mathbf{x}_k^\alpha = 0$ for all monomial $\mathbf{x}^\alpha$ with $|\alpha| = n$. Let $\mathbf{x}_k = (x_{k,1},\ldots,x_{k,d})$. Since $x_{k,i}\xi_k = \left[x_{k,i}\mathbf{x}_k^\beta\right]_{|\beta|\le n-1}$ and $x_i\mathbf{x}^\beta$ is a monomial of degree n, we deduce that $\sum_{k=1}^M c_k x_{k,i}\xi_k = 0$, $1 \le i \le d$. Multiplying the equation $\sum_{k=1}^M c_k\xi_k = 0$ by $x_{M,i}$ and then subtracting it from the equation $\sum_{k=1}^M c_k x_{k,i}\xi_k = 0$, we obtain

$$\sum_{k=1}^{M-1} c_k(x_{k,i} - x_{M,i})\xi_k = 0, \quad 1 \le i \le d.$$

Since $\xi_1,\ldots,\xi_{M-1}$ are linearly independent, $c_k(x_{k,i} - x_{M,i}) = 0$, $1 \le k \le M-1$, $1 \le i \le d$. Since $c_1 \ne 0$, this gives $x_{1,i} = x_{M,i}$ for $1 \le i \le M-1$. Thus, $\mathbf{x}_1 = \mathbf{x}_M$, contradicting the assumption that points are pairwise distinct. Hence, X_n has full rank. This completes the proof. □

2.3 Existence of Positive Cubature Rules

The positive cubature rules always exist for a regular weight function. We state this theorem more generally for a Borel measure μ. Let $\mathrm{supp}(\mu)$ denote the closed support of the measure μ.

Theorem 2.3.1. *Let μ be a positive Borel measure on $\mathbb{R}^d$ such that, for $m \in \mathbb{N}$,*

$$\int_{\mathbb{R}^d} \|\mathbf{x}\|^k \mathrm{d}\mu(\mathbf{x}) < \infty, \qquad 0 \le k \le m.$$

Then μ admits a positive cubature rule of degree m with N nodes, where $N \le \dim \Pi_m^d$ and all nodes belong to $\mathrm{supp}(\mu)$.

This theorem is named after Tchakaloff, who initiated its study. We give a proof later under the assumption that $\mathrm{supp}(\mu)$ is compact.

Let $K = \mathrm{supp}(\mu)$. Without losing generality, we assume $0 \in K$ since, otherwise, we can consider a translation of μ. Let Y be a vector function defined by

$$\mathsf{Y}(\mathbf{x}) = \left\{\mathbf{x}^\alpha : |\alpha| \le m, \alpha \in \mathbb{N}_0^d\right\} \in \mathbb{R}^M, \qquad M = \binom{m+d}{d},$$

where the elements are ordered in the graded lexicographical order. Every polynomial in Π_m^d can be written as $P(\mathbf{x}) = \langle p, \mathsf{Y}(\mathbf{x})\rangle$, where $p \in \mathbb{R}^M$ and $\langle \cdot,\cdot \rangle$ denotes the Euclidean inner product in $\mathbb{R}^M$. Let $\mathcal{U}(K)$ be the space spanned by M-dimensional vectors of the form $\mathsf{Y}(\mathbf{x})$ for $\mathbf{x} \in K$ and let $N = \dim \mathcal{U}(K)$. Then, $N \le M$ and the strict inequality $N < M$ holds only if K is a subset of a hypersurface of degree m. In particular, $\mathcal{U}(K)$ can be identified with the Euclidean space $\mathbb{R}^N$. Furthermore, there are N points $\mathbf{x}_1, \ldots, \mathbf{x}_N$ in K such that $\mathsf{Y}(\mathbf{x}_1), \ldots, \mathsf{Y}(\mathbf{x}_N)$ are linearly independent, which follows from Proposition 2.2.4 and leads to the existence of the cubature rule but without positivity.

Let $\mathcal{B}(K) = \{\mathsf{Y}(\mathbf{x}) : \mathbf{x} \in K\}$. Since K is compact, $\mathcal{B}(K)$ is compact in $\mathcal{U}(K)$. Moreover, $\mathcal{B}(K)$ contains N linearly independent vectors in $\mathcal{U}(K)$. We denote by $\mathrm{cone}(K)$ the convex cone in $\mathcal{U}(K)$, which consists of elements that can be written as linear combinations with nonnegative coefficients of at most N elements in $\mathcal{B}(K)$,

$$\mathrm{cone}(K) = \left\{c_1\mathbf{x}_1 + \cdots + c_N\mathbf{x}_N : c_j \ge 0, \quad \mathbf{x}_j \in \mathcal{B}(K), 1 \le j \le N\right\}.$$

Lemma 2.3.2 justifies the restriction on the number of elements in $\mathcal{B}(K)$.

Lemma 2.3.2. *Let $\mathbf{x}_1, \ldots, \mathbf{x}_p \in \mathcal{B}(K)$ and $\mathbf{x} = c_1\mathbf{x}_1 + \cdots + c_p\mathbf{x}_p$ with $c_1, \ldots, c_p$ positive. If $p > N$, then $\mathbf{x}$ can be written as a linear combination, with nonnegative coefficients, of at most N of the $\mathbf{x}_j$.*

Proof If $p > N$, then $\mathbf{x}_1, \dots, \mathbf{x}_p$ are linearly dependent. Hence, there exist real numbers $b_1, \dots, b_p$ so that $b_1\mathbf{x}_1 + \cdots b_p\mathbf{x}_p = 0$ and at least one of the b_j is positive. Thus, $\mu = \max_j b_j/c_j > 0$, and we can write

$$\mathbf{x} = h_1\mathbf{x}_1 + \cdots + h_p\mathbf{x}_p, \qquad h_j = \mu^{-1}(\mu c_j - b_j) \geq 0, \quad 1 \leq j \leq p,$$

which has at least one h_j being zero. Thus, $\mathbf{x}$ can be written as a linear combination, with nonnegative coefficients, of $p-1$ elements in $\mathcal{B}(K)$. The argument can be repeated until the number of elements is $\leq N$. This completes the proof. □

Lemma 2.3.3. *The set* $\operatorname{cone}(K)$ *is a closed convex cone in* $\mathcal{U}(K)$.

Proof If $\mathbf{x}_1, \mathbf{x}_2 \in \operatorname{cone}(K)$, then the lemma implies that $c_1\mathbf{x}_1 + c_2\mathbf{x}_2 \in \operatorname{cone}(K)$ for all $c_1 \geq 0$ and $c_2 \geq 0$. In particular, it shows that $\operatorname{cone}(K)$ is convex. To show that $\operatorname{cone}(K)$ is closed, let y_n be a sequence of elements in $\operatorname{cone}(K)$ so that

$$\mathbf{y}_n = c_{n,1}\mathbf{x}_{n,1} + \cdots c_{n,N}\mathbf{x}_{n,N}, \qquad c_{n,j} \geq 0, \quad \mathbf{x}_{n,j} \in \mathcal{B}(K), \quad 1 \leq j \leq N,$$

and $\mathbf{y}_n \to \mathbf{y}_0$ as $n \to \infty$. We need to show $\mathbf{y}_0 \in \operatorname{cone}(K)$. Let $\mathbf{z}$ be a vector in $\mathcal{B}(K)$, such that $\langle \mathbf{x}, \mathbf{z} \rangle > 0$ for all $x \in \mathcal{U}(K)$. We claim that $\inf_{\mathbf{x} \in \mathcal{B}(K)} \langle \mathbf{x}, \mathbf{z} \rangle > 0$. Indeed, assume otherwise, then there is a sequence $\mathbf{x}_n \in \mathcal{B}(K)$ such that $\langle \mathbf{x}_n, \mathbf{z} \rangle \to 0$. By compactness, there is a convergence subsequence $\mathbf{x}_{n_k} \to \mathbf{x}_0 \in \mathcal{B}(K)$ and $\langle \mathbf{x}_0, \mathbf{z} \rangle = 0$, which is a contradiction. Writing

$$\langle \mathbf{y}_n, \mathbf{z} \rangle = c_{n,1}\langle \mathbf{x}_{n,1}, \mathbf{z} \rangle + \cdots + c_{n,N}\langle \mathbf{x}_{n,N}, \mathbf{z} \rangle \geq c_{n,j}\langle \mathbf{x}_{n,j}, \mathbf{z} \rangle \geq c_{n,j} \inf_{\mathbf{x} \in \mathcal{B}(K)} \langle \mathbf{x}, \mathbf{z} \rangle,$$

we conclude that $0 \leq c_{n,j} \leq \langle \mathbf{y}_n, \mathbf{z} \rangle / \inf_{\mathbf{x} \in \mathcal{B}(K)} \langle \mathbf{x}, \mathbf{z} \rangle$. Since $\mathbf{y}_n$ converges, it is bounded, so $\langle \mathbf{y}_n, \mathbf{z} \rangle$ is bounded by the Cauchy inequality. Thus, $\{c_{n,j}\}_n$ is bounded for each j. As a consequence, there is a sequence of integers n_k such that $c_{n_k,j} \to c_j$ and $\mathbf{x}_{n_k,j} \to \mathbf{x}_j$, and $\mathbf{y}_0 = c_1\mathbf{x}_1 + \cdots c_N\mathbf{x}_N$. This shows $\mathbf{y}_0 \in \operatorname{cone}(K)$ and completes the proof. □

Lemma 2.3.4. *If a vector* $\mathbf{z} \in \mathcal{U}(K)$ *satisfies* $\langle \mathbf{y}, \mathbf{z} \rangle \geq 0$ *for all* $\mathbf{y} \in \mathcal{U}(K)$ *that satisfies* $\langle \mathbf{y}, \mathrm{Y}(\mathbf{x}) \rangle \geq 0$ *for* $\mathbf{x} \in K$, *then* $\mathbf{z} \in \operatorname{cone}(K)$.

Proof Assume $\mathbf{z} \in \mathcal{U}(K) \setminus \operatorname{cone}(K)$. Since $\operatorname{cone}(K)$ is closed and convex, by Minkowski's separation theorem, there exists a hyperplane H in $\mathbb{R}^{N-1}$ that separates $\operatorname{cone}(K)$ and $\mathbf{z}$, and we can assume H passes through the origin since the original is the vertex of the cone. Let $\mathbf{y}$ be a point on the same side of H as $\operatorname{cone}(K)$. Let $\ell_{\mathbf{y}}$ be the line that passes through $\mathbf{y}$ and is perpendicular to H. Then,

$$\langle \mathbf{z}, \mathbf{y} \rangle < 0 \quad \text{and} \quad \langle \mathrm{Y}(\mathbf{x}), \mathbf{y} \rangle \geq 0, \quad \forall \mathrm{Y}(\mathbf{x}) \in \operatorname{cone}(K),$$

since the angle between $\mathbf{y}$ and $\mathbf{z}$ is $>90^\circ$ and, for any $\mathbf{x} \in \mathrm{cone}(K)$, the angle between $\mathbf{y}$ and $\mathbf{x}$ is $\le 90^\circ$. This, however, contradicts the assumption on $\mathbf{z}$. □

We are ready to prove the existence of a positive cubature rule on the compact set K.

Proof of Theorem 2.3.1 for compact K Let $\mathbf{z} = \int_K \mathrm{Y}(\mathbf{x}) \mathrm{d}\mu(\mathbf{x})$. Then, $\mathbf{z} \in \mathbb{R}^M$, and the components of $\mathbf{z}$ are the moments of the measure μ of degree up to m. It is clear that $\mathbf{z} \in \mathcal{U}(K)$. If $\mathbf{y} \in \mathcal{U}(K)$ satisfies $\langle \mathbf{y}, \mathrm{Y}(\mathbf{x}) \rangle \ge 0$ for all $\mathbf{x} \in K$, then

$$\langle \mathbf{y}, \mathbf{z} \rangle = \int_K \langle \mathbf{y}, \mathrm{Y}(\mathbf{x}) \rangle \mathrm{d}\mu(\mathbf{x}) \ge 0.$$

Hence, by Lemma 2.3.4, $\mathbf{z} \in \mathrm{cone}(K)$, which means that $\mathbf{z}$ can be written as a linear combination, with nonnegative coefficients, of $\mathrm{Y}(\mathbf{x}_1), \ldots, \mathrm{Y}(\mathbf{x}_N)$ for some $\mathbf{x}_1, \ldots \mathbf{x}_N$ in K. This implies, in particular, the identity

$$\int_K P(\mathbf{x}) \mathrm{d}\mu(\mathbf{x}) = c_1 P(\mathbf{x}_1) + \cdots + c_N P(\mathbf{x}_N), \qquad c_j \ge 0, \tag{2.7}$$

for all polynomials given by $P(\mathbf{x}) = \langle \mathbf{p}, \mathrm{Y}(\mathbf{x}) \rangle$ with $\mathbf{p} \in \mathcal{U}(K)$. These polynomials are of degree at most m. We now show that this identity holds for all polynomials in Π_m^d, which gives our positive cubature after ignoring terms with zero coefficients. Let $f \in \Pi_m^d$. We can write $f = P + Q$ on K, where $P = \langle \mathbf{p}, \mathrm{Y}(\mathbf{x}) \rangle$ with $\mathbf{p} \in \mathcal{U}(K)$ and $Q = \langle \mathbf{q}, \mathrm{Y}(\mathbf{x}) \rangle$ with $\mathbf{q} \perp \mathcal{U}(K)$. It follows that $f(\mathbf{x}) = \langle P, \mathrm{Y}(\mathbf{x}) \rangle = P(\mathbf{x})$ for $\mathbf{x} \in K$. Furthermore, since $\mathbf{x}_j \in K$, we also have $f(\mathbf{x}_j) = P(\mathbf{x}_j)$. Thus, (2.7) holds for all polynomials in Π_m^d. This completes the proof. □

2.4 Product-Type Cubature Rules on Regular Domains

On some regular domains, cubature rules can be constructed via quadrature rules of one variable. We give several examples of what we shall call the *cubature rule of product type*. In each example, the integral on the domain can be written as a d-layer of integrals of one variable so that a cubature rule can be derived as a product of d quadrature rules, each for an integral of one variable in the d-layer. To have maximal precision, a preferred choice of quadrature rule is the Gauss quadrature. If w is a weight function on $\mathbb{R}$, we denote the Gauss quadrature rule for w by

$$\int_{\mathbb{R}} f(t) w(t) \mathrm{d}t = \sum_{k=1}^{n} \lambda_k(w) f(t_k(w)), \qquad \forall f \in \Pi_{2n-1},$$

to emphasize the dependence of the nodes and the weight on w. A cubature rule of product type of degree $2n-1$ uses $N = n^d$ number of nodes, which is excessive and way above the minimal number of nodes. Nevertheless, they are among a handful of cubature families that can be constructed explicitly for large n.

2.4.1 Product Cubature Rules on a Cubic Domain

A cubic domain is a product of intervals,

$$\Omega = [a_1, b_1] \times \cdots \times [a_d, b_d] \subset \mathbb{R}^d,$$

where $-\infty \le a_i < b_i \le \infty$. In such a domain, we consider the product weight function

$$W(\mathbf{x}) = w_1(x_1) w_2(x_2) \cdots w_d(x_d), \qquad x_i \in (a_i, b_i).$$

Then, the product Gauss cubature rule on the cubic domain is given by

$$\int_\Omega f(\mathbf{x}) W(\mathbf{x}) \mathrm{d}\mathbf{x} = \sum_{k_1=1}^{n} \cdots \sum_{k_d=1}^{n} \prod_{i=1}^{d} \lambda_{k_i}(w_i) f\big(t_{k_1}(w_1), \ldots, t_{k_d}(w_d)\big). \tag{2.8}$$

This cubature rule is exact for the polynomial space $\Pi_{2n-1} \times \cdots \times \Pi_{2n-1}$, which consists of polynomials that are of degree at most $2n-1$ in each variable. For our definition of precision in terms of total degree, the product cubature rule is of degree $2n-1$ but uses n^d number of nodes. In contrast, if the Gauss cubature rule of degree $2n-1$ were to exist, its number of points would be

$$\binom{n+d-1}{d} = \frac{n^d}{d!}\big(1 + O(n^{-1})\big).$$

Thus, the number of nodes of a product Gauss cubature rule is way more than the lower bound in (2.2). Moreover, it is even more than the rough upper bound given in Theorem 2.3.1 for large d.

2.4.2 Product-Type Cubature Rules on the Simplex

Since an affine transformation preserves the degree of polynomials and polynomial subspaces, we work with the simplex Δ^d defined by

$$\Delta^d = \left\{ \mathbf{x} \in \mathbb{R}^d : x_1 \ge 0, \ldots, x_d \ge 0, \ |\mathbf{x}| := x_1 + \cdots + x_d \le 1 \right\}. \tag{2.9}$$

For $\boldsymbol{\kappa} = (\kappa_1, \ldots, \kappa_{d+1})$, the Jacobi weight function $W_{\boldsymbol{\kappa}}$ on the simplex is defined by

$$W_{\boldsymbol{\kappa}}(\mathbf{x}) = x_1^{\kappa_1} \cdots x_d^{\kappa_d} (1 - |x|)^{\kappa_{d+1}}, \quad \mathbf{x} \in \Delta^d, \quad \kappa_j > -1, \quad 1 \le j \le d+1. \tag{2.10}$$

The normalization constant for the Jacobi weight is denoted by

$$\mathbf{c}_{\boldsymbol{\kappa}} = \left[\int_{\Delta^d} W_{\boldsymbol{\kappa}}(\mathbf{x})\mathrm{d}\mathbf{x}\right]^{-1} = \frac{\Gamma(|\boldsymbol{\kappa}|+d+1)}{\Gamma(\kappa_1+1)\ldots\Gamma(\kappa_{d+1}+1)}.$$

We derive cubature rules of the product type for the integral with the Jacobi weight on the simplex by using the Gaussian quadrature rule for the Jacobi weight on $[0,1]$,

$$u_{\alpha,\beta}(s) = s^\alpha(1-s)^\beta, \qquad \alpha,\beta > -1, \quad 0 < s < 1.$$

Let $-1 < t_{1,n} < t_{2,n} < \cdots < t_{n,n} < 1$ be the zeros of polynomial $P_n^{(\alpha,\beta)}(t)$ and set $s_{j,n}^{(\alpha,\beta)} = \left(1 + t_{j,n}^{(\alpha,\beta)}\right)/2$. Then, the Gauss quadrature rule for $u_{\alpha,\beta}$ is given by

$$c_{\alpha,\beta}\int_0^1 f(t)s^\alpha(1-s)^\beta \mathrm{d}s = \sum_{j=1}^n \lambda_{j,n}^{(\alpha,\beta)} f\left(s_{j,n}^{(\alpha,\beta)}\right), \qquad \forall f \in \Pi_{2n-1}. \tag{2.11}$$

where $c_{\alpha,\beta}$ is the normalization constant given in (1.12), and the weight $\lambda_{j,n}^{(\alpha,\beta)}$ can be given explicitly by

$$\lambda_{j,n}^{(\alpha,\beta)} = \frac{(\alpha+1)_n(\beta+1)_n(n+\alpha+\beta+1)}{(\alpha+\beta+2)_n\left(1-(t_{j,n}^{(\alpha,\beta)})^2\right)\left[P_n^{(\alpha,\beta)\prime}\left(t_{j,n}^{(\alpha,\beta)}\right)\right]^2}. \tag{2.12}$$

Proposition 2.4.1. *For $d \geq 2$, a product Gauss cubature rule of degree $2n-1$ for $W_{\boldsymbol{\kappa}}$ on the simplex Δ^d is given by*

$$\mathbf{c}_{\boldsymbol{\kappa}}\int_{\Delta^d} f(\mathbf{x})W_{\boldsymbol{\kappa}}(\mathbf{x})\mathrm{d}\mathbf{x} = \sum_{j_1=1}^n \cdots \sum_{j_d=1}^n \lambda_{j_1,n}^{(\kappa_1,\beta_1)}\ldots\lambda_{j_d,n}^{(\kappa_d,\beta_d)} f(\mathbf{x}_{\mathbf{j},n}), \tag{2.13}$$

where $\beta_j = \kappa_{j+1} + \cdots + \kappa_{d+1} + d - j$ for $1 \leq j \leq d$ so that $\beta_d = \kappa_{d+1}$, and

$$\mathbf{x}_{\mathbf{j},n} = \left(s_{j_1,n}^{(\kappa_1,\beta_1)}, \left(1-s_{j_1,n}^{(\kappa_1,\beta_1)}\right)s_{j_2,n}^{(\kappa_2,\beta_2)}, \ldots, \left(1-s_{j_1,n}^{(\kappa_1,\beta_1)}\right)\cdots\left(1-s_{j_{d-1},n}^{(\kappa_{d-1},\beta_d)}\right)s_{j_d,n}^{(\kappa_d,\beta_d)}\right).$$

Proof If $\mathbf{x}' \in \Delta^{d-1}$, then $\mathbf{x} = (x_1, (1-x_1)\mathbf{x}') \in \Delta^d$. We start with the identity

$$\int_{\Delta^d} f(\mathbf{x})\mathrm{d}\mathbf{x} = \int_0^1 \int_{\Delta^{d-1}} f(x_1,(1-x_1)\mathbf{x}')d\mathbf{x}'(1-x_1)^{d-1}\mathrm{d}x_1.$$

Setting $\boldsymbol{\kappa}_1 = (\kappa_2,\ldots,\kappa_{d+1})$, so that $|\boldsymbol{\kappa}_1| = |\boldsymbol{\kappa}| - \kappa_1$, it is easy to verify

$$W_{\boldsymbol{\kappa}}\left(x_1,(1-x_1)\mathbf{x}'\right) = W_{\boldsymbol{\kappa}_1}(\mathbf{x}')x_1^{\kappa_1}(1-x_1)^{|\boldsymbol{\kappa}_1|} \quad \text{and} \quad \mathbf{c}_{\boldsymbol{\kappa}} = c_{\kappa_1,\beta_1}\mathbf{c}_{\boldsymbol{\kappa}_1},$$

so that we can write the integral over the simplex as

$$\mathbf{c}_{\boldsymbol{\kappa}}\int_{\Delta^d} f(\mathbf{x})W_{\boldsymbol{\kappa}}(\mathbf{x})\mathrm{d}\mathbf{x} = c_{\kappa_1,|\boldsymbol{\kappa}_1|+d-1}\int_0^1\left[\mathbf{c}_{\boldsymbol{\kappa}_1}\int_{\Delta^{d-1}} f\left(x_1,(1-x_1)\mathbf{x}'\right)W_{\boldsymbol{\kappa}'}(\mathbf{x}')\mathrm{d}\mathbf{x}'\right] \times x_1^{\kappa_1}(1-x_1)^{|\boldsymbol{\kappa}_1|+d-1}\mathrm{d}x_1. \tag{2.14}$$

For $\boldsymbol{\alpha} \in \mathbb{N}_0^d$, write $\boldsymbol{\alpha}_1 = (\alpha_2, \dots, \alpha_d)$; then, the monomial $m_\alpha(\mathbf{x}) = \mathbf{x}^\alpha$ satisfies

$$m_{\boldsymbol{\alpha}}(x_1, (1-x_1)\mathbf{x}') = x_1^{\alpha_1}(1-x_1)^{|\boldsymbol{\alpha}_1|} m_{\boldsymbol{\alpha}_1}(\mathbf{x}').$$

In particular, if $|\alpha| = n$, then $m_{\boldsymbol{\alpha}}$ is of degree n and $m_{\boldsymbol{\alpha}_1}$ is of degree $n - \alpha_1 \le n$. As a consequence of these considerations, we can deduce the cubature of product type by induction and applying the Gauss quadrature rule in each step. □

The cubature rule (2.13) is positive, and all its nodes are inside Δ^d. However, its nodes are not evenly distributed in the simplex. There are more nodes toward the hyperplane $x_1 = 0$ and an accumulation at the origin. The weight function $W_{\boldsymbol{\kappa}}$ is invariant under simultaneous permutation of the elements of $\boldsymbol{\kappa} = (\kappa_1, \dots, \kappa_d)$ and $(x_1, \dots, x_d, 1 - |x|)$. Hence, any permutation of the $\boldsymbol{\kappa}$ and $\mathbf{x}_{\mathbf{j},n}$ gives another product type Gauss cubature rule.

2.4.3 Product-Type Cubature Rules on the Unit Ball

On the unit ball $\mathbb{B}^d = \{\mathbf{x} \in \mathbb{R}^d : \|\mathbf{x}\| \le 1\}$, the classical weight function is an analog of the Gegenbauer weight defined by

$$W_\mu(\mathbf{x}) = (1 - \|\mathbf{x}\|^2)^{\mu - \frac{1}{2}}, \qquad \mu > -\tfrac{1}{2}.$$

Its normalization constant on the ball is given by

$$\mathbf{c}_{\mu,d} = \left[\int_{\mathbb{B}^d} W_\mu(\mathbf{x}) \mathrm{d}\mathbf{x}\right]^{-1} = \frac{2}{\omega_d} \frac{\Gamma(\mu + \frac{d+1}{2})}{\Gamma(\frac{d}{2})\Gamma(\mu + \frac{1}{2})} = \frac{\Gamma(\mu + \frac{d+1}{2})}{\pi^{d/2}\Gamma(\mu + \frac{1}{2})},$$

where ω_d denotes the surface area of the unit sphere $\mathbb{S}^{d-1}$. To derive a product-type cubature rule for W_μ, we use the Gauss quadrature for the Gegenbauer weight $w_\mu(t) = (1-t^2)^{\mu - \frac{1}{2}}$ on $[-1, 1]$. Let $t_1^{(\mu)} < t_2^{(\mu)} < \cdots < t_n^{(\mu)}$ be zeros of the Gegenbauer polynomial C_n^μ. Then, the Gauss quadrature for w_μ is given by

$$c_\mu \int_{-1}^{1} f(t) w_\mu(t) \mathrm{d}t = \sum_{j=1}^{n} \lambda_{j,n}^{(\mu)} f\left(t_{j,n}^{(\mu)}\right), \quad \forall f \in \Pi_{2n-1}, \tag{2.15}$$

where $\lambda_{k,n}^{(\mu)}$ is given by (2.12) with $\alpha = \beta = \lambda - \frac{1}{2}$, which simplifies to

$$\lambda_{j,n}^{(\mu)} = \frac{(2\mu + 1)_n}{(n + 2\mu)\left(1 - (t_{j,n}^{(\mu)})^2\right)\left[C_{n-1}^{\mu+1}(t_{j,n}^{(\mu)})\right]^2}.$$

Since w_μ is even, the zeros of C_n^λ are symmetric and satisfy $t_{j,n}^{(\mu)} = t_{n-j,n}^{(\mu)}$ for $1 \le j \le n$.

Proposition 2.4.2. *For $d \ge 2$, a product Gauss cubature rule of degree $2n-1$ for W_μ on the unit ball $\mathbb{B}^d$ is given by*

$$\mathbf{c}_{\mu,d} \int_{\mathbb{B}^d} f(x) W_\mu(\mathbf{x}) d\mathbf{x} = \sum_{j_1=1}^{n} \cdots \sum_{j_d=1}^{n} \lambda_{j_1,n}^{(\mu_1)} \cdots \lambda_{j_d,n}^{(\mu_d)} f(\mathbf{x}_{\mathbf{j},n}), \tag{2.16}$$

where $\mu_j = \mu + \frac{d-j}{2}$ for $1 \le j \le d$, so that $\mu_d = \mu$, and

$$\mathbf{x}_{\mathbf{j},n} = \left(t_{j_1,n}^{(\mu_1)}, \sqrt{1-\left(t_{j_1,n}^{(\mu_1)}\right)^2} t_{j_2,n}^{(\mu_2)}, \ldots, \sqrt{1-\left(t_{j_1,n}^{(\mu_1)}\right)^2} \cdots \sqrt{1-\left(t_{j_{d-1},n}^{(\mu_{d-1})}\right)^2} t_{j_d,n}^{(\mu_d)} \right).$$

Proof If $\mathbf{x}' \in \mathbb{B}^{d-1}$, then $\left(x_1, \sqrt{1-x_1^2}\mathbf{x}'\right) \in \mathbb{B}^d$. We use the identity

$$\int_{\mathbb{B}^d} f(\mathbf{x}) W(\mathbf{x}) d\mathbf{x} = \int_{-1}^{1} \int_{\mathbb{B}^{d-1}} f\left(x_1, \sqrt{1-x_1^2}\mathbf{x}'\right) (1-x_1^2)^{\frac{d-1}{2}} d\mathbf{x}' dx_1, \tag{2.17}$$

the relation $\mathbf{c}_{\mu,\mathrm{d}} = c_{\mu+\frac{d-1}{2}} \mathbf{c}_{\mu,d-1}$, where c_μ is given in (1.17), as well as

$$W_\mu\left(x_1, \sqrt{1-x_1^2}\mathbf{x}'\right) = (1-x_1^2)^{\mu-\frac{1}{2}} W_\mu(\mathbf{x}') = w_\mu(x_1) W_\mu(\mathbf{x}')$$

to write the integral over $\mathbb{B}^d$ as, with $\mu_1 = \mu + \frac{d-1}{2}$,

$$\begin{aligned} &\mathbf{c}_{\mu,d} \int_{\mathbb{B}^d} f(\mathbf{x}) W_\mu(\mathbf{x}) d\mathbf{x} \\ &\quad = c_{\mu_1} \int_{-1}^{1} \left[\mathbf{c}_{\mu,d-1} \int_{\mathbb{B}^{d-1}} f\left(x_1, \sqrt{1-x_1^2}\mathbf{x}'\right) d\mathbf{x}' \right] w_{\mu_1}(x_1) dx_1. \end{aligned}$$

This identity allows us to apply induction on d. Assume that (2.16) has been established for W_μ on $\mathbb{B}^{d-1}$. For $\boldsymbol{\alpha} \in \mathbb{N}_0^d$ and $|\boldsymbol{\alpha}| \le n$, the monomial $m_{\boldsymbol{\alpha}}(\mathbf{x}) = \mathbf{x}^{\boldsymbol{\alpha}}$ for $\mathbf{x} \in \mathbb{R}^d$ satisfies

$$m_{\boldsymbol{\alpha}}\left(x_1, \sqrt{1-x_1^2}\mathbf{x}'\right) = x_1^{\alpha_1} (1-x_1^2)^{\frac{|\boldsymbol{\alpha}_1|}{2}} m_{\boldsymbol{\alpha}_1}(\mathbf{x}'), \qquad \mathbf{x}' \in \mathbb{B}^{d-1},$$

where $\boldsymbol{\alpha}_1 = (\alpha_2, \ldots, \alpha_d)$. If $\boldsymbol{\alpha}_1$ has an odd component, then the cubature rule for $\mathbb{B}^{d-1}$ is zero and so is the integral over $\mathbb{B}^{d-1}$. Hence, we only need to establish (2.16) for $\mathbb{B}^d$ for those $m_{\boldsymbol{\alpha}}$ with even α_j for $2 \le j \le d$, which satisfy, in particular, that $|\boldsymbol{\alpha}_1|$ is even, so that its x_1 factor, $x_1^{\alpha_1}(1-x_1^2)^{\frac{|\boldsymbol{\alpha}_1|}{2}}$, is a polynomial of degree $\alpha_1 + |\boldsymbol{\alpha}_1| = |\boldsymbol{\alpha}|$. Thus, applying the Gauss cubature for w_{μ_1} completes the proof. □

Any permutation of the variables and the indices of the right-hand side of (2.16) gives another product-type cubature rule. These cubature rules of degree $2n-1$ use n^d nodes, but the distribution of its nodes does not reflect the symmetry of the unit ball. Another cubature rule can be derived using the spherical polar coordinates. We first consider the case $d = 2$.

For integral over the unit disk $\mathbb{B}^2$, we use polar coordinates to write

$$\int_{\mathbb{B}^2} f(x_1,x_2)\mathrm{d}x_1\mathrm{d}x_2 = \int_0^1 r\int_0^{2\pi} f(r\cos\theta, r\sin\theta)\mathrm{d}\theta\mathrm{d}r.$$

This allows us to obtain a cubature rule from the well-known quadrature rule

$$\frac{1}{2\pi}\int_0^{2\pi} f(\theta)\mathrm{d}\theta = \frac{1}{n}\sum_{j=0}^{n-1} f\left(\frac{2\pi j}{n}\right), \quad \forall f\in\Pi_n(\mathbb{S}^1), \tag{2.18}$$

for trigonometric polynomials in

$$\Pi_n(\mathbb{S}^1) := \operatorname{span}\{1,\cos\theta,\sin\theta,\ldots,\cos n\theta,\sin n\theta\}.$$

Proposition 2.4.3. *Let n be a positive even integer and let $s_k = s_{k,\frac{n}{2}}^{(0,\mu-\frac12)}$, $1\le k\le\frac{n}{2}$, be the zeros of the Jacobi polynomial $P_{\frac{n}{2}}^{(0,\mu-\frac12)}(2s-1)$. Then,*

$$\mathbf{c}_2\int_{\mathbb{B}^2} f(\mathbf{x})W_\mu(\mathbf{x})\mathrm{d}\mathbf{x} = \frac{1}{2n-1}\sum_{j=0}^{2n-2}\sum_{k=1}^{n/2}\lambda_{k,\frac{n}{2}}^{(0,\mu-\frac12)} f\left(\sqrt{s_k}\cos\tfrac{2\pi j}{2n-1}, \sqrt{s_k}\sin\tfrac{2\pi j}{2n-1}\right)$$

for all $f\in\Pi_{2n-1}^2$, where $\mathbf{c}_2 = \frac{2\mu+1}{2\pi}$ and $\lambda_{j,\frac{n}{2}}^{(0,\mu-\frac12)}$ is given by (2.12).

Proof In polar coordinates $x_1 = r\cos\theta$ and $x_2 = r\sin\theta$, the space $\mathcal{P}_k^2$ of homogeneous polynomials of degree k has a basis

$$\left\{r^k\cos(k-2j)\theta : 0\le 2j\le k\right\}\cup\left\{r^k\sin(k-2j)\theta : 0\le 2j\le k-1\right\}.$$

The union of the basis for $0\le k\le n$ gives a basis of Π_n^2. If $f\in\Pi_{2n-1}^2$, we obtain

$$\frac{1}{2\pi}\int_{\mathbb{B}^2} f(\mathbf{x})W_\mu(\mathbf{x})\mathrm{d}\mathbf{x} = \frac{1}{2n-1}\sum_{j=0}^{2n-2}\int_0^1 f\left(r\cos\tfrac{2\pi j}{2n-1}, r\sin\tfrac{2\pi j}{2n-1}\right) r(1-r^2)^{\mu-\frac12}\mathrm{d}r,$$

and both sides become zero for $f(\mathbf{x}) = r^k\cos(k-2j)$ and $r^k\sin(k-2j)$ if $k\ne 2j$. Thus, we only need to discretize the integral for $\mathrm{d}r$ in the right-hand side for $g_j(r^2) = r^{2j}$ for $0\le 2j\le 2n-2$ or $0\le j\le n-1$. Changing variable $v = r^2$ shows that

$$\int_0^1 g_j(r^2)r(1-r^2)^{\mu-\frac12}\mathrm{d}r = \frac12\int_0^1 g_j(v)(1-v)^{\mu-\frac12}\mathrm{d}v,$$

which we discretize by the Gauss quadrature rule of degree $2\frac{n}{2}-1$ for the Jacobi weight function $(1-v)^{\mu-\frac12}$ on $[0,1]$. This completes the proof. □

This cubature rule on the disk of degree $2n-1$ has $n^2-\frac{n}{2}$ nodes, slightly less than n^2 but still excessive. The distribution of the nodes, however, is gravitated toward the center of the disk undesirably. It is possible to find cubature rules with nodes better distributed, but they are known only for small or moderate n.

There is an analog construction for the cubature rules on the unit ball $\mathbb{B}^d$ for $d > 2$ that, based on the spherical polar coordinates, use cubature rules on the spheres, which we discuss in the Subsection 2.4.4.

2.4.4 Cubature Rules on the Unit Sphere and the Unit Ball

For the unit sphere $\mathbb{S}^{d-1}$, we consider $\mathcal{I}(f) = \int_{\mathbb{S}^{d-1}} f(\xi)\mathrm{d}\sigma(\xi)$, where $d\sigma$ denotes the surface measure. Let $\Pi_n(\mathbb{S}^{d-1})$ denote the space of polynomials of d-variables restricted on the unit sphere $\mathbb{S}^{d-1}$. Since any polynomial of the form $(1-\|\mathbf{x}\|^2)P(\mathbf{x})$ will be zero on $\mathbb{S}^{d-1}$, the dimension of the space is

$$\dim \Pi_n(\mathbb{S}^{d-1}) = \dim \mathcal{P}_n^d + \dim \mathcal{P}_{n-1}^d = \binom{n+d-1}{n} + \binom{n+d-2}{n}. \qquad (2.19)$$

A cubature rule $\mathsf{C}(f)$ for $\mathcal{I}(f)$ has all its nodes on $\mathbb{S}^{d-1}$, and it is of degree m if it is exact for all polynomials in $\Pi_m(\mathbb{S}^{d-1})$. For $d = 2$, the cubature rule in (2.18) is akin to the Gauss quadrature since it uses n points and is of degree n and $\dim \Pi_n(\mathbb{S}^{d-1}) = 2n+1$.

For $d > 2$, we can derive a cubature rule on the sphere by using the identity

$$\int_{\mathbb{S}^{d-1}} f(\xi)\mathrm{d}\sigma = \int_{\mathbb{B}^{d-1}} \left[f\left(\mathbf{x}, \sqrt{1-\|\mathbf{x}\|^2}\right) + f\left(\mathbf{x}, -\sqrt{1-\|\mathbf{x}\|^2}\right)\right] \frac{\mathrm{d}\mathbf{x}}{\sqrt{1-\|\mathbf{x}\|^2}}. \qquad (2.20)$$

This identity follows from making a change of variable $\xi = (\sqrt{1-t^2}\xi', t)$, where $\xi' \in \mathbb{S}^{d-2}$ and $-1 \le t \le 1$, so that $\mathrm{d}\sigma_d(\xi) = (1-t^2)^{\frac{d-3}{2}}\mathrm{d}t\,\mathrm{d}\sigma_{d-1}(\xi')$ in the left-hand side integral, followed by writing the integral over $[-1,1]$ as a sum of two integrals over $[0,1]$ and changing variables $t = \sqrt{1-r^2}$.

Since $f(\mathbf{x},t) + f(\mathbf{x},-t)$ is even in the variable t, it follows that if f is a polynomial of degree n, then so is $f\left(\mathbf{x}, \sqrt{1-\|\mathbf{x}\|^2}\right) + f\left(\mathbf{x}, -\sqrt{1-\|\mathbf{x}\|^2}\right)$. Hence, applying the cubature rule (2.16) with $\mu = 0$, and d replaced by $d-1$, to the right-hand side of (2.20) leads to a cubature rule on the unit sphere $\mathbb{S}^{d-1}$. Let us write the zeros of C_n^μ as

$$t_{j,n}^{(\mu)} = \cos\theta_{j,n}^{(\mu)}, \quad \theta_{j,n} \in (0,\pi), \quad 1 \le j \le n.$$

If $\mu = 0$, the Gegenbauer polynomial degenerates to the Chebyshev polynomials of the first kind so that $\theta_{j,n}^{(0)} = \frac{(j-\frac{1}{2})\pi}{n}$. In particular, it follows that

$$\left\{t_{j,n}^{(0)} : 1 \le j \le n\right\} \bigcup \left\{-t_{l,n}^{(0)} : 1 \le j \le n\right\} = \left\{\cos\tfrac{(2j-1)\pi}{2n} : 1 \le j \le 2n\right\}.$$

Moreover, $\lambda_{j,n}^{(0)} = \frac{\pi}{n}$. Comparing with the nodes of the cubature rule (2.16) with

d replaced by $d-1$, we define

$$\xi_{\mathbf{j},n} = \left(\cos\theta_{j_1,n}^{\left(\frac{d-2}{2}\right)}, \sin\theta_{j_1,n}^{\left(\frac{d-2}{2}\right)}\cos\theta_{j_2,n}^{\left(\frac{d-3}{2}\right)}, \ldots, \sin\theta_{j_1,n}^{\left(\frac{d-2}{2}\right)}\cdots\sin\theta_{j_{d-2},n}^{\left(\frac{1}{2}\right)}\cos\frac{(2j_1-1)\pi}{2n}\right).$$

Theorem 2.4.4. *For $n \in \mathbb{N}$, the cubature formula*

$$\frac{1}{\omega_d}\int_{\mathbb{S}^{d-1}} f(x)d\sigma = \frac{\pi}{n}\sum_{j_1=1}^{2n}\sum_{j_2=1}^{n}\cdots\sum_{j_{d-1}=1}^{n}\prod_{i=2}^{d-1}\lambda_{i,n}^{\left(\frac{i-1}{2}\right)} f(\xi_{\mathbf{j},n}) \tag{2.21}$$

is exact for all $f \in \Pi_{2n-1}(\mathbb{S}^{d-1})$, and it uses $2n^{d-1}$ nodes.

This cubature rule can also be derived from writing the integral over $\mathbb{S}^{d-1}$ as an integral over $[0,\pi]$ and $\mathbb{S}^{d-2}$ and use induction.

For the integral with a radial weight function $W(\mathbf{x}) = w(\|\mathbf{x}\|)$ on a domain $\mathbb{B}_R^d = \{\mathbf{x} : \|\mathbf{x}\| \le R\}$, where $0 < R \le +\infty$, we can use the spherical polar coordinates, $\mathbf{x} = r\xi$, $r > 0$ and $\xi \in \mathbb{S}^{d-1}$, to write

$$\int_{\mathbb{B}_R^d} f(\mathbf{x})W(\mathbf{x})d\mathbf{x} = \int_0^R r^{d-1}\int_{\mathbb{S}^{d-1}} f(r\xi)d\sigma(\xi)w(r)dr.$$

As in the case of $d = 2$, we consider only the case of n being an even integer. Applying the cubature rule of (2.21) on the sphere and then applying a quadrature rule for $g(r^2)$ for polynomials g of degree $n-1$ for $r^{d-1}w(r)$ on $[0,R]$, which can be deduced from the Gauss quadrature rule for $r^{d-2}2w(\sqrt{r})$ on $[0,\sqrt{R}]$,

$$\frac{1}{2}\int_0^{\sqrt{R}} g(r)r^{d-2}2w\left(\sqrt{r}\right)dr = \sum_{k=1}^{n/2}\lambda_k(w)g(t_k(w)), \quad \deg g \le n-1,$$

as in the case of $d = 2$.

Theorem 2.4.5. *For an even positive integer n, the cubature formula*

$$\frac{1}{\omega_d}\int_{\mathbb{B}_R^d} f(\mathbf{x})w(\|\mathbf{x}\|)d\mathbf{x} = \frac{\pi}{n}\sum_{k=1}^{n/2}\lambda_k(w)\sum_{j_1=1}^{2n}\sum_{j_2=1}^{n}\cdots\sum_{j_{d-1}=1}^{n}\prod_{i=2}^{d-1}\lambda_{i,n}^{\left(\frac{i-1}{2}\right)} f\left(\sqrt{t_k(w)}\xi_{\mathbf{j},n}\right),$$

is of degree $2n-1$, and it uses n^d nodes.

The nodes of this cubature rule are distributed on $n/2$ spheres that are clustered toward the origin so that the nodes are accumulated toward the origin.

2.5 Cubature Rules Invariant under a Finite Group

If a domain Ω is invariant under a finite group, we can consider cubature rules invariant under the same group. Such cubature rules have much simpler structures.

Let G be a finite subgroup of the orthogonal group $O(d)$ acting on $\mathbb{R}^d$. For $\mathbf{x} \in \mathbb{R}^d$ and $\tau \in G$, we denote by $\mathbf{x}\tau$ the action of τ on $\mathbf{x}$. A domain Ω in $\mathbb{R}^d$ is called *invariant* under G if $\mathbf{x} \in \Omega$ implies $\mathbf{x}\tau \in \Omega$ for every $\tau \in G$. For a function f defined on $\mathbb{R}^d$ and $\tau \in G$, we define $\tau f(\mathbf{x}) := f(\mathbf{x}\tau)$. The function f is called invariant under G if $\tau f = f$ for all $\tau \in G$. If both the domain Ω and the weight function W are invariant under G, then so is the integral $\mathcal{I}(f) = \int_\Omega f(\mathbf{x})W(\mathbf{x})d\mathbf{x}$ since G is a subgroup of $O(d)$.

Definition 2.5.1. Let Ω and the weight function W be invariant under the subgroup G of $O(d)$. A cubature rule $\mathsf{C}(f)$ on a domain Ω is invariant under the group G if $\mathsf{C}(\tau f) = \mathsf{C}(f)$ for all $\tau \in G$.

Let $\mathbf{x}G$ be the orbit of $\mathbf{x} \in \mathbb{R}^d$, $\mathbf{x}G := \{\mathbf{x}\tau : \tau \in G\}$, and let $|\mathbf{x}G|$ denote the cardinality of the orbit $\mathbf{x}G$. An invariant cubature rule $\mathsf{C}(f)$ is of the form

$$\mathsf{C}(f) = \sum_{k=1}^{M} \lambda_k \frac{1}{|\mathbf{x}_k G|} \sum_{\mathbf{y} \in \mathbf{x}_k G} f(\mathbf{y}), \qquad \mathbf{x}_k \in \mathbb{R}^d, \quad \lambda_k \in \mathbb{R}.$$

The number of nodes of $\mathsf{C}(f)$ is equal to $\sum_{k=1}^{M} |\mathbf{x}_k G|$. We only need to consider invariant polynomials to establish such a cubature rule. Define

$$\Pi_n^G = \left\{ f \in \Pi_n^d : f \text{ invariant under G} \right\}.$$

Theorem 2.5.2 (Sobolev theorem). *Let Ω and the integral $\mathcal{I}$ over Ω be invariant under the group G. If the cubature rule $\mathsf{C}(f)$ is invariant under G, then $\mathsf{C}(f)$ is of degree m if and only if $\mathsf{C}(f) = \mathcal{I}(f)$ for $f \in \Pi_m^G$.*

Proof We only need to prove the sufficiency since Π_m^G is a subset of Π_m^d. Assume that $\mathsf{C}(f)$ is invariant and $\mathsf{C}(f) = \mathcal{I}(f)$ for all $f \in \Pi_m^G$. Let $|G|$ denote the order of the group G and let

$$f_G = \frac{1}{|G|} \sum_{\tau \in G} \tau f, \qquad f \in \Pi_m^d.$$

It follows readily that f_G is an element of Π_m^G. For $f \in \Pi_n^d$, we obtain

$$\begin{aligned} \mathsf{C}(f) &= \mathsf{C}(f_G) = \int_\Omega f_G(\mathbf{x})W(\mathbf{x})d\mathbf{x} \\ &= \frac{1}{|G|} \sum_{\tau \in G} \int_\Omega f(\mathbf{x})W\left(\mathbf{x}\tau^{-1}\right)d\mathbf{x} = \int_\Omega f(\mathbf{x})W(\mathbf{x})d\mathbf{x} = \mathcal{I}(f), \end{aligned}$$

where the second line follows from the invariance of Ω and W under G. □

This theorem is particularly useful for finding cubature rules numerically. The existence of a cubature rule of degree m is equivalent to a solution of the system of equations

$$\sum_{k=1}^{N} c_\alpha P_\alpha(\mathbf{x}_k) = \mathcal{I}(P_\alpha), \qquad |\alpha| \le m,$$

where $\{P_\alpha : |\alpha| \le m\}$ is a basis of Π_m^d, which is the system of moment equations (2.1) if $P_\alpha(\mathbf{x}) = \mathbf{x}^\alpha$. If both $\mathcal{I}(f)$ and $\mathsf{C}(f)$ are invariant under a group G, then we only need to consider invariant polynomials and solve the system of equations for a basis of Π_m^G. The latter has a much smaller cardinality than that of Π_m. Later, we consider several examples.

A. Invariant cubature rule on $[-1,1]^d$. The group $\mathbb{Z}_2^d$ of sign changes is perhaps the simplest example. A weight function invariant under $\mathbb{Z}^2$ can be written as $W(\mathbf{x}) = w\left(x_1^2, \ldots, x_d^2\right)$. Its invariant cubature rule of degree $2n-1$ on $[-1,1]^d$ takes the form

$$\int_{[-1,1]^d} f(\mathbf{x}) W(\mathbf{x}) d\mathbf{x} = \sum_{k=1}^{M} \frac{\lambda_k}{2^{\sigma_k}} \sum_{\boldsymbol{\varepsilon} \in \{-1,1\}^d} f(\varepsilon_1 x_{k,1}, \ldots, \varepsilon_d x_{k,d}), \quad f \in \Pi_{2n-1}^d,$$

where the inner sum is over all possible sign changes and σ_k denotes the number of nonzero elements of $\mathbf{x}_k = (x_{k,1}, \ldots, x_{k,d})$, so that $2^{\sigma_k} = |\mathbf{x}_k \mathbb{Z}_2^{d+1}|$. This cubature rule has $\sum_{k=1}^{M} 2^{\sigma_k}$ nodes. To find λ_k and $\mathbf{x}_k$, we need only to consider polynomials that are even in every variable. Moreover,

$$\Pi_{2n-2}^G = \Pi_{2n-1}^G = \left\{p\left(x_1^2, \ldots, x_d^2\right) : p \in \Pi_{n-1}^d\right\},$$

so that we only need to establish the cubature rule for polynomials in Π_{2n-1}^G, which has the dimension $\dim \Pi_{2n-2}^G = \dim \Pi_{n-1}^d$.

Another group is the symmetric group $\mathcal{S}_d$ of d elements, which acts on $x \in \mathbb{R}^d$ by permuting the components of $\mathbf{x}$. Let W be a weight function invariant under $\mathcal{S}_d$. For example, $W(\mathbf{x}) = w(x_1 + \cdots + x_d)$ or $W(\mathbf{x}) = w(x_1) \cdots w(x_d)$ for a weight function w of one variable. Then, the invariant cubature rule of degree m takes the form

$$\int_{[-1,1]^d} f(\mathbf{x}) W(\mathbf{x}) d\mathbf{x} = \sum_{k=1}^{M} \frac{\lambda_k}{|\mathbf{x}_k \mathcal{S}_d|} \sum_{\mathbf{x} \in \mathbf{x}_k \mathcal{S}_d} f(\mathbf{x}), \quad f \in \Pi_m^d,$$

where the inner sum on the right-hand side is over distinct permutations of $\mathbf{x}_k \tau$ for $\tau \in \mathcal{S}_d$. This cubature rule has $N = \sum_{k=1}^{M} |\mathbf{x}_k \mathcal{S}_d|$ number of nodes. The polynomials invariant under S_d are the symmetric polynomials. Hence, to establish this cubature rule, we only need to work with symmetric polynomials.

B. Invariant cubature rule on the simplex. The simplex Δ^d is invariant under the symmetric group $\mathcal{S}_{d+1}$ of permutations, acting on $(x_1,\dots,x_d,1-|\mathbf{x}|)$ for $\mathbf{x}\in\Delta^d$. For convenience, we consider homogeneous coordinates on the simplex and, instead of Δ^d, we work with $\hat{\Delta}^{d+1}$ defined as

$$\hat{\Delta}^{d+1}=\left\{\mathbf{y}=(y_1,\dots,y_{d+1})\in\mathbb{R}^{d+1}:y_1\ge 0,\ y_{d+1}\ge 0,\ |\mathbf{y}|=1\right\},$$

with $|\mathbf{y}|=y_1+\cdots+y_{d+1}$, of which Δ^d is the special parametrization with $y_{d+1}=1-y_1-\cdots-y_d$. For $\kappa>-1$, the Jacobi weight function $W_{\boldsymbol{\kappa}}$ with $\boldsymbol{\kappa}=(\kappa,\kappa,\dots,\kappa)\in\mathbb{R}^{d+1}$ becomes

$$W_\kappa(\mathbf{y})=y_1^\kappa\cdots y_{d+1}^\kappa=(y_1\cdots y_{d+1})^\kappa,$$

which is invariant under $\mathcal{S}_{d+1}$, so is the integral $\int_{\hat{\Delta}^{d+1}}f(\mathbf{y})W_\kappa(\mathbf{y})d\mathbf{y}$. We consider the invariant cubature rule for this integral.

For $\mathbf{y}\in\mathbb{R}^{d+1}$, let $\mathbf{x}\mathcal{S}_{d+1}=\{\mathbf{y}\tau:\tau\in\mathcal{S}_{d+1}\}$ denote the orbit of $\mathbf{y}$ under $\mathcal{S}_{d+1}$ and $|\mathbf{y}\mathcal{S}_{d+1}|$ denote the cardinality of the orbit. Then, the invariant cubature rule of degree m on $\hat{\Delta}^{d+1}$ is of the form

$$\int_{\hat{\Delta}^{d+1}}f(\mathbf{y})W_{\boldsymbol{\kappa}}(\mathbf{y})d\mathbf{y}=\sum_{k=1}^{M}\frac{\lambda_k}{|\mathbf{y}_k\mathcal{S}_{d+1}|}\sum_{\mathbf{y}\in\mathbf{y}_k\mathcal{S}_{d+1}}f(\mathbf{y}),\qquad f\in\Pi_m^d. \tag{2.22}$$

The number of nodes of this cubature rule equals $\sum_{k=1}^{M}|\mathbf{y}_k\mathcal{S}_{d+1}|$, and its value depends on the locations of $\mathbf{y}_k$. For example, the vertex $\mathbf{e}_1=(1,0,\dots,0)$ has $|\mathbf{e}_1\mathcal{S}_{d+1}|=d+1$, whereas the center of the simplex $\mathbf{c}=(\frac{1}{d+1},\dots,\frac{1}{d+1})$ has $|\mathbf{c}\mathcal{S}_{d+1}|$ $=1$. These cubature rules will be discussed further in Chapter 11.

C. Invariant cubature rule on the unit ball $\mathbb{B}^d$. The unit ball of $\mathbb{R}^d$ is invariant under the rotation group $O(d)$, so that if W is a weight function invariant under a subgroup of $O(d)$, then so is the integral $\mathcal{I}(f)=\int_{\mathbb{B}^d}f(\mathbf{x})W(\mathbf{x})d\mathbf{x}$ and we can consider invariant cubature rules under the subgroup.

As an application of Theorem 2.5.2, we consider cubature rules invariant under the group $\mathbb{Z}_2^d$ on $\mathbb{B}^d$. The weight function invariant under $\mathbb{Z}_2^d$ is necessarily the form $W(x_1^2,\dots,x_d^2)$.

Theorem 2.5.3. *Let W be a weight function on the simplex Δ^d and define $W_{\mathbb{B}}(x)=W(x_1^2,\dots,x_d^2)$. Every cubature rule $\mathsf{C}(f)$ of degree n for W,*

$$\int_{\Delta^d}f(\mathbf{x})W(\mathbf{x})\frac{d\mathbf{x}}{\sqrt{x_1\cdots x_d}}=\sum_{j=1}^{M}\lambda_j f(\mathbf{x}_j),\quad f\in\Pi_n^d, \tag{2.23}$$

that has all its nodes $\mathbf{x}_j=(x_{j,1},\dots,x_{j,d})$ in Δ^d corresponds to a cubature rule *of degree* $2n+1$ *for the weight* $W^{\mathbb{B}}(\mathbf{x}) = W(x_1^2,\dots,x_d^2)$ *on the unit ball*

$$\int_{\mathbb{B}^d} f(\mathbf{x}) W^{\mathbb{B}}(\mathbf{x}) d\mathbf{x} = \sum_{j=1}^{M} \frac{\lambda_j}{2^{\sigma_j}} \sum_{\varepsilon\in\{-1,1\}^d} f\left(\varepsilon_1\sqrt{x_{j,1}},\dots,\varepsilon_d\sqrt{x_{j,d}}\right), \quad f\in\Pi_{2n+1}^d, \tag{2.24}$$

where $\sigma_j = |\mathbf{x}_j \mathbb{Z}_2^d|$, *and the correspondence is reciprocal.*

Proof By Theorem 2.5.2, the invariant cubature rule (2.24) on the unit ball holds if it works for polynomials invariant under $\mathbb{Z}_2^d$, which are polynomials of the form $P(x_1^2,\dots,x_d^2)$ for $P\in\Pi_n^d$. Changing variable $y_i = x_i^2$ for $1\le i\le d$, it follows readily that

$$\int_{\mathbb{B}^d} g(x_1^2,\dots,x_d^2) W^{\mathbb{B}}(\mathbf{x}) d\mathbf{x} = \int_{\Delta^d} g(x_1,\dots,x_d) W(\mathbf{x}) \frac{d\mathbf{x}}{\sqrt{x_1\cdots x_d}}. \tag{2.25}$$

Consequently, applying the invariant cubature (2.24) on $f(x_1^2,\dots,x_d^2)$ gives (2.23). Furthermore, if (2.23) holds, then the identity (2.25) shows that (2.24) holds for $\mathbb{Z}_2^d$ invariant polynomials of degree $2n-2$. Moreover, if f is a monomial of degree $2n-1$, then both sides of (2.24) are zero by parity. Hence, by Theorem 2.5.2, the invariant cubature rule (2.24) holds for all polynomials of degree up to $2n-1$. □

For $\boldsymbol{\kappa} = (\kappa_1,\dots,\kappa_{d+1})$ with $\kappa_i > -1$, we define the weight function on $\mathbb{B}^d$ by

$$W_{\boldsymbol{\kappa}}^{\mathbb{B}}(\mathbf{x}) = \prod_{i=1}^{d} |x_i|^{2\kappa_i+1}(1-\|\mathbf{x}\|^2)^{\kappa_{d+1}}, \quad x\in\mathbb{B}^d, \tag{2.26}$$

which is invariant under $\mathbb{Z}_2^d$. The invariant cubature rule (2.24) for $W_{\boldsymbol{\kappa}}^{\mathbb{B}}$ corresponds to the cubature rule (2.23) on the simplex for the weight function

$$\frac{W_{\boldsymbol{\kappa}}^{\mathbb{B}}(\sqrt{x_1},\dots,\sqrt{x_d})}{\sqrt{x_1\cdots x_d}} = \prod_{i=1}^{d} |x_i|^{\kappa_i}(1-|\mathbf{x}|)^{\kappa_{d+1}} = W_{\boldsymbol{\kappa}}(\mathbf{x}), \quad \mathbf{x}\in\Delta^d, \tag{2.27}$$

which is the Jacobi weight on Δ^d. In particular, the product cubature rule (2.13) of degree $2n-1$ for $W_{\mathbf{k}}$ leads to a $\mathbb{Z}_2^d$ invariant cubature rule for $W_{\boldsymbol{\kappa}}^{\mathbb{B}}$ on $\mathbb{B}^d$ of degree $4n-1$. Moreover, since the zeros of $P_n^{(\alpha,\beta)}(2s-1)$ are in the interior of $(0,1)$, all nodes of the cubature rule (2.13) are in the interior of Δ^d, and each node corresponds to 2^d nodes for (2.24). Consequently, the invariant cubature rule of degree $4n-1$ has $4n^2 = (2n)^2$ number of nodes, which is the same as the number of nodes of the product-type cubature rule of degree $4n-1 = 2(2n)-1$. This is, however, not surprising since the proof of the product-type cubature rule (2.16) for W_μ can be extended to the weight function $W_{\boldsymbol{\kappa}}^{\mathbb{B}}$.

D. Relation between cubature rules on $\mathbb{B}^d$ and $\mathbb{S}^d$. In Subsection 2.4.4, the cubature rule (2.21) of product type on the unit sphere is derived from the cubature rule (2.16) of product type on the unit ball. This connection can be explored one step further. Let H be a weight function defined on $\mathbb{S}^d$ such that H is symmetric with respect to the last variable; that is,

$$H(\mathbf{x}',x_{d+1}) = H(\mathbf{x}',-x_{d+1}) \quad \text{for} \quad (\mathbf{x}',x_{d+1}) \in \mathbb{S}^d, \quad \mathbf{x}' \in \mathbb{B}^d.$$

Theorem 2.5.4. *Let H be a weight function defined on $\mathbb{S}^d$ as earlier. Every cubature rule of degree m on the unit ball $\mathbb{B}^d$*

$$\int_{\mathbb{B}^d} f(\mathbf{x}) \frac{H(\mathbf{x},\sqrt{1-\|\mathbf{x}\|^2})}{\sqrt{1-\|\mathbf{x}\|^2}} \mathrm{d}\mathbf{x} = \sum_{k=1}^{N} \lambda_k f(\mathbf{x}_k), \quad \mathbf{x}_k \in \mathbb{B}^d, \quad f \in \Pi_m^d, \tag{2.28}$$

that has all its nodes in $\mathbb{B}^d$ is equivalent to the cubature rule of degree m on the unit sphere $\mathbb{S}^d$

$$\int_{\mathbb{S}^d} g(\xi)H(\xi)\mathrm{d}\sigma(\xi) = \sum_{k=1}^{N} \lambda_k \left[g(\mathbf{x}_k, x_{k,d+1}) + g(\mathbf{x}_k, -x_{k,d+1}) \right], \quad g \in \Pi_m(\mathbb{S}^d), \tag{2.29}$$

where $x_{k,d+1} = \sqrt{1-\|\mathbf{x}_k|^2}$ so that $(\mathbf{x}_k, x_{k,d+1}) \in \mathbb{S}^d$.

Proof Let $\xi = (\mathbf{x}, x_{d+1}) \in \mathbb{S}^d$ with $\mathbf{x} \in \mathbb{B}^d$. Each polynomial P_m of degree m can be written as $\sum_{k=0}^{n} p_{m-k}(\mathbf{x})x_{d+1}^k$, where $p_{m-k} \in \Pi_{m-k}^d$. For $(\mathbf{x}, x_{d+1}) \in \mathbb{S}^d$, the monomial $x_{d+1}^{2k} = (1-\|\mathbf{x}\|^2)^k$ is a polynomial in $\mathbf{x}$, whereas $x_{d+1}^{2k+1} = x_{d+1}(1-\|\mathbf{x}\|^2)^k$ is x_{d+1} multiple of a polynomial in $\mathbf{x}$. Hence, $P(\xi)$ can be written as $P_m(\xi) = p_m(\mathbf{x}) + x_{d+1}q_{m-1}(\mathbf{x})$. It follows that

$$\Pi_m(\mathbb{S}^d) = \Pi_m^d[\mathbf{x}] \cup x_{d+1}\Pi_{m-1}^d[\mathbf{x}], \qquad \xi = (\mathbf{x}, x_{d+1}),$$

where we write $\Pi_m^d[\mathbf{x}]$ instead of Π_m^d to emphasize that it is the space in variable $\mathbf{x}$. Since the cubature rule (2.29) is invariant under the sign change of the last variable of ξ, or $\mathbb{Z}_2$ acting on the ξ_{d+1}, by Theorem 2.5.2, it has degree m if and only if it holds for those polynomials in $\Pi_m(\mathbb{S}^d)$ that are invariant under the same group, which is Π_m^d in the variable $\mathbf{x}$, that is, $\Pi_m^d[\mathbf{x}]$. For $f \in \Pi_n^d$, the cubature rule (2.29) is equivalent to (2.28) by (2.20). This completes the proof. □

E. Cubature rules on the simplex Δ^d and the sphere $\mathbb{S}^d$. Combining Theorem 2.5.3 and Theorem 2.5.4 gives a relation between cubature rules on the simplex Δ^d and those on the unit sphere $\mathbb{S}^d$.

Theorem 2.5.5. *Let* w *be a weight function on* Δ^d. *Every cubature rule* $\mathsf{C}(f)$ *of degree* n *on the simplex,*

$$\int_{\Delta^d} f(\mathbf{x}) \frac{w(\mathbf{x}\sqrt{1-|\mathbf{x}|}}{\sqrt{x_1\cdots x_d(1-|x|)}} \mathrm{d}\mathbf{x} = \sum_{j=1}^{M} \lambda_j f(\mathbf{x}_j), \quad f \in \Pi_n^d, \tag{2.30}$$

that has all its nodes $\mathbf{x}_j = (x_{j,1}, \ldots, x_{j,d})$ *in* Δ^d *is equivalent to a cubature rule of degree* $2n+1$ *on the unit sphere*

$$\int_{\mathbb{S}^d} g(\xi) \mathsf{w}(\xi_1^2, \ldots, \xi_{d+1}^2) \mathrm{d}\sigma(\xi) \tag{2.31}$$

$$= \sum_{j=1}^{M} \frac{\lambda_j}{2^{\sigma_j}} \sum_{\varepsilon \in \{-1,1\}^{d+1}} g\left(\varepsilon_1 \sqrt{x_{j,1}}, \ldots, \varepsilon_d \sqrt{x_{j,d+1}}\right), \quad f \in \Pi_{2n+1}(\mathbb{S}^d),$$

where $\mathbf{x}_j = (x_{j,1}, \ldots, x_{j,d})$, $x_{j,d+1} = \sqrt{1 - \|\mathbf{x}_j\|^2}$, *and* $\sigma_j = |\mathbf{x}_j \mathbb{Z}_2^{d+1}|$.

In particular, an invariant cubature rule on the simplex Δ^d is equivalent to a cubature rule on the unit sphere $\mathbb{S}^d$ invariant under the octahedral group. The latter is defined as a semidirect product of the permutation group $\mathcal{S}_{d+1}$ and $\mathbb{Z}_2^{d+1}$.

Corollary 2.5.6. *If the cubature rule* (2.30) *is invariant under the permutation group of* Δ^d, *then it is equivalent to the cubature rule* (2.31) *on* $\mathbb{S}^d$ *invariant under the octahedral group.*

2.6 Constructing Cubature Rules Numerically

In practical applications, such as FEM or hp-FEM, it is desirable to use cubature rules of an assumed degree with as few nodes as possible. As mentioned in the Preface, many cubature rules of lower to moderate degrees used in applications are found by solving the system of moment equations since a cubature rule $\mathsf{C}_m(f)$ is equivalent to (2.1), which we restate as

$$\sum_{k=1}^{N} \lambda_k \mathbf{x}_k^\alpha = \mathcal{I}(\mathbf{x}^\alpha), \qquad |\alpha| \le m, \quad \alpha \in \mathbb{N}_0^d.$$

To keep the number of nodes small, we assume both λ_k and $\mathbf{x}_k \in \mathbb{R}^d$ as variables and, consequently, the system of moment equations is highly nonlinear. Moreover, it has $(d+1)N$ variables and $\dim \Pi_m^d = \binom{m+d}{d}$ equations.

The nonlinear systems of equations are solved mostly numerically by brute force and usually under the presumption that the number of variables is no less than the number of equations, that is,

$$(d+1)N \geq \binom{m+d}{d}, \quad \text{or} \quad N \geq \frac{1}{d+1}\binom{m+d}{d}. \tag{2.32}$$

Moreover, it is often necessary to assume that the number of variables equals the number of equations. For example, a system of N equations and N variables can be solved by, say, Newton's iterative method. This requires choosing m so that $\frac{1}{d+1}\binom{m+d}{d}$ is an integer or adjusting the number of variables by imposing certain restrictions on nodes judiciously, such as putting some nodes on the boundary of the domain. Solving the nonlinear system of equations is a challenging problem and has limitations on the size of the system. For example, choosing initial points for Newton's method is difficult if the system is large. Hence, the approach is limited to finding cubature rules of lower degrees. One way to reduce the size of the system is to consider invariant cubature rules when the integral is invariant under a finite group G, for which we only need to consider, by the Sobolev theorem, invariant polynomials in Π_m^G, so that the size of the system of moment equations is reduced to a fraction. Most cubature rules of moderate degrees found numerically are invariant.

The approach outlined earlier has been utilized in numerous works for constructing cubature rules of lower to moderate degrees on specific domains. Choosing cubature rules on the triangle and simplex as a specimen, we describe details and various aspects of the approach in Chapter 11.

As mentioned in the Preface, our main goal is to understand minimal cubature rules, which have the smallest number of nodes among cubature rules of the same degree. The cubature rules found numerically, however, may not be minimal. Indeed, per the presumption that the number of variables is no less than the number of equations, the cubature rules constructed numerically have the number of nodes satisfying (2.32). In particular, if $m = 2n-1$, then it shows that

$$N \geq \frac{1}{d+1}\binom{2n-1+d}{d} = \frac{2^d}{(d+1)!}n^d + \mathcal{O}\left(n^{d-1}\right).$$

However, if a Gauss cubature rule of degree $2n-1$ exists, then its number of nodes is substantially smaller and satisfies

$$N = \dim \Pi_{n-1}^d = \binom{n+d-1}{d} = \frac{1}{d!}n^d + \mathcal{O}\left(n^{d-1}\right).$$

This shows that the presumption on the number of variables does not hold for the system of moment equations corresponding to the Gauss cubature rule. In particular, the discussion calls into question the minimality of the number of nodes of cubature rules found numerically.

By its nature, numerical calculations could produce cubature rules of lower to moderate degrees but do not help our theoretical understanding of cubature rules in general.

2.7 Notes and Further Results

There are few books on the theory of cubature rules. The book [146] by A. Stroud is classical and remains a standard reference, and it also contains a list of product-type cubature rules of high dimensions. The book [50] by H. Engels deals with both quadrature and cubature rules and covers a variety of related topics, but its coverage of cubature rules is on the lighter side. The book [115] by Mysovskikh is a comprehensive work devoted entirely to cubature rules, and it discusses the Gauss cubature and minimal cubature rules. The book is in Russian, and there is no English translation.

There are also two books by S. Sobolev [143, 144], the second one coauthored with V. Vaskevich, on cubature rules. Both cover invariant cubature under a finite subgroup based on Theorem 2.5.2, which was first proved by Sobolev himself in [142], but the central theme of the two books is the construction of cubature rules that minimize the norm of the error functional in a given class of functions and contain mostly the research of Sobolev and his school on approximate multivariate integration, which does not pertain to our interests here.

The lower bound (2.2) is classical and already appeared in [146]. The Gauss cubature rule was discussed and characterized in [115]. The proof of Proposition 2.2.6 first appeared in [114].

The existence of positive cubature rules stated in Theorem 2.3.1 was first proved by Tchakaloff in [153] under the assumption that the measures are compactly supported and absolutely continuous with respect to the Lebesgue measure. The theorem was generated in [113] and, more recently, in [34, 129], to its current version but with the additional assumption of the finiteness of the $m+1$ moments. The final statement in Theorem 2.3.1 was proved in [8].

The product-type cubature rules for the special domains have already appeared in [146]. For the unit sphere, the unit ball, and the standard simplex, these cubature rules are the only ones known explicitly in high degrees; thus, even though their nodes are not well separated and too numerous, these cubature rules are used when cubature rules of high degrees are needed. For these and several other regular domains, positive cubature rules for well-separated nodes and sharp estimates of the cubature weights exist; more precisely, if a well-separated set of data with cardinality about n^d is given on, say, the ball $\mathbb{B}^d$, then there is a positive cubature rule of degree cn for some positive $c > 0$, where the cubature weights satisfy sharp estimates from above and below. While such cubature rules are useful tools for analysis based on the Fourier orthogonal expansions on the regular domains, they are far less so for numerical calculation since their cubature weights are difficult to determine, and neither is the constant c of cn. The first result for such positive cubature rules is established on the unit sphere in [103] and the unit ball in [127]. See [35, 182] and extensive references therein for other domains.

The study of invariant cubature rules was initiated by S. Sobolev when he proved Theorem 2.5.2 in [142], which has since been used extensively for finding cubature rules numerically, especially for the unit sphere. The close relations between cubature rules on the unit sphere, the unit ball, and the standard simplex are observed in [171, 172] and summarized in [175].

Constructing cubature rules numerically has been carried out extensively for various specific domains, such as the square, triangle, disk, and their higher-dimensional analogs, mostly in two and three dimensions. We refer to the extensive survey of [26]. A list of known cubature rules over several standard regions is given in [30] and amended in [27]. A digital collection is documented in [28] and available on the webpage therein. A more updated online respiratory of cubature rules is [136], where they are compiled in Python.

3
Orthogonal Polynomials of Several Variables

The study of cubature rules with few nodes relies heavily on understanding the zeros of orthogonal polynomials. A concise account of orthogonal polynomials in several variables is given in this chapter. After a brief introduction in Section 3.1, the structure of orthogonal polynomials in terms of the three-term relation is presented in Section 3.2. The common zeros of orthogonal polynomials of the same degree are discussed in Section 3.3. Several examples are given in Section 3.4. Section 3.5 consists of two lemmas, needed in the next chapter, on the multi-index Hankel matrix.

3.1 Orthogonal Polynomials

Let Ω be a domain with a nonempty interior in $\mathbb{R}^d$ and let W be a regular weight function on Ω. We define an inner product by

$$\langle f, g \rangle = \int_{\Omega} f(\mathbf{x}) g(\mathbf{x}) W(\mathbf{x}) \mathrm{d}\mathbf{x} = \mathcal{I}(fg).$$

A polynomial $P \in \Pi_n^d$ is called an orthogonal polynomial with respect to W if $\langle P, Q \rangle = 0$ for all $Q \in \Pi_{n-1}^d$. More generally, we can define orthogonality under a linear functional $\mathcal{L}$ by defining $\langle f, g \rangle = \mathcal{L}(fg)$, where we require that $\mathcal{L}$ is *square positive* in the sense that $\mathcal{L}(p^2) > 0$ for all nontrivial polynomials. It should be noted that our definition does not require the orthogonality of polynomials of the same degree. For $n = 0, 1, 2, \ldots$, we denote by $\mathcal{V}_n^d$ the space of orthogonal polynomials of degree n,

$$\mathcal{V}_n^d := \left\{ P \in \Pi_n^d \mid \langle P, Q \rangle = 0, \quad \forall Q \in \Pi_{n-1}^d \right\}. \tag{3.1}$$

Since Ω has a nonempty interior, the dimension of $\mathcal{V}_n^d$ is the same as that of $\mathcal{P}_n^d$, the space of homogeneous polynomials of degree n in d variables. We introduce the notation

$$r_n^d := \dim \mathcal{V}_n^d = \binom{n+d-1}{n}$$

and write r_n for r_n^d when no confusion is likely. A basis $\{P_\alpha : |\alpha| = n\}$ of $\mathcal{V}_n^d$ may not be orthogonal. It is orthogonal if $\langle P_\alpha, P_\beta \rangle = 0$ for $|\alpha| = |\beta|$ and $\alpha \neq \beta$, and it is orthonormal if $\langle P_\alpha, P_\alpha \rangle = 1$ in addition.

Given an inner product, a sequence of orthogonal polynomials can be obtained from the moments $\mu_\alpha = \int_\Omega \mathbf{x}^\alpha W(\mathbf{x}) d\mathbf{x}$, $\alpha \in \mathbb{N}_0^d$. In the graded lexicographical order, monomials in $\mathcal{P}_n^d$ precede those in $\mathcal{P}_{n+1}^d$ and those in $\mathcal{P}_n^d$ are ordered lexicographically. For $\mathbf{x} \in \mathbb{R}^d$, we use the notation $\mathbf{x}^n$ to denote the set of monomials of degree n,

$$\mathbf{x}^n := \{\mathbf{x}^\alpha : |\alpha| = n\}, \qquad n = 0, 1, 2, \ldots,$$

and we regard $\mathbf{x}^n$ as a column vector in $\mathbb{R}^{r_n}$, where the monomials are ordered lexicographically. For $n = 0, 1, \ldots$, let M_n be the moment matrix

$$M_{n,d} = \left[\mu_{\alpha+\beta}\right]_{|\alpha|\le n, |\beta| \le n}. \tag{3.2}$$

The matrix is of size $N \times N$ with $N = \dim \Pi_n^d$ and it is positive definite since $\mathbf{c}^\mathsf{T} M_{n,d} \mathbf{c} = \int_\Omega (\mathbf{c}^\mathsf{T} \mathbf{x}^n)^2 W(\mathbf{x}) d\mathbf{x}$ for any vector $\mathbf{c} \in \mathbb{R}^N$ and

$$\Delta_{n,d} = \det M_{n,d} > 0.$$

For $\alpha \in \mathbb{N}^d$, we further denote by $\boldsymbol{\mu}_{\alpha,k}$ the row vector of $\mathbb{R}^{r_k}$ defined by

$$\boldsymbol{\mu}_{k,\alpha} = [\mu_{\alpha+\beta}]_{|\beta| \le n}, \qquad k = 0, 1, 2, \ldots.$$

The verification of the following proposition is straightforward.

Proposition 3.1.1. *For $\alpha \in \mathbb{N}_0^d$ and $|\alpha| = n$, the polynomial*

$$P_\alpha(\mathbf{x}) := \frac{1}{\Delta_{n-1,d}} \det \left[\begin{array}{c|c} M_{n-1,d} & \begin{array}{c} 1 \\ \mathbf{x} \\ \vdots \\ \mathbf{x}^{n-1} \end{array} \\ \hline \boldsymbol{\mu}_{0,\alpha} \boldsymbol{\mu}_{1,\alpha} \cdots \boldsymbol{\mu}_{n-1,\alpha} & \mathbf{x}^\alpha \end{array}\right] \tag{3.3}$$

is an orthogonal polynomial of degree n with respect to W. Moreover,

$$P_\alpha(\mathbf{x}) = \mathbf{x}^\alpha + R_\alpha(\mathbf{x}), \qquad R_\alpha \in \Pi_{n-1}^d,$$

so that $\{P_\alpha : |\alpha| = n, \alpha \in \mathbb{N}_0^d\}$ is a basis of the space $\mathcal{V}_n^d$.

The orthogonal polynomial P_α^n is called *monic* because it has a single leading term x^α. In general, a basis of $\mathcal{V}_n^d$ can be conveniently denoted by $\mathbb{P}_n = \{P_\alpha\}_{|\alpha|=n}$ with the understanding that its elements are arranged according to the lexicographical order $\alpha(1),\ldots,\alpha(r_n^d)$. We will also regard $\mathbb{P}_n$ as a vector

$$\mathbb{P}_n = \left[P_\alpha\right]_{|\alpha|=n} = \left[P_{\alpha(1)},\ldots,P_{\alpha(r_n)}\right]^{\mathsf{T}}, \tag{3.4}$$

where the superscript T means that $\mathbb{P}_n$ is a column vector. This is convenient since, for example, the orthonormality of a basis is simply $\mathcal{I}\left(\mathbb{P}_n\mathbb{P}_m^{\mathsf{T}}\right) = \delta_{m,n}I_{r_n}$, where I_k denotes the identity matrix of size $k \times k$.

The space $\mathcal{V}_n^d$ can have many different bases. Indeed, if $\mathbb{P}_n$ denotes the basis of monic orthogonal polynomials of degree n and Q_n is an invertible matrix of $r_n \times r_n$, then $Q_n\mathbb{P}_n$ is also a basis of $\mathcal{V}_n^d$. In particular, if $\mathbb{P}_n$ is an orthonormal basis, then $O_n\mathbb{P}_n$ is also an orthonormal basis whenever O_n is an orthogonal matrix of size r_n. Each component of $\mathbb{P}_n$ is a polynomial of degree n. Hence, it can be written in terms of vectors of monomials as

$$\mathbb{P}_n(\mathbf{x}) = G_n\mathbf{x}^n + G_{n,n-1}\mathbf{x}^{n-1} + \cdots + G_{n,1}\mathbf{x} + G_{n,0},$$

where $G_{n,k}$ is a matrix of $r_n^d \times r_k^d$ and the matrix G_n is invertible and called the leading coefficient of $\mathbb{P}_n$; $G_n = I_{r_n}$ if and only if $\mathbb{P}_n$ is the monic basis of $\mathcal{V}_n^d$.

The plethora of bases is one of the main obstacles to studying orthogonal polynomials of more than one variable. As one way to bypass the obstacle, some of the notions and results for orthogonal polynomials of one variable can be extended to several variables by stating them in terms of the bases $\mathbb{P}_n$ collectively instead of in individual components. More precisely, we adopt a vector-matrix notation for extending structural relations of orthogonal polynomials to several variables in Section 3.2.

3.2 Structural Relations of Orthogonal Polynomials

The structural relations in this section are based on an extension of the three-term relation of one variable, which holds in the vector-matrix notation. These relations are well documented in [47]. We shall indicate the proof but will not provide full details.

We start with the three-term relations between the orthogonal subspaces, which hold for any basis of $\mathcal{V}_n^d$ but are of the simplest form when stated in terms of orthonormal bases.

Theorem 3.2.1. *Let $\mathbb{P}_m$ denote an orthonormal basis of $\mathcal{V}_m^d$, $m = 0,1,\ldots$ and let*

$\mathbb{P}_{-1}(\mathbf{x}) := 0$. *Then, there exist unique matrices* $A_{m,i} : r_m^d \times r_{m+1}^d$ *and* $B_{m,i} : r_m^d \times r_m^d$, *such that*

$$x_i \mathbb{P}_m(\mathbf{x}) = A_{m,i}\mathbb{P}_{m+1}(\mathbf{x}) + B_{m,i}\mathbb{P}_m(\mathbf{x}) + A_{m-1,i}^{\mathsf{T}}\mathbb{P}_{m-1}(\mathbf{x}), \quad 1 \le i \le d, \tag{3.5}$$

where $A_{m,i} = \mathcal{I}(x_i \mathbb{P}_m \mathbb{P}_{m+1}^{\mathsf{T}})$ *and* $B_{m,i} = \mathcal{I}(x_i \mathbb{P}_m \mathbb{P}_m^{\mathsf{T}})$. *Moreover,*

$$\operatorname{rank} A_{m,i} = \operatorname{rank} r_m^d \quad \textit{and} \quad \operatorname{rank} A_m = (A_{m,1}^{\mathsf{T}}, \ldots, A_{m,d}^{\mathsf{T}})^{\mathsf{T}} = r_{m+1}^d, \tag{3.6}$$

The proof of (3.5) follows from expanding $x_i\mathbb{P}_n$ in the basis $\mathbb{P}_0, \mathbb{P}_1, \ldots, \mathbb{P}_{n+1}$ of Π_{n+1}^d, writing the coefficients as matrices and applying the orthogonality, just like in the case of one variable. The proof of (3.6) follows from comparing the leading coefficients of the three-term relations. Recall that G_n denotes the leading coefficient of $\mathbb{P}_n$. By (3.5), comparing the coefficients of $\mathbf{x}^{d+1}$, we see that $A_{n,i}$ can be written as

$$A_{n,i} = G_n L_{n,i} G_{n+1}^{-1}, \qquad 1 \le i \le d, \tag{3.7}$$

where the matrix $L_{n,i} : r_n^d \times r_n^{d+1}$ is defined uniquely by

$$L_{n,i}\mathbf{x}^{n+1} = x_i \mathbf{x}^n, \qquad 1 \le i \le d. \tag{3.8}$$

For $d = 2$, the matrices $L_{n,i}$ are given in terms of the identity matrix I_n by

$$L_{n,1} = [I_{n+1}|0] \qquad \text{and} \qquad L_{n,2} = [0|I_{n+1}],$$

where 0 stands for a column vector of zeros.

As an analog of the Favard theorem of one variable, the three-term relations and the rank conditions characterize orthogonality.

Theorem 3.2.2. *Let* $\{\mathbb{P}_n\}_{n=0}^{\infty} = \{P_\alpha^n : |\alpha| = n, n \in \mathbb{N}_0\}$, $\mathbb{P}_0 = 1$, *be an arbitrary sequence in* Π^d *such that* $\{P_\alpha^m : |\alpha| \le m \le n\}$ *spans* Π_n^d *for each n. Then, the following statements are equivalent.*

1. *There exists a positive definite linear functional* $\mathcal{L}$ *that makes* $\{\mathbb{P}_n\}_{n=0}^{\infty}$ *an orthonormal basis for* Π^d *in the sense that* $\mathcal{L}(\mathbb{P}_n \mathbb{P}_m^{\mathsf{T}}) = 0$ *for* $n \ne m$.
2. *For* $m \ge 0$, $1 \le i \le d$, *there exist matrices* $A_{m,i}$ *and* $B_{m,i}$ *such that* (3.5) *and* (3.6) *are satisfied.*

Let $\mathbf{P}_n(\cdot,\cdot)$ denote the reproducing kernel of $\mathcal{V}_n^d$ and let $\mathbf{K}_n(\cdot,\cdot)$ denote the reproducing kernel of Π_n^d in $L^2(W)$. These kernels are uniquely defined by their reproducing property; for example,

$$\int_\Omega f(\mathbf{y})\mathbf{P}_n(\mathbf{x},\mathbf{y})W(\mathbf{y})d\mathbf{y} = f(\mathbf{x}), \qquad \forall f \in \mathcal{V}_n^d.$$

If $\mathbb{P}_m$ is an orthonormal basis of $\mathcal{V}_m^d$ for $0 \le m \le n$, then we can write

$$\mathbf{P}_n(\mathbf{x},\mathbf{y}) = \mathbb{P}_n^{\mathsf{T}}(\mathbf{x})\mathbb{P}_n(\mathbf{y}) \quad \text{and} \quad \mathbf{K}_n(\mathbf{x},\mathbf{y}) = \sum_{m=0}^{n} \mathbf{P}_m(\mathbf{x},\mathbf{y}).$$

An immediate consequence of the three-term relations (3.5) leads to an analog of the Christoffel–Darboux formula, which is proved just like in one variable, but there are d such formulas.

Theorem 3.2.3. *Let $\mathbb{P}_k$ be orthonormal polynomials for $\mathcal{V}_k^d$, $0 \le k \le n+1$. Then, for $\mathbf{x},\mathbf{y} \in \mathbb{R}^d$,*

$$\mathbf{K}_n(\mathbf{x},\mathbf{y}) = \frac{\mathbb{P}_{n+1}(\mathbf{x})^{\mathsf{T}} A_{n,i}^{\mathsf{T}} \mathbb{P}_n(\mathbf{y}) - \mathbb{P}_n(\mathbf{x})^{\mathsf{T}} A_{n,i} \mathbb{P}_{n+1}(\mathbf{y})}{x_i - y_i}, \quad 1 \le i \le d, \tag{3.9}$$

where $\mathbf{x} = (x_1,\ldots,x_d)$ and $\mathbf{y} = (y_1,\ldots,y_d)$ and $x_i \ne y_i$. Moreover,

$$\mathbf{K}_n(\mathbf{x},\mathbf{x}) = \partial_i \mathbb{P}_{n+1}(\mathbf{x})^{\mathsf{T}} A_{n,i}^{\mathsf{T}} \mathbb{P}_n(\mathbf{x}) - \partial_i \mathbb{P}_n(\mathbf{x})^{\mathsf{T}} A_{n,i} \mathbb{P}_{n+1}(\mathbf{x}), \quad 1 \le i \le d, \tag{3.10}$$

where ∂_i denotes the partial derivative with respect to x_i.

The three-term relations also lead to a recursive formula that relies on a general inverse of A_m. Let A be a matrix of size $k \times j$ with $k \ge j$ and $\operatorname{rank} A = j$. We define a unique generalized inverse, denoted by A^+, of A in terms of the singular decomposition

$$A = P \begin{bmatrix} \Lambda \\ \mathrm{O} \end{bmatrix} Q, \quad \text{so that} \quad A^+ = Q^{-1} \left[\Lambda^{-1} \ \mathrm{O}^{\mathsf{T}} \right] P^{-1}, \tag{3.11}$$

where $P : k \times k$ and $Q : j \times j$ are invertible matrices, $\Lambda : j \times j$ is an invertible diagonal matrix, and O is the zero matrix of $(k-j) \times j$. Then, $A^+ A = I_j$. Notice that the identity $A^+ A = I_j$ remains true if we replace O^{T} in A^+ with any nonzero matrix; there could be many general inverses. Now, let $D_m^{\mathsf{T}} = (D_{m,1}^{\mathsf{T}}, \ldots, D_{m,d}^{\mathsf{T}})$, where $D_{m,i}^{\mathsf{T}} : r_{m+1}^d \times r_m^d$, be A_m^+ or any general inverse of A_m so that

$$D_m^{\mathsf{T}} A_m = \sum_{i=1}^{d} D_{m,i}^{\mathsf{T}} A_{m,i} = I_{r_{m+1}^d}. \tag{3.12}$$

Then, the three-term relation (3.5) implies

$$\mathbb{P}_{m+1} = \sum_{i=1}^{d} x_i D_{m,i}^{\mathsf{T}} \mathbb{P}_m + E_m \mathbb{P}_m + F_m \mathbb{P}_{m-1}, \tag{3.13}$$

where the matrices E_m and F_m are defined by

$$E_m := -\sum_{i=1}^{d} D_{m,i}^{\mathsf{T}} B_{m,i} \quad \text{and} \quad F_m := -\sum_{i=1}^{d} D_{m,i}^{\mathsf{T}} A_{m-1,i}^{\mathsf{T}}.$$

These relations can be used to deduce several generic properties of orthogonal polynomials in several variables.

Definition 3.2.4. The domain $\Omega \in \mathbb{R}^d$ is called centrally symmetric if it is symmetric to the origin. The function f defined on Ω is called centrally symmetric if Ω is centrally symmetric and $f(-\mathbf{x}) = f(\mathbf{x})$ for all $\mathbf{x} \in \Omega$.

If W is centrally symmetric on Ω, changing variables $\mathbf{x} \mapsto -\mathbf{x}$ shows that $\int_\Omega \mathbf{x}^\alpha W(\mathbf{x}) d\mathbf{x} = 0$ whenever $|\alpha|$ is an odd integer. More generally, we call a linear functional $\mathcal{L}$ centrally symmetric if

$$\mathcal{L}(\mathbf{x}^\alpha) = 0, \quad |\alpha| = \text{odd integer}, \quad \alpha \in \mathbb{N}^d.$$

The two notions are equivalent if $\mathcal{L}$ is the integral $\mathcal{I}f = \int_\Omega f(\mathbf{x})W(\mathbf{x})d\mathbf{x}$. With the help of the recursive relation (3.13), we can deduce the following theorem:

Theorem 3.2.5. *A positive definite linear functional $\mathcal{L}$ is centrally symmetric if and only if $B_{n,i} = 0$ for all $n \in \mathbb{N}_0$ and $1 \le i \le d$. Furthermore, if $\mathcal{L}$ is centrally symmetric, then an orthogonal polynomial of degree n is a sum of monomials of even degrees if n is even and a sum of monomials of odd degrees if n is odd.*

The three-term relations also lead to a matrix representation of the product operator $X_i : Y \mapsto x_i Y$ for $1 \le i \le d$, which gives, as an analog of the Jacobi matrix in one variable, a family of block tridiagonal matrices J_i, $1 \le i \le d$, defined via the coefficients of (3.5),

$$J_i = \begin{bmatrix} B_{0,i} & A_{0,i} & & \bigcirc \\ A_{0,i}^{\mathsf{T}} & B_{1,i} & A_{1,i} & \\ & A_{1,i}^{\mathsf{T}} & B_{2,i} & \ddots \\ \bigcirc & & \ddots & \ddots \end{bmatrix}, \qquad 1 \le i \le d, \tag{3.14}$$

called *block Jacobi matrices*. The entries of the J_i are matrices of increasing sizes going down the main diagonal. Since the product operators X_i and X_j commute, the corresponding Jacobi matrices must also commute. Writing the commuting relations $J_i J_j = J_j J_i$, $1 \le i, j \le d$, entrywise shows that the coefficient matrices of the three-term relations (3.5) satisfy a set of commutativity relations: for $1 \le i, j \le d$ and $k \ge 0$,

$$\begin{aligned} A_{k,i}A_{k+1,j} &= A_{k,j}A_{k+1,i}, \\ A_{k,i}B_{k+1,j} + B_{k,i}A_{k,j} &= B_{k,j}A_{k,i} + A_{k,j}B_{k+1,i}, \\ A_{k-1,i}^{\mathsf{T}}A_{k-1,j} + B_{k,i}B_{k,j} + A_{k,i}A_{k,j}^{\mathsf{T}} &= A_{k-1,j}^{\mathsf{T}}A_{k-1,i} + B_{k,j}B_{k,i} + A_{k,j}A_{k,i}^{\mathsf{T}}, \end{aligned} \tag{3.15}$$

where $A_{-1,i} := 0$. These can be formally verified by computing $\mathcal{L}(x_i x_j \mathbb{P}_k \mathbb{P}_{k\pm1}^{\mathsf{T}})$ in two different ways, applying the three-term relations for $x_i \mathbb{P}_n$ and $x_j \mathbb{P}_{k\pm1}$ and then reversing the role of x_i and x_j.

3.3 Common Zeros of Orthogonal Polynomials

The set of zeros of a polynomial in more than one variable is an algebraic variety and, in particular, can be an algebraic surface of one dimension lower. For example, zeros of an algebraic polynomial of degree n in two variables could be an algebraic curve of degree n on the plane. For studying cubature rules, we need to consider common zeros of a set of polynomials, as defined in Definition 2.2.5. In particular, if $\mathbf{x}$ is a common zero of the set $\mathbb{P}_n$, we call it a zero of $\mathbb{P}_n$.

The structural matrices can be utilized to characterize the common zeros of orthogonal polynomials in $\mathcal{V}_n^d$ or zeros of a basis $\mathbb{P}_n$ of $\mathcal{V}_n^d$. We need the truncated block Jacobi matrices defined by

$$J_{n,i} := \begin{bmatrix} B_{0,i} & A_{0,i} & & & \bigcirc \\ A_{0,i}^{\mathsf{T}} & B_{1,i} & A_{1,i} & & \\ & \ddots & \ddots & \ddots & \\ & & A_{n-3,i}^{\mathsf{T}} & B_{n-2,i} & A_{n-2,i} \\ \bigcirc & & & A_{n-2,i}^{\mathsf{T}} & B_{n-1,i} \end{bmatrix}, \quad 1 \le i \le d,$$

where $A_{n,i}$ and $B_{n,i}$ are coefficients of (3.5). The matrix $J_{n,i}$ is of the size $N \times N$ with $N = \dim \Pi_{n-1}^d$. For $d = 1$, the characteristic polynomial of J_n is an orthogonal polynomial $p_n(w)$ of degree n by (1.8), so that the eigenvalues of J_n are zeros of $p_n(w)$. Although the former is no longer feasible, the latter has an analog in several variables.

For $d > 1$, we call an element $\lambda \in \mathbb{R}^d$ *joint eigenvalue* of $J_{n,1}, \ldots, J_{n,d}$ if there is a vector $\boldsymbol{\xi} \ne 0$, $\boldsymbol{\xi} \in \mathbb{R}^N$, such that $J_{n,i}\boldsymbol{\xi} = \lambda_i \boldsymbol{\xi}$ for $i = 1, \ldots, d$; the vector $\boldsymbol{\xi}$ is called a *joint eigenvector.*

Theorem 3.3.1. *A point $\lambda \in \mathbb{R}^d$ is a zero of $\mathbb{P}_n$ if and only if it is a joint eigenvalue of $J_{n,1}, \ldots, J_{n,d}$; moreover, a joint eigenvector of the joint eigenvalue λ is $(\mathbb{P}_0^{\mathsf{T}}(\lambda), \ldots, \mathbb{P}_{n-1}^{\mathsf{T}}(\lambda))^{\mathsf{T}}$.*

Proof If $\mathbb{P}_n(\lambda) = 0$, then the relation (3.5) with $m = 0, 1, \ldots, n-1$ gives

$$\begin{aligned} B_{0,i}\xi_0 + A_{0,i}\xi_1 &= \lambda_i \xi_0, \\ A_{k-1,i}^{\mathsf{T}}\xi_{k-1} + B_{k,i}\xi_k + A_{k,i}\xi_{k+1} &= \lambda_i \xi_k, \quad 1 \le k \le n-2, \\ A_{n-2,i}^{\mathsf{T}}\xi_{n-2} + B_{n-1,i}\xi_{n-1} &= \lambda_i \xi_{n-1}, \end{aligned} \tag{3.16}$$

where $\xi_j = \mathbb{P}_j(\lambda)$ and $1 \le i \le d$. Written in matrix form, these are precisely

$$J_{n,i}\xi = \lambda_i \xi, \qquad \xi = [\mathbb{P}_0^{\mathsf{T}}(\lambda), \ldots, \mathbb{P}_{n-1}^{\mathsf{T}}(\lambda)]^{\mathsf{T}},$$

which shows that λ is the eigenvalue of $J_{n,1}, \ldots, J_{n,d}$ with joint eigenvector ξ. In the other direction, suppose that $\lambda = (\lambda_1, \ldots, \lambda_d)$ is a joint eigenvalue of $J_{n,i}$

with a joint eigenvector ξ. Let us write $\xi = (\xi_0^{\mathsf{T}}, \xi_1^{\mathsf{T}}, \ldots, \xi_{n-1}^{\mathsf{T}})^{\mathsf{T}}$, where $\xi_j \in \mathbb{R}^{r_j}$. Then, $J_{n-1,i}\xi = \lambda_i \xi$ becomes (3.16). We first claim that ξ_0 cannot be zero. Indeed, if $\xi_0 = 0$, then the first equation of (3.16) shows $A_{0,i}\xi_1 = 0$ so that $A_0\xi_1 = 0$, and thus $\xi_1 = 0$ by (3.6). Starting from $\xi_0 = \xi_1 = 0$, we deduce inductively by (3.16) and (3.6) that $A_k\xi_k = 0$ for $k = 2, 3, \ldots n$, so that $\xi = 0$, which however is a contradiction to ξ being an eigenvector. Thus, $\xi_0 \neq 0$, and we can normalize ξ so that $\xi_0 = \mathbb{P}_0(\lambda)$. We further define $\xi_n = 0$. Let $\zeta_j = \xi_j - \mathbb{P}_j(\lambda)$. Then, $\{\zeta_j\}_{j=0}^n$ satisfies the three-term relations (3.5) for $m = 0, 1, \ldots, n-1$. Since the three-term relations determine the sequence recursively by (3.13) and $\zeta_0 = 0$, we conclude that $\zeta_k = 0$ for $1 \le k \le n$. In particular, $\zeta_n = \mathbb{P}_n(\lambda) = 0$. The proof is complete. □

Corollary 3.3.2. *The orthogonal polynomials $\mathbb{P}_n$ have at most $\dim \Pi_{n-1}^d$ common zeros, and all zeros are real and simple. Moreover, polynomials $\mathbb{P}_n$ and $\mathbb{P}_{n-1}$ do not have common zeros.*

Proof Since the zeros of $\mathbb{P}_n$ are eigenvalues of $J_{n,1}, \ldots, J_{n,d}$, the number of zeros is at most $N = \dim \Pi_{n-1}^d$ because of the size of the matrices, and all zeros are real since $J_{n,i}$ are symmetric. Moreover, if ξ is a zero of $\mathbb{P}_n$, then by $\mathbf{K}_{n-1}(\xi,\xi) \ge \mathbf{K}_0(\xi,\xi) > 1$ and (3.10), we conclude that $\mathbb{P}_{n-1}(\xi) \neq 0$, so that it has no common zero with $\mathbb{P}_n$, and $\partial_i \mathbb{P}_n(\xi) \neq 0$, so that ξ is simple. □

Our next theorem characterizes when $\mathbb{P}_n$ can have a maximal number of zeros.

Theorem 3.3.3. *For $n = 1, 2, \ldots$, $\mathbb{P}_n$ has $N = \dim \Pi_{n-1}^d$ zeros if and only if*

$$A_{n-1,i}A_{n-1,j}^{\mathsf{T}} = A_{n-1,j}A_{n-1,i}^{\mathsf{T}}, \qquad 1 \le i, j \le d. \tag{3.17}$$

Proof By Theorem 3.3.1, $\mathbb{P}_n$ has N zeros is the same as that the matrices $J_{n,1}, \ldots, J_{n,d}$ have N joint eigenvalues. Since $J_{n,i}$ is of size $N \times N$, this means that these matrices can be simultaneously diagonalized by an invertible matrix, which holds if and only if the matrices commute (cf. [69, p. 52]). Since the commutativity relations (3.15) come from $J_iJ_j = J_jJ_i$, a quick verification shows that $J_{n,i}J_{n,j} = J_{n,j}J_{n,i}$ holds if and only if (3.17) holds. □

Since the matrix multiplication is not commutable, the condition (3.17) cannot hold in general. Our next result shows that it does not hold if the integral is centrally symmetric. Let $\mathcal{M}_{m,n}$ denote the set of matrices of size $m \times n$. It is known that the matrix product satisfies

$$\operatorname{rank}(AB) \ge \operatorname{rank} A + \operatorname{rank} B - k, \qquad A \in \mathcal{M}_{n,k}, \quad B \in \mathcal{M}_{k,m}. \tag{3.18}$$

Proposition 3.3.4. *For $d = 2$, a necessary condition for* (3.17) *to hold is*

$$\operatorname{rank}[B_{n,1}, B_{n,2}] = 2 \quad \textit{where} \quad [B_{n,1}, B_{n,2}] := B_{n,1}B_{n,2} - B_{n,2}B_{n,1}. \tag{3.19}$$

In particular, Eq. (3.17) *fails to hold if the integral is centrally symmetric.*

Proof Assume (3.17). Then, the third identity of (3.15) can be written as

$$[B_{n,1}, B_{n,2}] = A_{n-1,2}^{\mathsf{T}}A_{n-1,1} - A_{n-1,1}^{\mathsf{T}}A_{n-1,2} = S_n(A^{\mathsf{T}})A_{n-1}, \tag{3.20}$$

where $S_n(A^{\mathsf{T}}) := [-A_{n-1,2}^{\mathsf{T}}\ A_{n-1,1}^{\mathsf{T}}] \in \mathcal{M}_{n+1,2n}$ since $r_n^d = n+1$ for $d = 2$. Using (3.7), it is easy to see that $\operatorname{rank}(S_n(A^{\mathsf{T}}) = n+1$. Hence, by (3.18) and (3.6), we obtain

$$\operatorname{rank}[B_{n,1}, B_{n,2}] \ge n+1+n+1-2n = 2.$$

To prove the equality holds, we observe that the first identity of (3.15) with $k = n-1$ can be written as $S_n(A)A_n = 0$, where $S_n(A) = [-A_{n-1,2}\ A_{n-1,1}] \in \mathcal{M}_{n,2(n+2)}$. Since $\operatorname{rank} A_n = n+2$, it follows that the columns of A_n consist of a basis for the null space of $S_n(A)$. Furthermore, Eq. (3.17) can be written as $S_n(A)T_n(A) = 0$, where $T_n(A) = \begin{bmatrix} A_{n-1,1}^{\mathsf{T}} \\ A_{n-1,2}^{\mathsf{T}} \end{bmatrix} \in \mathcal{M}_{2(n+1),n}$, which shows that the columns of $T_n(A)$ belong to the null space of $S_n(A)$. Since $\operatorname{rank} T_n(A) = n$, there is a matrix $M \in \mathcal{M}_{n+2,n}$ such that $T_n(A) = A_nM$ and $\operatorname{rank} M = n$. By (3.20) with n replaced by $n+1$, we obtain

$$\begin{aligned} [B_{n+1,1}, B_{n+2,2}]M &= S_{n+1}(A^{\mathsf{T}})A_nM \\ &= S_{n+1}(A^{\mathsf{T}})T_n(A) = A_{n,1}^{\mathsf{T}}A_{n-1,2}^{\mathsf{T}} - A_{n,2}^{\mathsf{T}}A_{n-1,1}^{\mathsf{T}} = 0, \end{aligned}$$

where the last step uses the first identity of (3.15). Hence, it follows that

$$\operatorname{rank}[B_{n+1,1}, B_{n+2,2}] \le n+2-\operatorname{rank} M = 2.$$

This proves (3.19). In the centrally symmetric setting, $B_{ni} = 0$ by Theorem 3.2.5 so that (3.17) cannot hold. □

The proposition and its proof can be extended to higher dimensions but require much more involved notations. Another proof that (3.17) fails for centrally symmetric weight function will follow from an improved lower bound for the number of nodes in Section 5.2, but not the rank condition (3.19).

3.4 Examples of Orthogonal Polynomials of Several Variables

We give several examples of orthogonal polynomials that are orthogonal for classical weight functions on regular domains. Most of these examples can be given in terms of orthogonal polynomials of one variable. They are distinct by the weight functions and the domains of their orthogonality. The simplest case is the product weight functions of one variable on a product domain in Section 3.4.1, followed by a family of a wrapt product of weight functions in two variables that include the unit disk, triangle, and parabolic biangle as special cases in Section 3.4.2. The next family consists of rotationally invariant weight functions and domains, with classical orthogonal polynomials on the unit ball as an example, for which an orthogonal basis can be given with the help of spherical harmonics, as discussed in Section 3.4.3. Another basis of orthogonal polynomials on the unit ball, given in Cartesian coordinates, and on the simplex is given in Section 3.4.4. Finally, the section is closed with a family of weight functions derived from symmetry.

3.4.1 Product Weight Functions

For $i = 1,2,\ldots,d$, let w_i be a weight function on $[a_i,b_i]$. Define the product weight function W by

$$W(\mathbf{x}) = w_1(x_1)\cdots w_d(x_d), \qquad \mathbf{x} \in [a_1,b_1]\times\cdots\times[a_d,b_d].$$

We denote by $\mathcal{V}_n^d(W)$ the space of orthogonal polynomials of degree n associated with W. Let $p_n(w)$ be the orthonormal polynomial with respect to w. Then, an orthonormal basis for the space $\mathcal{V}_n^d(W)$ is given by

$$P_\alpha(\mathbf{x}) = p_{\alpha_1}(w_1;x_1)\cdots p_{\alpha_d}(w_d;x_d), \quad |\alpha| = n,\, \alpha \in \mathbb{N}_0^d.$$

For $d = 2$, we write $\alpha = (n-k,k)$ for $0 \le k \le n$, and the basis is given by

$$P_k^n(\mathbf{x}) = p_{n-k}(w_1;x_1)p_k(w_2;x_2), \qquad 0 \le k \le n.$$

Let $a_n(w)$ and $b_n(w)$ be the coefficients of the three-term relation of $p_n(w)$ as in (1.6). For $d = 2$, $\mathbb{P}_n = (P_{0,n},P_{1,n},\ldots,P_{n,n})^\mathsf{T}$ and the three-term relations are

$$x_i\mathbb{P}_n = A_{n,i}\mathbb{P}_{n+1} + B_{n,i}\mathbb{P}_n + A_{n-1,i}^\mathsf{T}\mathbb{P}_{n-1}, \quad i = 1,2,$$

where the coefficient matrices $A_{n,i}$ of size $(n+1)\times(n+2)$ and $B_{n,i}$ of size

$(n+1)\times(n+1)$ are given by

$$A_{n,1} = \begin{bmatrix} a_n(w_1) & & \bigcirc & 0 \\ & \ddots & & \vdots \\ \bigcirc & & a_0(w_1) & 0 \end{bmatrix} \quad \text{and} \quad B_{n,1} = \begin{bmatrix} b_n(w_1) & & \bigcirc \\ & \ddots & \\ \bigcirc & & b_0(w_1) \end{bmatrix},$$

$$A_{n,2} = \begin{bmatrix} 0 & a_0(w_2) & & \bigcirc \\ \vdots & & \ddots & \\ 0 & \bigcirc & & a_n(w_2) \end{bmatrix} \quad \text{and} \quad B_{n,2} = \begin{bmatrix} b_0(w_2) & & \bigcirc \\ & \ddots & \\ \bigcirc & & b_n(w_2) \end{bmatrix}.$$

If w is centrally symmetric, then $b_k(w) = 0$; hence, $B_{n,i} = 0$.

3.4.2 Wrapt Product Weight Functions of Two Variables

Let w_1 and w_2 be weight functions defined on the interval (a,b) and (c,d), respectively. We consider the weight function in two variables

$$W(x_1,x_2) = w_1(x_1)w_2\left(\frac{x_2}{\rho(x_1)}\right) \tag{3.21}$$

on the domain $R = \{(x_1,x_2) : a < x_1 < b,\, c\rho(x_1) < x_2 < d\rho(x_1)\}$, where ρ is a positive function on (a,b) specified in the following cases.

Case I. ρ is a polynomial of degree 1;

Case II. ρ is the square root of a nonnegative quadratic polynomial, and w_2 is an even function on $(-c,c)$ with $d = -c$.

More specifically, we assume that ρ is given by

$$\rho(t) = \begin{cases} \rho_0 t + \rho_1, & \text{in Case I} \\ (\rho_0 t^2 + \rho_1 t + \rho_2)^{\frac{1}{2}}, & \text{in Case II} \end{cases}. \tag{3.22}$$

For a fixed $k \in \mathbb{N}_0$, let $\{q_{n,k}\}_{n=0}^{\infty}$ denote the system of orthonormal polynomials with respect to the weight function $\rho^{2k+1}(t)w_1(t)$. Define

$$P_k^n(\mathbf{x}) = q_{n-k,k}(x)(\rho(x_1))^k p_k\left(w_2; \frac{x_2}{\rho(x_1)}\right), \quad 0 \le k \le n. \tag{3.23}$$

The assumption on ρ shows that P_k^n is a polynomial of degree n in $\mathbf{x} = (x_1,x_2)$. Furthermore, a change of variables in the integral over R shows that $\{P_k^n : 0 \le k \le n\}$ is an orthogonal basis of $\mathcal{V}_n^2(W)$. Several examples of this class of polynomials will be given before the end of this subsection.

The three-term relation (3.5) for $\mathbb{P}_n = (P_0^n,\ldots,P_n^n)^{\mathsf{T}}$ defined by (3.23) can be determined from the coefficients $a_n(w_2)$ and $b_n(w_2)$ of the three-term relation for $p_n(w_2)$, and the three-term relation for $q_{k,n}$, for each fixed k,

$$tq_{n,k}(t) = a_{n,k}q_{n+1,k}(t) + b_{n,k}q_{n,k}(t) + a_{n-1,k}q_{n-1,k}(t),$$

where we assume $q_{0,k}(t) = 1$ and $q_{-1,k}(t) = 0$. We write $q_{n,k}(t) = \gamma_{n,k}t^n + \cdots$, so that $\gamma_{n,k}$ is the leading coefficient of $q_{n,k}$. For ρ_0 given in (3.22), we define

$$\tau = \begin{cases} \rho_0^2 & \text{in Case I} \\ \rho_0, & \text{in Case II} \end{cases}. \tag{3.24}$$

Then, it is easy to verify that the coefficient matrices of (3.5) are given by

$$A_{n,1} = \begin{bmatrix} a_{n,0} & & & \bigcirc & 0 \\ & a_{n-1,1} & & & 0 \\ & & \ddots & & \vdots \\ \bigcirc & & & a_{0,n} & 0 \end{bmatrix}, \qquad B_{n,1} = \begin{bmatrix} b_{n,0} & & & \bigcirc \\ & b_{n-1,1} & & \\ & & \ddots & \\ \bigcirc & & & b_{0,n} \end{bmatrix},$$

and

$$A_{n,2} = \begin{bmatrix} e_{0,n} & d_{0,n} & & & \bigcirc & 0 \\ c_{1,n} & e_{1,n} & d_{1,n} & & & \vdots \\ & \ddots & \ddots & \ddots & & \vdots \\ & & c_{n-1,n} & e_{n-1,n} & d_{n-1,n} & 0 \\ \bigcirc & & & c_{n,n} & e_{n,n} & d_{n,n} \end{bmatrix},$$

where

$$c_{k,n} = \tau a_{k-1}(w_2)\frac{\gamma_{n-k,k}}{\gamma_{n-k+2,k-1}} \quad \text{and} \quad d_{k,n} = a_k(w_2)\frac{\gamma_{n-k,k}}{\gamma_{n-k,k+1}}, \tag{3.25}$$

$$e_{k,n} = b_k(w_2)\int_a^b q_{n-k,k}(t)q_{n+1-k,k}(t)(\rho(t))^{2k+2}w_1(t)\mathrm{d}t,$$

and, finally,

$$B_{n,2} = \begin{bmatrix} f_{0,n} & g_{0,n} & & & \bigcirc \\ g_{0,n} & f_{1,n} & g_{1,n} & & \\ & \ddots & \ddots & \ddots & \\ & & g_{n-2,n} & f_{n-1,n} & g_{n-1,n} \\ \bigcirc & & & g_{n-1,n} & f_{n,n} \end{bmatrix},$$

where

$$f_{k,n} = b_k(w_2)(\rho_1 b_{n-k,k} + \rho_2),$$
$$g_{k,n} = a_k(w_2)\int_a^b q_{n-k,k}(t)q_{n-k-1,k+1}(t)[\rho(t)]^{2k+3}w_2(t)\mathrm{d}t.$$

In particular, $e_{k,n} = f_{k,n} = 0$ in Case II since $b_k(w_2) = 0$ in Case II.

The class of orthogonal polynomials (3.23) consists of several classical families of orthogonal polynomials. In particular, the product weight function for $d = 2$ is the case when $\rho(t) = 1$. We give three other examples, in which we state P_k^n without normalizing it to orthonormal.

Jacobi polynomials on the disk. Let $w_1(t) = w_2(t) = (1-t^2)^{\mu-1/2}$ on $[-1,1]$ and $\rho(t) = (1-t^2)^{1/2}$. Then, the weight function (3.21) becomes

$$W_\mu(\mathbf{x}) = (1-x_1^2-x_2^2)^{\mu-1/2}, \quad \mu > -1/2,$$

on the unit disk $\mathbb{B}^2 = \{\mathbf{x} : x_1^2+x_2^2 \le 1\}$, and the orthogonal polynomials P_k^n in (3.23) become

$$P_k^n(\mathbf{x}) = C_{n-k}^{k+\mu+\frac{1}{2}}(x_1)(1-x_1^2)^{\frac{k}{2}} C_k^\mu\left(\frac{x_2}{\sqrt{1-x_1^2}}\right), \tag{3.26}$$

which holds under limit $\lim_{\mu\to 0}\mu^{-1}C_k^\mu(t) = (2/k)T_k(t)$ if $\mu = 0$.

Jacobi polynomials on the triangle. Let $w_1(t) = t^\alpha(1-t)^{\beta+\gamma}$ and $w_2(t) = t^\beta(1-t)^\gamma$, both defined on the interval $(0,1)$, and $\rho(t) = 1-t$. Then, the weight function (3.21) becomes

$$W_{\alpha,\beta,\gamma}(\mathbf{x}) = x_1^\alpha x_2^\beta(1-x_1-x_2)^\gamma, \quad \alpha,\beta,\gamma > -1, \tag{3.27}$$

on the triangle $\Delta^2 = \{\mathbf{x} : x_1 \ge 0, x+2 \ge 0, 1-x_1-x_2 \ge 0\}$, and the orthogonal polynomials P_k^n in (3.23) become

$$P_k^n(\mathbf{x}) = P_{n-k}^{(2k+\beta+\gamma+1,\alpha)}(2x_1-1)(1-x_1)^k P_k^{(\gamma,\beta)}\left(\frac{2x_2}{1-x_1}-1\right). \tag{3.28}$$

Jacobi polynomials on the parabolic domain. Let $w_1(x) = x^a(1-x)^b$ on $[0,1]$, $w_2(x) = (1-x^2)^a$ on $[-1,1]$, and $\rho(x) = \sqrt{x}$. Then, the weight function (3.21) becomes

$$W_{a,b}(\mathbf{x}) = (1-x_1)^b(x_1-x_2^2)^a, \qquad x_2^2 < x_1 < 1, \quad \alpha,\beta > -1.$$

The domain $R = \{\mathbf{x} : x_2^2 < x_1 < 1\}$ is bounded by a straight line and a parabola. The orthogonal polynomials P_k^n in (3.23) become

$$P_k^n(\mathbf{x}) = p_{n-k}^{(a,b+k+1/2)}(2x_1-1)x_1^{k/2}p_k^{(b,b)}\left(\frac{x_2}{\sqrt{x_1}}\right). \tag{3.29}$$

3.4.3 Radial Symmetric Weight Function and Spherical Harmonics

A radial symmetric weight function is of the form $W(x) = w(\|x\|)$, where w is a nonnegative weight function on $\mathbb{R}_+$, and it is invariant under rotations. For the classical weight function $W_\mu(\mathbf{x}) = (1 - \|\mathbf{x}\|^2)^{\mu-\frac{1}{2}}$ on the unit ball $\mathbb{B}^d$, $w(t) = (1-t^2)^{\mu-\frac{1}{2}}\chi_{[0,1]}(t)$, where χ_E denotes the characteristic function of the set E.

For $d = 2$, an orthogonal basis for $\mathcal{V}_n^2(W)$ can be given in terms of the polar coordinates

$$\mathbf{x} = (x_1, x_2) = (r\cos\theta, r\sin\theta), \quad 0 \le \theta \le 2\pi \quad \text{and} \quad r \ge 0.$$

Proposition 3.4.1. *For $k \ge 0$, let $p_m^{(k)}$ denote the orthogonal polynomial of degree m with respect to the weight function $r^{k+1}w(r)$ on $\mathbb{R}_+$. Define*

$$\begin{aligned} P_{j,1}(\mathbf{x}) &= p_{2j}^{(2n-4j)}(r) r^{n-2j}\cos(n-2j)\theta, \\ P_{j,2}(\mathbf{x}) &= p_{2j}^{(2n-4j)}(r) r^{n-2j}\sin(n-2j)\theta. \end{aligned} \tag{3.30}$$

Then, $\{P_{j,1} : 0 \le j \le n/2\} \cup \{P_{j,2} : 1 \le j \le n/2\}$ is an orthogonal basis of $\mathcal{V}_n^2(W)$ for $W(\mathbf{x}) = w(\|\mathbf{x}\|)$.

Using polar coordinates and the orthogonality of $\sin m\theta$ and $\cos m\theta$ on the unit circle, the proof follows from a straightforward calculation. In particular, for $w_\mu(t) = (1-t^2)^{\mu-\frac{1}{2}}$, the polynomial $p_m^{(k)}$ can be given in terms of the Jacobi polynomials, so that we obtain another orthogonal basis for $\mathcal{V}_n^2(W_\mu)$.

Proposition 3.4.2. *Define in polar coordinates $(x,y) = (r\cos\theta, r\sin\theta)$,*

$$\begin{aligned} P_{j,1}(\mathbf{x}) &= P_j^{(\mu-\frac{1}{2}, n-2j)}(2r^2-1) r^{n-2j}\cos(n-2j)\theta, \\ P_{j,2}(\mathbf{x}) &= P_j^{(\mu-\frac{1}{2}, n-2j)}(2r^2-1) r^{n-2j}\sin(n-2j)\theta. \end{aligned} \tag{3.31}$$

Then, $\{P_{j,1} : 0 \le j \le \frac{n}{2}\} \cup \{P_{j,2} : 0 \le j < \frac{n}{2}\}$ is an orthogonal basis of $\mathcal{V}_n^2(W_\mu)$.

Let us normalize the basis (3.31) and denote the orthonormal basis by $\hat{P}_{j,1}$ and $\hat{P}_{j,2}$. Defining the vector $\mathbb{P}_n$ by $\mathbb{P}_{2m} = \left[\hat{P}_{0,1}, \hat{P}_{0,2}, \ldots, \hat{P}_{m-1,1}, \hat{P}_{m-1,2}, \hat{P}_{m,1}\right]^{\mathsf{T}}$ and $\mathbb{P}_{2m-1} = \left[\hat{P}_{0,1}, \hat{P}_{0,2}, \ldots, \ldots, \hat{P}_{m,-11}, \hat{P}_{m-1,2}\right]^{\mathsf{T}}$. Then, the matrices in $A_{n,i}$ in the three-term relation (3.5) are of a checkerboard pattern

$$A_{n,1} = \begin{bmatrix} * & 0 & * & \cdots \\ 0 & * & 0 & \cdots \\ * & 0 & * & \cdots \\ \cdots & \cdots & \cdots & \cdots \end{bmatrix} \quad \text{and} \quad A_{n,2} = \begin{bmatrix} 0 & * & * & \cdots \\ * & 0 & 0 & \cdots \\ 0 & * & * & \cdots \\ \cdots & \cdots & \cdots & \cdots \end{bmatrix}.$$

We can extend the basis (3.4.10) to radial symmetric weight function in more than two variables. This requires spherical harmonics that are analogs of trigonometric polynomials on the unit sphere.

Harmonic polynomials are polynomial solutions of $\Delta Y = 0$, where Δ is the Laplace operator. Spherical harmonics are the restriction of homogeneous harmonic polynomials. Let $\mathcal{H}_n^d$ denote the space of spherical harmonics of d variables. Since $\Delta : \mathcal{P}_n^d \mapsto \mathcal{P}_{n-2}^d$, it follows that

$$\dim \mathcal{H}_n^d = \dim \mathcal{P}_n^d - \mathcal{P}_{n-2}^d = \binom{n+d-1}{n} - \binom{n+d-3}{n-2}. \tag{3.32}$$

The spherical harmonics are orthogonal for the inner product in the space $L^2(\mathbb{S}^{d-1})$,

$$\langle f, g \rangle_{\mathbb{S}^{d-1}} = \frac{1}{\omega_d} \int_{\mathbb{S}^{d-1}} f(\xi) g(\xi) \mathrm{d}\sigma(\xi),$$

where $\mathrm{d}\sigma$ and ω_d denote the surface measure and the surface area of $\mathbb{S}^{d-1}$.

Theorem 3.4.3. *If $Y_n \in \mathcal{H}_n^d$, $Y_m \in \mathcal{H}_m^d$, and $n \neq m$, then $\langle Y_n, Y_m \rangle_{\mathbb{S}^{d-1}} = 0$.*

Since Y_n is homogeneous, $Y_n(r\xi) = r^n Y(\xi)$ in the spherical polar coordinates $x = r\xi$, $r > 0$ and $\xi \in \mathbb{S}^{d-1}$, so that $\frac{\mathrm{d}Y_n}{\mathrm{d}r}(x') = nY_n(x')$. The orthogonality follows from Green's identity

$$\begin{aligned}(n-m) \int_{\mathbb{S}^{d-1}} Y_n Y_m \mathrm{d}\sigma &= \int_{\mathbb{S}^{d-1}} \left(Y_m \frac{\mathrm{d}Y_n}{\mathrm{d}r} - Y_n \frac{\mathrm{d}Y_m}{\mathrm{d}r} \right) \mathrm{d}\sigma \\ &= \int_{\mathbb{B}^d} (Y_m \Delta Y_n - Y_n \Delta Y_m) \, \mathrm{d}\mathbf{x} = 0.\end{aligned}$$

For $d = 2$, $\{\cos n\theta, \sin n\theta\}$ is an orthogonal basis for $\mathcal{H}_n^2$ in the spherical coordinates $(\cos\theta, \sin\theta) \in \mathbb{S}^1$ or $0 \le \theta < 2\pi$. For $d > 2$, an orthogonal basis of $\mathcal{H}_n^d$ can be given explicitly via the Gegenbauer polynomials in spherical coordinates. The latter is given by

$$\begin{cases} x_1 = r \sin\theta_{d-1} \ldots \sin\theta_2 \sin\theta_1, \\ x_2 = r \sin\theta_{d-1} \ldots \sin\theta_2 \cos\theta_1, \\ \ldots \\ x_{d-1} = r \sin\theta_{d-1} \cos\theta_{d-2}, \\ x_d = r \cos\theta_{d-1}, \end{cases} \tag{3.33}$$

where $r \ge 0$, $0 \le \theta_1 \le 2\pi$, $0 \le \theta_i \le \pi$ for $i = 2, \ldots, d-1$. The spherical coordinates can be written as $\mathbf{x} = (\xi \sin\theta_{d-1}, \cos\theta_{d-1}) \in \mathbb{S}^{d-1}$ with $\xi \in \mathbb{S}^{d-2}$. Let $\mathrm{d}\sigma = \mathrm{d}\sigma_d$ be the surface measure on $\mathbb{S}^{d-1}$. Then, $\mathrm{d}\sigma_d(\mathbf{x}) = (\sin\theta_{d-1})^{d-2} \mathrm{d}\theta_{d-1} \mathrm{d}\sigma_{d-1}(\xi)$, and it follows by induction that

$$\mathrm{d}\sigma = \mathrm{d}\sigma_d = \prod_{j=1}^{d-2} \left(\sin\theta_{d-j}\right)^{d-j-1} \mathrm{d}\theta_{d-1}\ldots\mathrm{d}\theta_2\mathrm{d}\theta_1$$

in the spherical coordinates (3.33). Furthermore, we can write

$$\int_{\mathbb{S}^{d-1}} f(\mathbf{x})\mathrm{d}\sigma_d(\mathbf{x}) = \int_0^{\pi}\int_{\mathbb{S}^{d-2}} f(\xi\sin\theta,\cos\theta)\mathrm{d}\sigma_{d-1}(\xi)(\sin\theta)^{d-2}\mathrm{d}\theta. \tag{3.34}$$

This allows us to work recursively in dimension and leads to an orthogonal basis for spherical harmonics in terms of the Gegenbauer polynomials.

Theorem 3.4.4. *For $d>2$ and $\mathbf{v}\in\mathbb{N}_0^d$, define*

$$Y_{\mathbf{v}}(x) := [h_{\mathbf{v}}]^{-1} r^{|\mathbf{v}|} g_{\mathbf{v}}(\theta_1)\prod_{j=1}^{d-2}(\sin\theta_{d-j})^{|\mathbf{v}^{j+1}|} C_{\nu_j}^{\lambda_j}(\cos\theta_{d-j}), \tag{3.35}$$

where $g_{\mathbf{v}}(\theta_1)=\cos\nu_{d-1}\theta_1$ for $\nu_d=0$, $\sin(\nu_{d-1}+1)\theta_1$ for $\nu_d=1$, $|\mathbf{v}^j|=\nu_j+\cdots+\nu_d$, $\lambda_j=|\mathbf{v}^{j+1}|+(d-j-1)/2$, and

$$[h_{\mathbf{v}}]^2 := b_{\mathbf{v}}\prod_{j=1}^{d-2}\frac{(2\lambda_j)_{\nu_j}(\frac{d-j}{2})_{|\mathbf{v}^{j+1}|}\lambda_j}{\nu_j!(\frac{d-j+1}{2})_{|\mathbf{v}^{j+1}|}(\nu_j+\lambda_j)},$$

in which $b_{\mathbf{v}}=2$ if $\nu_{d-1}+\nu_d>0$ and $b_{\mathbf{v}}=1$ otherwise. Then, $\{Y_{\mathbf{v}}:|\mathbf{v}|=n,\nu_d=0,1\}$ is an orthonormal basis of $\mathcal{H}_n^d$; that is, $\langle Y_{\mathbf{v}},Y_{\mathbf{v}'}\rangle_{\mathbb{S}^{d-1}}=\delta_{\mathbf{v},\mathbf{v}'}$.

We are ready to give a basis for orthogonal polynomials for the radial symmetric weight function $W(x)=w(\|\mathbf{x}\|)$. In terms of the spherical-polar coordinates $\mathbf{x}=r\xi$, $r\ge 0$ and $\xi\in\mathbb{S}^{d-1}$, we can write

$$\int_{\mathbb{R}^d} f(\mathbf{x})W(\mathbf{x})\mathrm{d}x = \int_0^{\infty} r^{d-1}\int_{\mathbb{S}^{d-1}} f(r\xi)\mathrm{d}\sigma(\xi)\mathrm{d}r,$$

so that an orthogonal basis for the space $\mathcal{V}_n^d(W)$ of orthogonal polynomials can be given in terms of spherical harmonics.

Theorem 3.4.5. *For $0\le j\le n/2$, let $\left\{Y_{\mathbf{v}}^{n-2j}:|\mathbf{v}|=n-2j,\nu_d=0,1,\mathbf{v}\in\mathbb{N}_0^d\right\}$ be an orthonormal basis for the space $\mathcal{H}_{n-2j}^d$ and let $p_m^{(2n-4j+d-1)}$ denote the orthogonal polynomials of degree m with respect to $w_{j,n}(t)=t^{2n-4j+d-1}w(t)$ on $\mathbb{R}_+$. Define*

$$P_{j,\mathbf{v}}^n(\mathbf{x}) = p_{2j}^{(2n-4j+d-1)}(\|\mathbf{x}\|)Y_{\mathbf{v}}^{n-2j}(\mathbf{x}).$$

Then, $\{P_{j,\nu}^n:|\mathbf{v}|=n-2j,\ \nu_d=0, 0\le j\le n/2\}$ is an orthogonal basis of $\mathcal{V}_n^d(W)$.

The verification is immediate upon using $Y_{\mathbf{v}}^{n-2j}(\mathbf{x}) = r^{n-2j} Y_{\mathbf{v}}^{n-2j}(\xi)$ and counting the number of elements in the basis.

For the classical weight function $W_\mu(\mathbf{x}) = (1-\|\mathbf{x}\|^2)^{\mu-\frac{1}{2}}$ on the unit ball $\mathbb{B}^d$, $w(t) = (1-t^2)^{\mu-\frac{1}{2}}$ so that $w_{j,n}(t) = t^{2n-4j+d-1}(1-t^2)^{\mu-\frac{1}{2}}$. The orthogonal polynomials of even degree for $w_{j,n}$ can be given in terms of the Jacobi polynomials, which leads to the following corollary.

Corollary 3.4.6. *For W_μ on the unit ball $\mathbb{B}^d$, the orthogonal basis for $\mathcal{V}_n^d(W_\mu)$ in Theorem 3.4.5 consists of*

$$P_{j,\mathbf{v}}^n(\mathbf{x}) = P_j^{(\mu-\frac{1}{2},n-2j+\frac{d-2}{2})}(2\|\mathbf{x}\|^2-1) Y_{\mathbf{v}}^{n-2j}(\mathbf{x}).$$

For the weight function W_μ on the unit ball, another orthogonal basis can be given in terms of the Cartesian coordinates using (2.17), as we shall see later.

3.4.4 Orthogonal Polynomials on the Ball and Simplex

For the Jacobi weight function $W_{\boldsymbol{\kappa}}$ on the simplex Δ^d in (2.10), an orthogonal basis for $\mathcal{V}_n^d(W_{\boldsymbol{\kappa}})$ can be given by a wrapt product of the Jacobi polynomials, as an extension of the basis in (3.28) on the triangle. We adopt the following notations: Associated with $x \in \mathbb{R}^d$, we define $\mathbf{x}_0 = 0$ and $\mathbf{x}_j = (x_1,\ldots,x_j)$ for $1 \le j \le d$. For $\boldsymbol{\kappa} \in \mathbb{R}^{d+1}$, let $|\boldsymbol{\kappa}^j| = \kappa_j + \cdots + \kappa_{d+1}$ for $1 \le j \le d+1$ and, for $\mathbf{v} \in \mathbb{N}^d$, let $|\mathbf{v}^j| = v_j + \cdots + v_d$ for $1 \le j \le d$ and $|\mathbf{v}^{d+1}| = 0$.

Theorem 3.4.7. *For $\mathbf{v} \in \mathbb{N}_0^d$ and $|\mathbf{v}| = n$, the polynomials*

$$P_{\mathbf{v}}(W_{\boldsymbol{\kappa}};\mathbf{x}) = [h_{\mathbf{v}}]^{-1} \prod_{j=1}^{d} (1-|\mathbf{x}_{j-1}|)^{v_j} P_{v_j}^{(a_j,\kappa_j)}\left(\frac{2x_j}{1-|\mathbf{x}_{j-1}|}-1\right),$$

where $a_j = 2|\mathbf{v}^{j+1}| + |\boldsymbol{\kappa}^{j+1}| + d - j$ and

$$[h_{\mathbf{v}}]^2 = \frac{1}{(|\boldsymbol{\kappa}|+d+1)_{2|\mathbf{v}|}} \prod_{j=1}^{d} \frac{(a_j+\kappa_j+1)_{2v_j}(a_j+1)_{v_j}(\kappa_j+1)_{v_j}}{(a_j+\kappa_j+1)_{v_j} v_j!},$$

consist of an orthonormal basis of $\mathcal{V}_n^d(W_{\boldsymbol{\kappa}})$.

This orthogonal basis is an extension of the basis (3.28) on the triangle, and it can be verified by using (2.14), which writes the integral over Δ^d as a wrapt product of the integral over $[0,1]$ and Δ^{d-1}.

The orthogonal polynomials on the simplex and those on the unit ball are closely related. Let $\mathcal{V}_n(W_{\boldsymbol{\kappa}},\Delta^d)$ be the space of orthogonal polynomials of degree n for the Jacobi weight on the simplex. Let $\mathbf{x}^2 = (x_1^2,\ldots,x_d^2)$ for $\mathbf{x} \in \Delta^d$ and define

$$W_{\kappa}^{\mathbb{B}}(\mathbf{x}) = \prod_{i=1}^{d} |x_i|^{2\kappa_i+1}(1-\|\mathbf{x}\|^2)^{\kappa_{d+1}} = |x_1|\cdots|x_d| W_{\kappa}(\mathbf{x}^2), \quad \mathbf{x}\in\mathbb{B}^d,$$

as in (2.26). The space $\mathcal{V}_n(W_{\kappa}^{\mathbb{B}},\mathbb{B}^d)$ of orthogonal polynomials for the weight function $W_{\kappa}^{\mathbb{B}}$ on the unit ball can be decomposed in terms of $\mathcal{V}_n(W_{\kappa+\alpha},\Delta^d)$ for $\alpha = (\varepsilon,0)$ with $\varepsilon\in\{0,1\}^d$.

Theorem 3.4.8. *Let $\psi:\mathbf{x}\mapsto\mathbf{x}^2$ for $\mathbf{x}\in\Delta^d$ and define*

$$\mathcal{V}_n(W_{\kappa},\Delta^d)\circ\psi := \left\{P\circ\psi : P\in\mathcal{V}_n(W_{\kappa},\Delta^d)\right\}.$$

Then, for $\kappa_i > -1$, $1\le i\le d+1$,

$$\mathcal{V}_{2n}\left(W_{\kappa}^{\mathbb{B}},\mathbb{B}^d\right) = \bigoplus_{\varepsilon\in\{0,1\}^d,|\varepsilon|\,\text{even}} x^{\varepsilon}\mathcal{V}_{n-\frac{1}{2}|\varepsilon|}\left(W_{\kappa+(\varepsilon,0)},\Delta^d\right)\circ\psi,$$

$$\mathcal{V}_{2n+1}\left(W_{\kappa}^{\mathbb{B}},\mathbb{B}^d\right) = \bigoplus_{\varepsilon\in\{0,1\}^d,|\varepsilon|\,\text{odd}} x^{\varepsilon}\mathcal{V}_{n-\frac{1}{2}(|\varepsilon|-1)}\left(W_{\kappa+(\varepsilon,0)},\Delta^d\right)\circ\psi.$$

The proof follows readily from the parity of polynomials and the integral identity

$$\int_{\mathbb{B}^d} f\left(x_1^2,\ldots x_d^2\right)\mathrm{d}\mathbf{x} = \int_{\Delta^d} f(x_1,\ldots x_d)\frac{\mathrm{d}\mathbf{x}}{\sqrt{x_1,\ldots,x_d}}.$$

The orthogonal basis for $\mathcal{V}_n^d(W_{\kappa}^{\mathbb{B}},\mathbb{B}^d)$ derived from the basis $\mathcal{V}_n(W_{\kappa};\Delta^d)$ of Theorem 3.4.7 can be written collectively in terms of the generalized Gegenbauer polynomials $C_n^{(\lambda,\mu)}$, defined in (1.19), as follows:

Theorem 3.4.9. *An orthonormal basis for $\mathcal{V}_n(W_{\kappa}^{\mathbb{B}},\mathbb{B}^d)$ consists of, for $|\nu| = n$,*

$$P_{\nu}\left(W_{\kappa}^{\mathbb{B}};\mathbf{x}\right) = [h_{\nu}]^{-1}\prod_{j=1}^{d}(1-\|\mathbf{x}_{j-1}\|^2)^{\frac{\nu_j}{2}} C_{\nu_j}^{(\lambda_j,\kappa_j+\frac{1}{2})}\left(\frac{x_j}{\sqrt{1-\|\mathbf{x}_{j-1}\|^2}}\right), \tag{3.36}$$

where $\lambda_j = |\nu^{j+1}| + |\kappa^{j+1}| + d - j + \frac{1}{2}$ and h_{ν} is as in Theorem 3.4.7.

For $\kappa_1 = \cdots = \kappa_d = -\frac{1}{2}$ and $\kappa_{d+1} = \mu - \frac{1}{2}$, Theorem 3.4.9 shows that an orthogonal basis for the classical weight function, W_{μ}, on the unit ball can be given in terms of orthogonal polynomials on the simplex. Since $C_n^{(\lambda,0)} = C_n^{\lambda}$ is the classical Gegenbauer polynomial, the orthonormal basis for $\mathcal{V}_n(W_{\mu},\mathbb{B}^d)$ consists of

$$P_{\nu}(W_{\mu};\mathbf{x}) := [h_{\nu}]^{-1}\prod_{j=1}^{d}(1-\|\mathbf{x}_{j-1}\|^2)^{\frac{\nu_j}{2}} C_{\nu_j}^{\lambda_j}\left(\frac{x_j}{\sqrt{1-\|\mathbf{x}_{j-1}\|^2}}\right), \tag{3.37}$$

where $\lambda_j = \mu + |\mathbf{v}^{j+1}| + \frac{d-j}{2}$ and $h_{\mathbf{v}}$ is given by

$$[h_{\mathbf{v}}]^2 = \frac{(\mu+\frac{d}{2})_n}{(\mu+\frac{d+1}{2})_n} \prod_{j=1}^{d} \frac{(\mu+\frac{d-j}{2})_{|\mathbf{v}^j|}(2\mu+2|\mathbf{v}^{j+1}|+d-j)_{v_j}}{(\mu+\frac{d-j+1}{2})_{|\mathbf{v}^j|} v_j!}, \tag{3.38}$$

and the basis holds under $\lim_{\lambda\to 0}\lambda^{-1}C_n^\lambda(x) = (2/n)T_n(x)$ whenever $\lambda_j = 0$.

For $d = 2$, this is the basis given in (3.26), and the orthogonality of (3.36) can also be verified by using (2.17). Comparing with Corollary 3.4.6, we obtain two distinct orthonormal bases for the classical orthogonal polynomials on the unit ball.

Along the same line of reasoning, we can also obtain orthogonal polynomials on the sphere from those on the unit ball and, as shown earlier, from the simplex as well. Associate with $\mathbf{x} \in \mathbb{B}^d$, we define $\mathbf{y} \in \mathbb{R}^{d+1}$ so that $\mathbf{y} = r\xi$ with $\xi = (\mathbf{x}, x_{d+1}) \in \mathbb{S}^d$ in the spherical polar coordinates, where $x_{d+1}^2 = 1 - \|\mathbf{x}\|^2$. Let $\mathbf{1} = (1,1,\ldots,1)$ and $\mathbf{e}_{d+1} = (0,0,\ldots,0,1) \in \mathbb{R}^{d+1}$. For $|\mathbf{v}| = n$ and $|\mathbf{v}'| = n-1$, define

$$Y_{\mathbf{v}}^{(1)}(\mathbf{y}) = r^n P_{\mathbf{v}}\left(W_{\boldsymbol{\kappa}-\frac{1}{2}\mathbf{1}}^{\mathbb{B}};\mathbf{x}\right) \quad \text{and} \quad Y_{\mathbf{v}'}^{(2)}(\mathbf{y}) = r^{n+1} x_{d+1} P_{\mathbf{v}'}\left(W_{\boldsymbol{\kappa}-\frac{1}{2}\mathbf{1}+\frac{1}{2}\mathbf{e}_{d+1}}^{\mathbb{B}};\mathbf{x}\right).$$

Taking into account the parity of $P_{\mathbf{v}}(W_{\kappa}^{\mathbb{B}})$, it is easy to see that both $Y_{\mathbf{v}}^{(1)}$ and $Y_{\mathbf{v}}^{(2)}$ are homogeneous polynomials in y and are of degree n.

Theorem 3.4.10. *The polynomials in $\{Y_{\mathbf{v}}^{(1)} : |\mathbf{v}| = n\} \cup \{Y_{\mathbf{v}}^{(2)} : |\mathbf{v}| = n-1\}$ consist of an orthogonal basis for the space $\mathcal{V}_n^{d+1}(\mathsf{W}_{\boldsymbol{\kappa}};\mathbb{S}^d)$ of homogeneous orthogonal polynomials with respect to the weight function*

$$\mathsf{W}_{\boldsymbol{\kappa}}(\mathbf{y}) = \prod_{i=1}^{d+1} |y_i|^{2\kappa_i}, \qquad \kappa_i \ge 0, \quad x \in \mathbb{S}^d.$$

The orthogonality for either $\{Y_{\mathbf{v}}^{(1)} : |\mathbf{v}| = n\}$ or $\{Y_{\mathbf{v}}^{(2)} : |\mathbf{v}| = n-1\}$ follows from the integral identity (2.20), and the orthogonality between $Y_{\mathbf{v}}^{(1)}$ and $Y_{\mathbf{v}'}^{(2)}$ follows from the different parity of the two families since $\mathsf{W}_{\boldsymbol{\kappa}}$ is even in each of its variables.

When $\kappa_i = 0$ for all i in this theorem, this shows that an orthogonal basis for the space $\mathcal{H}_n^{d+1}$ of spherical harmonics of $d+1$ variables can be deduced from the orthogonal polynomials of degree n for $W_{-\frac{1}{2}}(\mathbf{x}) = 1/\sqrt{1-\|\mathbf{x}\|^2}$ and orthogonal polynomials of degree $n-1$ for $W_{\frac{1}{2}}(\mathbf{x}) = \sqrt{1-\|\mathbf{x}\|^2}$. For $d = 1$, this spells the familiar relation between the spherical harmonics $\{\cos n\theta, \sin n\theta\}$ and the Chebyshev polynomials of the first and the second kind.

3.4.5 Orthogonal Polynomials Derived from Symmetry

These polynomials are deduced from product orthogonal polynomials by a change of variables via symmetric polynomials. We consider the case $d = 2$ first.

A. Polynomials of two variables. Let w be a nonnegative weight function defined on $\mathbb{R}$. The mapping $(x_1, x_2) \mapsto (u_1, u_2)$

$$u_1 = x_1 + x_2 \quad \text{and} \quad u_2 = x_1 x_2, \tag{3.39}$$

maps the domain $\triangle = \{(x_1, x_2) : x_1 < x_2, x_1, x_2 \in \operatorname{supp}(w)\}$ in $\mathbb{R}^2$ one-to-one and onto the domain Ω defined by

$$\Omega = \left\{(u_1, u_2) : u_1^2 \ge 4u_2, \ \tfrac{1}{2}\left(u_1 \pm \sqrt{u_1^2 - 4u_2}\right) \in \operatorname{supp}(w)\right\}. \tag{3.40}$$

For w supported on $[-1, 1]$, the domain $\triangle$ is a triangle and Ω is given by

$$\Omega = \left\{(u_1, u_2) : u_1^2 \ge 4u_2, 1 + u_2 \ge |u_1|, \quad -2 \le u_1 \le 2\right\}, \tag{3.41}$$

which is bounded by two lines and a parabola, as depicted in Figure 3.1.

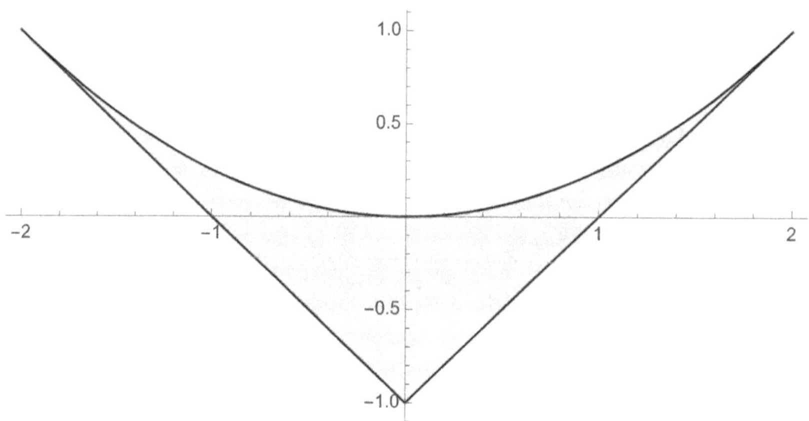

Figure 3.1 The domain Ω when w is supported on $[-1, 1]$.

For $\gamma \ge -\frac{1}{2}$, let W_γ be the weight function defined on Ω by

$$W_\gamma(u_1, u_2) = w(x_1)w(x_2)(u_1^2 - 4u_2)^\gamma, \quad (u_1, u_2) \in \Omega, \tag{3.42}$$

under the mapping (3.39). Since the map $u \mapsto x$ maps Ω to $\triangle$ and the latter is half of the product domain $\{\mathbf{x} \in \mathbb{R}^2 : x_1, x_2 \in \operatorname{supp}(w)\}$, and the Jacobian of the map is $|x_1 - x_2| = \sqrt{u_1^2 - 4u_2}$, we obtain

$$\int_\Omega f(\mathbf{u}) W_\gamma(\mathbf{u}) d\mathbf{u} = \frac{1}{2} \int_\triangle f(x_1 + x_2, x_1 x_2) w(x_1) w(x_2) |x_1 - x_2|^{2\gamma+1} d\mathbf{x}. \quad (3.43)$$

Let $\mathcal{V}_n(W_\gamma)$ denote the space of orthogonal polynomials of degree n with respect to W_γ on the domain Ω. An orthogonal basis can be given explicitly when $\gamma = \pm\frac{1}{2}$.

Proposition 3.4.11. *Let $p_n = p_n(w)$ be the orthonormal polynomial of degree n with respect to w. Under the mapping* (3.39)*, define*

$$P_k^{n,-\frac{1}{2}}(u_1, u_2) = \begin{cases} p_n(x_1) p_k(x_2) + p_n(x_2) p_k(x_1), & k < n, \\ \sqrt{2} p_n(x_1) p_n(x_2), & k = n. \end{cases} \quad (3.44)$$

Then, $\{P_k^{n,-\frac{1}{2}} : 0 \le k \le n\}$ is an orthonormal basis of $\mathcal{V}_n(W_{-\frac{1}{2}})$. Moreover, under the mapping (3.39)*, define*

$$P_k^{n,\frac{1}{2}}(u_1, u_2) = \frac{p_{n+1}(x_1) p_k(x_2) - p_{n+1}(x_2) p_k(x_1)}{x_1 - x_2}, \quad 0 \le k \le n. \quad (3.45)$$

Then, $\{P_k^{n,\frac{1}{2}} : 0 \le k \le n\}$ is an orthonormal basis of $\mathcal{V}_n(W_{\frac{1}{2}})$.

Proof First of all, the right-hand side of the polynomials in (3.44) is symmetric polynomials in (x_1, x_2). Since $x_1^n x_2^k = u_1^{n-k} u_2^k + q(u_1, u_2)$ for $0 \le k \le n$, where q is a polynomial of degree at most $n-1$, it is easy to see that $P_k^{n,-\frac{1}{2}}$ is a polynomial of degree n in (u_1, u_2). Similarly, $P_k^{n,-\frac{1}{2}}$ is a polynomial of degree n in (u_1, u_2). Making a change variable $u \mapsto x$ in (3.43), then the integrant in the right-hand side is symmetric in (x_1, x_2), so that we can write the integral over $\triangle$ as half of the integral over $\mathbb{R}^2$. If $\gamma = -\frac{1}{2}$, then the orthonormality of $P_k^{n,-\frac{1}{2}}$ follows immediately from (3.43) and the orthonormality of $p_n(w)$. If $\gamma = \frac{1}{2}$, the denominator $(x_1 - x_2)^2$ in $P_k^{n,\frac{1}{2}} P_j^{m,\frac{1}{2}}$ cancels out the $|x_1 - x_2|^2$ on the right-hand side of (3.43), so that the orthonormality of $P_k^{n,-\frac{1}{2}}$ follows again from the orthonormality of $p_n(w)$. □

The weight function W_γ and its domain Ω in the case that $p_n(w)$ is a classical orthogonal polynomial are given later.

Example 3.4.12. If $w(t) = (1-t)^\alpha (1+t)^\beta$ is the Jacobi weight function on $[-1, 1]$, then $p_n(w) = P_n^{(\alpha,\beta)}$ is the Jacobi polynomial, and the weight W_γ becomes

$$W_{\alpha,\beta,\gamma}(u_1, u_2) = (1 - u_1 + u_2)^\alpha (1 + u_1 + u_2)^\beta (u_1^2 - 4u_2)^\gamma, \quad \alpha, \beta > -1,$$

on the domain Ω given in (3.41) since $(1 \pm x_1)(1 \pm x_2) = 1 \pm u_1 + u_2$.

Example 3.4.13. If $w(t) = t^\alpha e^{-t}$ is the Laguerre weight function on $\mathbb{R}_+$, then $p_n(w) = L_n^\alpha$ is the Laguerre polynomial, and the weight W_γ becomes

$$W_{\alpha,\gamma}(u_1,u_2) = u_2^\alpha (u_1^2 - 4u_2)^\gamma e^{-u_1}, \quad \alpha > -1,$$

on the domain $\Omega = \{(u_1,u_2) : 0 \le 4u_2 \le u_1^2\}$.

Example 3.4.14. If $w(t) = e^{-t^2}$ is the Hermite weight function on $\mathbb{R}$, then $p_n(w) = H_n$ is the Hermite polynomial, and the weight W_γ becomes

$$W_\gamma(u_1,u_2) = (u_1^2 - 4u_2)^\gamma e^{-u_1^2 + 2u_2}$$

on the domain $\Omega = \{(u_1,u_2) : 4u_2 \le u_1^2,\ u_1 \in \mathbb{R}\}$.

B. Polynomials of more than two variables. We extend the abovementioned construction to d variables using elementary symmetric polynomials. A polynomial $f \in \Pi^d$ is called *symmetric* if f is invariant under any permutation of its variables. Let $\mathcal{S}_d$ be the symmetric group of d objects. A permutation of $\mathbf{x} \in \mathbb{R}^d$ is $\mathbf{x}\sigma = (x_{\sigma(1)},\ldots,x_{\sigma(d)})$ for some $\sigma \in \mathcal{S}_d$. A symmetric polynomial satisfies $f(\mathbf{x}) = f(\mathbf{x}\sigma)$ for all $\sigma \in \mathcal{S}_d$. In the following, we also denote by $\mathcal{S}_d$ the set of permutations of $\{1,2,\ldots,d\}$. Symmetric polynomials are indexed by partitions. A partition λ of d parts is an element in $\mathbb{N}_0^d$ such that $\lambda_1 \ge \lambda_2 \ge \cdots \ge \lambda_d$. Let Λ^d denote the set of such partitions,

$$\Lambda^d = \{\lambda \in \mathbb{N}_0^{d+1} : \lambda_1 \ge \lambda_2 \ge \cdots \ge \lambda_d\}.$$

Let w be a nonnegative measure on $\mathbb{R}$ and let $p_n = p_n(w)$ be the orthonormal polynomial of degree n with respect to w. We define two families of polynomials, denoted by $Q_\alpha^{\pm\frac{1}{2}}$, that are symmetric and orthogonal.

Proposition 3.4.15. *For $\alpha \in \Lambda^d$, define*

$$Q_\alpha^{-\frac{1}{2}}(\mathbf{x}) := \sum_{\beta \in \mathcal{S}_d} p_{\alpha_1}(x_{\beta_1}) \ldots p_{\alpha_d}(x_{\beta_d}). \tag{3.46}$$

Then, $\{Q_\alpha^{-\frac{1}{2}} : \alpha \in \Lambda^d\}$ is a family of symmetric polynomials and satisfies

$$\int_{\mathbb{R}^d} Q_\alpha^{-\frac{1}{2}}(\mathbf{x}) Q_\beta^{-\frac{1}{2}}(\mathbf{x}) \prod_{i=1}^d w(x_i) d\mathbf{x} = m_1! \ldots m_{d'}!\, \delta_{n,m}\, \delta_{\alpha,\beta}, \tag{3.47}$$

where d' is the number of distinct elements in α and m_i is the number of occurrences of the ith distinct element in α.

The polynomials $Q_\alpha^{-\frac{1}{2}}$ are symmetric. Their orthogonality follows from that of $p_n(w)$ by integrating the left-hand side of (3.47) one variable at a time. The second family is more involved.

Theorem 3.4.16. *For $\alpha \in \Lambda^d$ and $x \in \mathbb{R}^d$, define*

$$J_\alpha(\mathbf{x}) := \det\left[p_{\alpha_i+d-i}(x_j)\right]_{i,j=1}^d \quad \text{and} \quad J(\mathbf{x}) = \prod_{1\le i<j\le d}(x_i - x_j).$$

Then, the family of polynomials $\{Q_\alpha^{\frac{1}{2}}\}$ defined by

$$Q_\alpha^{\frac{1}{2}}(\mathbf{x}) = \frac{J_\alpha(\mathbf{x})}{J(\mathbf{x})}, \qquad \alpha \in \Lambda^d, \tag{3.48}$$

are symmetric and satisfy

$$\int_{\mathbb{R}^d} Q_\alpha^{\frac{1}{2}}(\mathbf{x})\, Q_\beta^{\frac{1}{2}}(\mathbf{x})\, [J(\mathbf{x})]^2 \prod_{i=1}^d w(x_i)\mathrm{d}\mathbf{x} = \delta_{\alpha,\beta}. \tag{3.49}$$

Proof First of all, since $J_\alpha(x)$ vanishes for $x_i = x_j$, $i \neq j$, $Q_\alpha^{\frac{1}{2}}$ is indeed a polynomial. Secondly, since J is the Vandermonde determinant, $J(\mathbf{x}) = \det(x_j^{\alpha_i})_{i,j=1}^d$, a permutation of variables changes rows of J_α and J simultaneously so that it does not change the value of their ratio. Thus, $Q_\alpha^{\frac{1}{2}}$ is indeed a symmetric polynomial. The orthogonality can be verified directly by writing the determinant as $\det[a_{i,j}]_{i,j=1}^d = \sum_{\sigma\in S_d} \operatorname{sign}(\sigma) a_{1,\sigma(1)} \cdots a_{d,\sigma(d)}$ or using the identity

$$\det\left[\int_{\mathbb{R}} f_i(t) g_i(t) w(t)\mathrm{d}t\right]_{i,j,=1}^d = \int_{\mathbb{R}^d} \det[f_i(t_j)]_{i,j,=1}^d \det[g_i(t_j)]_{i,j,=1}^d \prod_{j=1}^d w(t_j)\mathrm{d}t.$$

In either case, the orthogonality follows from that of $p_n(w)$. □

Our example of the Gauss cubature rules requires the corresponding orthogonal polynomials to have a maximal number of common zeros. Like in the case of $d = 2$, we derive them using $Q_\alpha^{\pm\frac{1}{2}}$ by a change of variables. Let e_k be the elementary symmetric polynomials defined by

$$e_k(\mathbf{x}) := \sum_{1\le i_1<\cdots<i_k\le d} x_{i_1} \ldots x_{i_k}, \qquad k = 1, 2, \ldots, d.$$

These polynomials generate the algebra of symmetric polynomials. Indeed, the fundamental theorem of symmetric polynomials states that every symmetric polynomial f can be uniquely represented as a polynomial of elementary symmetric polynomials. Let w be the weight function on $\mathbb{R}$. Let $\triangle^d$ be the domain

$$\triangle^d = \{\mathbf{x} \in \mathbb{R}^d \mid x_1 < x_2 < \cdots < x_d, x_i \in \operatorname{supp}(w)\}.$$

As an extension of (3.39) to d-variables, we consider the mapping $\mathbf{x} \in \triangle^d \mapsto u \in \Omega$, defined by

$$u_k = e_k(x_1, \dots x_d), \qquad k = 1, 2, \dots, d. \tag{3.50}$$

For $d = 2$, this agrees with (3.39). The domain Ω is the image of $\triangle^d$ under the mapping. For w defined on $[-1, 1]$, the domain is depicted in Figure 3.1 when $d = 2$. For $d = 3$, the domain is depicted in Figure 3.2. The mapping $\mathbf{x} \mapsto \mathbf{u}$ is

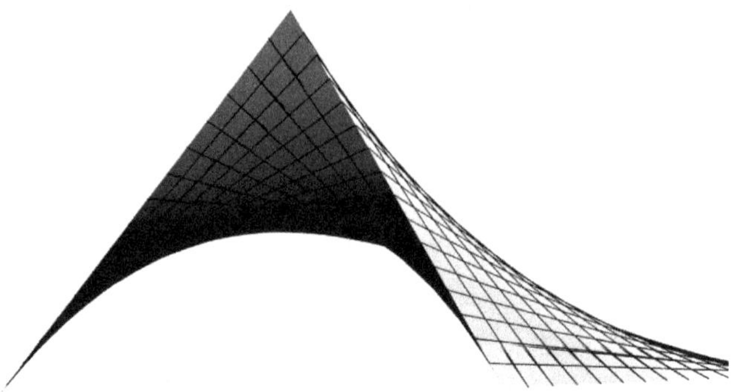

Figure 3.2 Ω for $d = 3$ and $\mathrm{supp}(w) = [-1, 1]$.

well defined since the Jacobian of the mapping satisfies

$$\det\left[\left(\frac{\partial e_i}{\partial x_j}\right)\right] = J(\mathbf{x}) = \prod_{1 \le i < j \le d} (x_i - x_j)^2.$$

Indeed, let us denote by $D(\mathbf{x})$ the determinant on the left-hand side. Then, $D(\mathbf{x}) = J(\mathbf{x})$ can be verified by using the identity

$$\begin{aligned} \frac{\partial e_i}{\partial x_j} &= e_{i-1}(x_1, \dots, x_{j-1}, x_{j+1}, \dots, x_d), \\ &= e_{i-1}(x_2, \dots, x_d) + (x_1 - x_j) e_{i-2}(x_2, \dots, x_{j-1}, x_{j+1}, \dots, x_d), \end{aligned}$$

where we define $e_{-1}(\mathbf{x}) = 0$ and $e_0(\mathbf{x}) = 1$, which allows us to extract $x_1 - x_i$ from the determinant by subtracting the first column from other columns to deduce $D(\mathbf{x}) = \prod_{i=2}^d (x_1 - x_i) D(x_2, \dots, x_{d-1})$, which shows $D(\mathbf{x}) = J(\mathbf{x})$. Let

$$\Delta(\mathbf{u}) = [J(\mathbf{x})]^2, \qquad u_k = e_k(\mathbf{x}), \quad 1 \le k \le d.$$

Under the mapping (3.50), the polynomials $Q_\alpha^{\pm\frac{1}{2}}(\mathbf{x})$ become polynomials in $\mathbf{u}$ variables, which are orthogonal with respect to the weight function $W_{\pm\frac{1}{2}}$ on Ω, where

$$W_{-\frac{1}{2}}(\mathbf{u}) = \prod_{i=1}^{d} w(x_i)[\Delta(\mathbf{u})]^{-\frac{1}{2}} \quad \text{and} \quad W_{\frac{1}{2}}(\mathbf{u}) = \prod_{i=1}^{d} w(x_i)[\Delta(\mathbf{u})]^{\frac{1}{2}}. \tag{3.51}$$

More precisely, an orthogonal basis for $\mathcal{V}_n(W_{\pm\frac{1}{2}})$, the space of orthogonal polynomials of degree n with respect to $W_{\pm\frac{1}{2}}$ on Ω, can be derived.

Proposition 3.4.17. *Let $Q_\alpha^{\pm\frac{1}{2}}$ be the symmetric orthogonal polynomials in* (3.47) *and* (3.49). *For $\alpha \in \Lambda^d$ and $n = \alpha_1 \ge \cdots \ge \alpha_d$, define*

$$P_\alpha^{n,-\frac{1}{2}}(\mathbf{u}) = Q_\alpha^{-\frac{1}{2}}(\mathbf{x}) \quad \text{and} \quad P_\alpha^{n,\frac{1}{2}}(u) = Q_\alpha^{\frac{1}{2}}(\mathbf{x}),$$

where $\mathbf{u} = \mathbf{u}(\mathbf{x})$ as in (3.50). *Then, $\mathbb{P}_n^{\pm\frac{1}{2}} = \left\{P_\alpha^{n,\pm\frac{1}{2}} : n = \alpha_1 \ge \cdots \ge \alpha_d \ge 0\right\}$ is an orthogonal basis of $\mathcal{V}_n(W_{\pm\frac{1}{2}})$.*

Proof For $n = \alpha_1 \ge \cdots \ge \alpha_d$, the polynomial $Q_\alpha^{-\frac{1}{2}}(\mathbf{x})$ is evidently a polynomial of degree n in each of its variables so that it is a symmetric polynomial of degree n. By the fundamental theorem of symmetric functions, it is a polynomial of degree n in $e_1, \ldots, e_d$. Moreover, it contains the term $[e_n(\mathbf{x})]^n$. Consequently, under the mapping (3.50), $P_\alpha^{-\frac{1}{2}}(\mathbf{u})$ is a polynomial of degree n in $\mathbf{u}$. The orthogonality follows immediately from the change of variable

$$\int_\Omega g(\mathbf{u})[\Delta(\mathbf{u})]^{-\frac{1}{2}} d\mathbf{u} = \int_{\mathcal{R}} g(\mathbf{u}(\mathbf{x}))d\mathbf{x} = \frac{1}{d!}\int_{\mathbb{R}^d} g(\mathbf{u}(\mathbf{x}))d\mathbf{x}, \tag{3.52}$$

where $\mathbf{u}(\mathbf{x}) = (e_1(\mathbf{x}), \ldots, e_d(\mathbf{x}))$ and the second equality follows since $\mathbf{u}(\mathbf{x})$ is symmetric, and the orthogonality of $Q_\alpha^{-\frac{1}{2}}(\mathbf{x})$. Moreover, let $N(n,d)$ be the cardinality of the index set,

$$N(n,d) = \#\left\{\alpha \in \mathbb{N}_0^d : n = \alpha_1 \ge \cdots, \ge \alpha_d\right\}.$$

A moment reflection shows that $N(n,d) = \sum_{k=0}^n N(k,d-1)$, so that $N(n,d) = \binom{n+d-1}{n}$. Consequently, we have shown that $\mathbb{P}^{-\frac{1}{2}}$ consists of an orthogonal basis of $\mathcal{V}_n(W_{-\frac{1}{2}})$.

A similar proof holds for $\mathbb{P}^{\frac{1}{2}}$. Indeed, for $n = \alpha_1 \ge \cdots \ge \alpha_d$, the nominator $J_\alpha(\mathbf{x})$ of $Q_\alpha^{-\frac{1}{2}}(\mathbf{x})$ is a polynomial of degree $n+d-1$ in each x_j and the denominator $J(\mathbf{x})$ is a polynomial of degree $d-1$ in x_j, so that $Q_\alpha^{\frac{1}{2}}(\mathbf{x})$ is a polynomial of degree n in each of its variables. Consequently, $P_\alpha^{\frac{1}{2}}(\mathbf{u})$ is a polynomial of degree n. The orthogonality of $P_\alpha^{\frac{1}{2}}(\mathbf{u})$ again follows that of $Q_\alpha^{\frac{1}{2}}(\mathbf{u})$ by the change of variables $\mathbf{x} \mapsto \mathbf{u}(\mathbf{x})$, where $[J(\mathbf{x})]^2$ in (3.49) becomes, when combining with the Jacobian, $[\Delta(\mathbf{u})]^{\frac{1}{2}}$ in $W_{\frac{1}{2}}$. □

3.5 Multi-index Hankel Matrix

A matrix $H : r_n^d \times r_m^d$ is called a multi-index Hankel matrix if it is of the form

$$H = (h_{\alpha+\beta})_{|\alpha|=n,|\beta|=n-1},$$

where the rows and columns are indexed in lexicographical order of $\{\alpha : |\alpha| = n\}$ and $\{\beta : |\beta| = n-1\}$, respectively. In particular, the moment matrix $M_{n,d}$ in (3.2) is a multi-index Hankel matrix. Let $L_{n,i}$, $1 \le i \le d$, be the matrices defined in (3.8). We establish two lemmas on these matrices that will be needed later.

Lemma 3.5.1. *A matrix $H_n : r_n^d \times r_{n-1}^d$ satisfies the equation*

$$L_{n,i} H_n L_{n-1,j}^{\mathsf{T}} = L_{n,j} H_n L_{n-1,i}^{\mathsf{T}}, \quad 1 \le i, j \le d, \tag{3.53}$$

if and only if it is a multi-index Hankel matrix.

Proof If H is a multi-index Hankel matrix, we define a linear functional $\mathcal{L}$ on the space of polynomials by $\mathcal{L}(\mathbf{x}^\alpha) = h_\alpha$. Recall that $\mathbf{x}^n$ is the column vector of $\{\mathbf{x}^\alpha : |\alpha| = n\}$. We can then write $H_n = \mathcal{L}(\mathbf{x}^n (\mathbf{x}^{n-1})^{\mathsf{T}})$, from which (3.53) follows readily. We now prove the other direction. Let $H_n = (h_{\alpha,\beta})_{|\alpha|=n,|\beta|=n-1}$ in (3.53). We need to prove

$$h_{\alpha,\beta} = h_{\alpha',\beta'}, \qquad \alpha+\beta = \alpha'+\beta', \tag{3.54}$$

where $|\alpha| = |\alpha'| = n$, $|\beta| = |\beta'| = n-1$. We call such (α,β) and (α',β') related.

First, we claim that we only need to prove (3.54) for the special case that α and α' differ at only two components, say, $\alpha_i \ne \alpha_i'$ and $\alpha_j \ne \alpha_j'$. Indeed, since $\alpha+\beta = \alpha'+\beta'$, this means that β and β' differ at only two components, $\beta_i \ne \beta_i'$ and $\beta_j \ne \beta_j'$, and we can always relate these two pairs by finite many intermediate pairs $(\alpha^{(k)},\beta^{(k)})$, where two consecutive pairs, say $(\alpha^{(k)},\beta^{(k)})$ and $(\alpha^{(k+1)},\beta^{(k+1)})$, are related and differ only at two components. Let

$$\mathcal{N}_n = \{\alpha \in \mathbb{N}_0^d : |\alpha| = n\} \quad \text{and} \quad \mathcal{N}_{n,i} = \{\alpha \in \mathbb{N}_0^d : |\alpha| = n\, \alpha_i > 0\}.$$

If (α,β) and (α',β') are related and differ only at their ith and jth components, we show that either

$$\alpha \in \mathcal{N}_{n,i}, \quad \beta \in \mathcal{N}_{n-1,j}, \quad \alpha' \in \mathcal{N}_{n,j}, \quad \beta' \in \mathcal{N}_{n-1,i}, \tag{3.55}$$

or (3.54) with i and j interchanged. Indeed, if $\alpha_i = 0$, then $\alpha_j \ne 0$, and from $|\alpha| = |\alpha'|$ or $\alpha_i + \alpha_j = \alpha_i' + \alpha_j'$, we have $\alpha_i' \ne 0$. Also, from $\alpha_i + \beta_i = \alpha_i' + \beta_i'$, we have $\beta_i \ne 0$. It then follows from $\beta_i + \beta_j = \beta_i' + \beta_j'$ that

$$\beta_j' = \beta_i + \beta_j - \beta_i' = \alpha_i' + \beta_j \ne 0.$$

The same proof works if $\alpha_i' \neq 0$. The case that α_i, α_j, α_i', and α_j' are all nonzero is easier. Thus, we have proved (3.55).

Suppose that (α,β) and (α',β') are related and differ only at their ith and jth components and that (3.55) holds. We only need to prove (3.54) for these two pairs. We note that $L_{n,i}$ maps $\mathcal{N}_n$ to $\mathcal{N}_{n,i}$. Therefore, using (3.55), the identity (3.53) implies

$$(h_{\alpha,\beta})_{\alpha\in\mathcal{N}_{n,i},\beta\in\mathcal{N}_{n-1,j}} = (h_{\alpha,\beta})_{\alpha\in\mathcal{N}_{n,j},\beta\in\mathcal{N}_{n-1,i}}.$$

Let $(\alpha_i) = (\alpha_1\ldots,\alpha_{i-1},\alpha_{i+1},\ldots,\alpha_d) \in \mathbb{N}_0^d$. Then, the abovementioned equation is equivalent to

$$(h_{(\alpha_i),(\beta_j)})_{\alpha\in\mathcal{N}_{n-1},\beta\in\mathcal{N}_{n-2}} = (h_{(\alpha_j),(\beta_i)})_{\alpha\in\mathcal{N}_{n-1},\beta\in\mathcal{N}_{n-2}},$$

in particular, $h_{(\alpha_i),(\beta_j)} = h_{(\alpha_j),(\beta_i)}$. Since (α_i) and (α_j) differ at only two components, α_j and β_j can be taken as depending on α_i and β_i, respectively. If we let $\alpha_i + \beta_i = k$, then we have

$$h_{\alpha,\beta}|_{\alpha_i=1,\beta_i=k-1} = h_{\alpha,\beta}|_{\alpha_i=2,\beta_i=k-2} = \cdots = h_{\alpha,\beta}|_{\alpha_i=k-1,\beta_i=1},$$

where $h_{\alpha,\beta}|_{\alpha_i=a,\beta_i=b}$ means that α_j and β_j are assigned accordingly and other elements of α and β are unchanged in these equations. In particular, we have proved (3.54) for these two pairs (α,β) and (α',β'). □

Lemma 3.5.2. *A matrix $H_n : r_n^d \times r_n^d$ satisfies the equation*

$$L_{n,i}H_nL_{n,j}^{\mathsf{T}} = L_{n,j}H_nL_{n,i}^{\mathsf{T}}, \quad 1 \le i,j \le d,$$

if and only if it is a multi-index Hankel matrix.

The proof is similar to Lemma 3.5.1, and we shall leave out the details.

3.6 Notes and Further Results

The need for orthogonal polynomials in studying cubature rules starts with the Gauss cubature and early works on minimal cubature rules [146, 50, 115]. Although specific examples of orthogonal polynomials of more than one variable appeared early in literature (see, e.g. chapter 12 of [51]), the theory of orthogonal polynomials is developed much later than its counterpart in one variable. In particular, the structural relations in (Section 3.2), which will play an essential role in the next two chapters, were developed only in the early 1990; see, for example, [161, 164, 165], and the references in [170]. Our main reference for this chapter is [47], the first book devoted to orthogonal polynomials in several variables.

Characterizing common zeros of orthogonal polynomials in terms of structural matrices in the three-term relation in Theorem 3.3.3 is the harbinger of what will come in later chapters and is important for studying cubature rules. This was established in [164], where the block Jacobi matrices were introduced. For $d = 2$, this theorem appeared in [115] but with a complicated condition equivalent to (3.17) but not explicitly.

The wrapt product of weight functions and its orthogonal polynomials first appeared in [85] and in [2] for some special cases, and they were unearthed and systematically discussed in [81]. The explicit formulas for the three-term relation appeared in [166].

Classical orthogonal polynomials in several variables possess many special properties, for which we refer to [47, chapter 2] for $d = 2$ and [47, chapter 5] for several variables and also to [35, chapter 1] for spherical harmonics.

For $d = 2$, the orthogonal polynomials in Subsection 3.4.5 appeared in [79] and [81]. For $d > 2$, they are studied in [12] in connection with Gauss cubature rules. The polynomials $Q_\alpha^{\frac{1}{2}}$ in (3.48) are the generalized Schur polynomials associated with the orthonormal system $\{p_n\}$. Their orthogonality is established in Theorem 3.4.16, and they are of interest in combinatorics and symmetric functions [98, 99].

4
Gauss Cubature Rules

We study cubature rules with the number of nodes attaining the lower bound (2.2) and define them formally as follows.

Definition 4.0.1. A cubature rule of degree $2n-1$ or $2n-2$ is called Gauss cubature if it has $\dim \Pi_{n-1}^d$ nodes; that is, if it attains the lower bound (2.2).

In the definition, we extend the name Gauss to cubature rules of degree $2n-2$ in contrast to the common practice for Gauss quadrature rules. A characterization of Gauss cubature rules is given in Section 4.1. A family of examples of these cubature rules is given in Section 4.2, based on cubature rules for symmetric polynomials. In Section 4.3, we show that the Gauss cubature rules of degree $2n-2$ exist for product Chebyshev weight on a square, even though these rules do not exist for degree $2n-1$. The existence of the Gauss cubature rules of degree $2n-1$ requires the matrices $A_{n-1,i}A_{n-1,j}^{\mathsf{T}} - A_{n-1,j}A_{n-1,i}^{\mathsf{T}}$ be zero. In Section 4.4, we show that these matrices are nonzero for the centrally symmetric weight functions and wrapt product weight functions by establishing a quantitative result on the rank of these matrices; hence, the Gauss cubature rules do not exist for these weight functions.

4.1 Characterization of Gauss Cubature Rules

Let W be a regular weight function on a domain Ω in $\mathbb{R}^d$ and Ω has a nonempty interior in $\mathbb{R}^d$. We again denote by $\mathcal{I}(f)$ the integral $\int_\Omega f(\mathbf{x})W(\mathbf{x})\mathrm{d}\mathbf{x}$ and, more generally, a positive linear functional well-defined for all polynomials in d variables. By Definition 4.0.1, a cubature rule $\mathsf{C}(f)$ is called a Gauss cubature rule for $\mathcal{I}$ if $\mathcal{I}(f) = \mathsf{C}(f)$ for $f \in \Pi_m^d$ and $\mathsf{C}(f)$ has $N = \dim \Pi_{n-1}^d$ nodes for $m = 2n-1$ or $2n-2$.

As an analog of quadrature rules, the existence of a Gauss cubature rule of degree $2n-1$ is characterized by the zeros of orthogonal polynomials $\mathbb{P}_n$ of $\mathcal{V}_n^d$, and a Gauss cubature rule of degree $2n-2$ is characterized by the zeros of quasi-orthogonal polynomials of degree n that are orthogonal to polynomials of degree $n-2$. In the vector notation, the quasi-orthogonal polynomials are of the form

$$\mathbb{Q}_n(\mathbf{x}) = \mathbb{P}_n(\mathbf{x}) + \Gamma \mathbb{P}_{n-1}(\mathbf{x}),$$

where Γ is a real matrix of the size $r_n^d \times r_{n-1}^d$, which includes $\mathbb{P}_n$ if $\Gamma = 0$. The components of $\mathbb{Q}_n$ are orthogonal to polynomials of degree at most $n-2$.

The zeros of $\mathbb{P}_n$ or $\mathbb{Q}_n$ are common zeros of the set of polynomials, as defined in Definition 2.2.5. It is called *simple* if at least one polynomial in the set has a partial derivative that does not vanish on the zero.

Theorem 4.1.1. *Let $\mathcal{I}$ be a positive definite linear functional and let $\mathbb{P}_n$ denote a basis of orthogonal polynomials of degree n with respect to $\mathcal{I}$. For $n \ge 2$,*

(i) *$\mathcal{I}$ admits a Gauss cubature rule of degree $2n-1$ if and only if $\mathbb{P}_n$ has $N = \dim \Pi_{n-1}^d$ real and simple zeros.*
(ii) *$\mathcal{I}$ admits a Gauss cubature rule of degree $2n-2$ if and only if there is a matrix $\Gamma : r_n^d \times r_{n-1}^d$ such that $\mathbb{P}_n + \Gamma \mathbb{P}_{n-1}$ has $N = \dim \Pi_{n-1}^d$ real and simple zeros.*

Proof We only need to prove (ii) since (i) follows from setting $\Gamma = 0$ in (ii) as seen in the proof. First, we assume that a Gauss cubature rule exists and it has $\mathbf{x}_k$, $1 \le k \le N = \dim \Pi_{n-1}^d$, as its nodes. Since $\mathcal{I}$ is positive definite and the cubature rule is of degree $2n-2$, there is no nonzero polynomial Q of degree less than n that vanishes on all nodes since otherwise $0 = \mathsf{C}_{2n-2}(Q^2) = \mathcal{L}(Q^2) > 0$. Hence, by Lemma 2.2.6, there exist monic polynomials Q_α^n, $|\alpha| = n$, that have $\{\mathbf{x}_k : 1 \le k \le N\}$ as their common zeros. Since the cubature rule is of the degree $2n-2$, $\mathcal{I}(Q_a^n R) = \mathsf{C}_{2n-2}(Q_\alpha^n R) = 0$ for all $R \in \Pi_{n-2}^2$. Hence, each Q_α is a quasi-orthogonal polynomial of degree n. Consequently, we can write $\mathbb{Q}_n = \mathbb{P}_n + \Gamma \mathbb{P}_{n-1}$ for some matrix Γ. This proves the sufficient direction.

We now prove the necessary direction. Suppose that $\mathbb{Q}_n = \mathbb{P}_n + \Gamma \mathbb{P}_{n-1}$ has N real and simple zeros, denoted by $\mathbf{x}_1, \mathbf{x}_2, \ldots, \mathbf{x}_N$, and we can assume that $\mathbb{P}_n$ consists of monic orthogonal polynomials. By Proposition 2.2.4, there is an interpolatory cubature rule of degree $n-1$,

$$\int_\Omega f(\mathbf{x}) W(\mathbf{x}) d\mathbf{x} = \sum_{k=1}^N \lambda_k f(\mathbf{x}_k), \qquad \forall f \in \Pi_{n-1}^d.$$

Write $\mathbb{Q}_n = \{Q_\alpha : |\alpha| = n\}$. Since Q_α vanish on $\mathbf{x}_k$, $1 \le k \le N$, and Q_α are quasi-orthogonal, both sides of the displayed equation hold for $\mathbb{Q}_n$, so that the cubature rule holds for all $f \in \Pi_n^d$. Now, let f be a polynomial of degree $2n-2$. We claim that there exist polynomials q_α and r such that

$$f(\mathbf{x}) = \sum_{|\alpha|=n} q_\alpha(\mathbf{x})Q_\alpha(\mathbf{x}) + r(\mathbf{x}), \qquad q_\alpha \in \Pi_{n-2}^d, \quad r \in \Pi_{n-1}^d.$$

Indeed, by linearity, it is sufficient to consider $f(x) = \mathbf{x}^\beta$ for $|\beta| = m$ and $n \le m \le 2n-2$. The case $m = n$ is trivial since $Q_\alpha(\mathbf{x}) = \mathbf{x}^\alpha + R_\alpha(x)$ for $|\alpha| = n$. Assume the claim has been established for monomials of degree $m \ge n$. If $|\beta| = m+1$, then we can write $\beta = \alpha + \gamma$ for some α and γ such that $|\alpha| = n$ and $|\gamma| = m+1-n$. Using $\mathbf{x}^\alpha = P_\alpha(x) - R_\alpha(x)$, we can write $\mathbf{x}^\beta = \mathbf{x}^\alpha\mathbf{x}^\gamma = \mathbf{x}^\gamma P_\alpha(\mathbf{x}) - \mathbf{x}^\gamma R_\alpha(\mathbf{x})$. Since $\mathbf{x}^\gamma R_\alpha(\mathbf{x})$ is of degree $|\alpha| - 1 + |\gamma| = m$, it can be written as $\sum_{|\nu|=n} q_\nu(\mathbf{x})Q_\nu(\mathbf{x}) + r(\mathbf{x})$ by the induction hypothesis. Putting these together proves the claim. In particular, we see that $f(\mathbf{x}_k) = r(\mathbf{x}_k)$ for $1 \le k \le N$. Thus, by the orthogonality of Q_α, we obtain

$$\begin{aligned}\int_\Omega f(\mathbf{x})W(\mathbf{x})d\mathbf{x} &= \int_\Omega r(\mathbf{x})W(\mathbf{x})d\mathbf{x} \\ &= \sum_{k=1}^N \lambda_k r(\mathbf{x}_k) = \sum_{k=1}^N \lambda_k f(\mathbf{x}_k)\end{aligned}$$

for all $f \in \Pi_{2n-2}^d$, where the second step follows from applying the interpolatory cubature rule on $r \in \Pi_{n-1}^d$. Now, let ℓ_j be the fundamental interpolation polynomial of degree $n-1$, determined by $\ell_j(\mathbf{x}_k) = \delta_{k,j}$. Since the cubature rule is of degree $2n-2$, it follows that

$$\int_\Omega [\ell_j(\mathbf{x})]^2 W(\mathbf{x})d\mathbf{x} = \sum_{k=1}^N \lambda_k [\ell_j(\mathbf{x}_k)]^2 = \lambda_j.$$

In particular, $\lambda_j > 0$, and the cubature rule is positive. □

The common zeros of orthogonal polynomials $\mathbb{P}_n$ are characterized by Theorem 3.3.3, which is proved by using the Jacobi matrices. We need an analog for the quasi-orthogonal polynomials $\mathbb{Q}_n = \mathbb{P}_n + \Gamma\mathbb{P}_{n-1}$, which relies on the following modification of the Jacobi matrices,

$$J_{n,i}(\Gamma) := \begin{bmatrix} B_{0,i} & A_{0,i} & & & \bigcirc \\ A_{0,i}^\top & B_{1,i} & A_{1,i} & & \\ & \ddots & \ddots & \ddots & \\ & & A_{n-3,i}^\top & B_{n-2,i} & A_{n-2,i} \\ \bigcirc & & & A_{n-2,i}^\top & B_{n-1,i} - A_{n-1,i}\Gamma \end{bmatrix}, \qquad 1 \le i \le d,$$

which reduces to the Jacobi matrix $J_{n,i}$ when $\Gamma = 0$. We retain the name of the joint eigenvalue and eigenvector for these modified Jacobi matrices.

Theorem 4.1.2. *A point $\lambda \in \mathbb{R}^d$ is a zero of $\mathbb{Q}_n$ if and only if it is a joint eigenvalue of $J_{n,1}, \ldots, J_{n,d}$, and a joint eigenvector of λ is $(\mathbb{P}_0^{\mathsf{T}}(\lambda), \ldots, \mathbb{P}_{n-1}^{\mathsf{T}}(\lambda))^{\mathsf{T}}$.*

Proof We observe that the $m = n-1$ equation of the three-term relations (3.5) can be rewritten as

$$A_{n-2,i}^{\mathsf{T}}\mathbb{P}_{n-2}(\lambda) + (B_{n-1,i} - A_{n-1,i}\Gamma)\mathbb{P}_{n-1}(\lambda) + A_{n-1,i}\mathbb{Q}_n(\lambda) = \lambda_i\mathbb{P}_{n-1}(\lambda).$$

Keeping this in mind and with the modification of replacing $\mathbb{P}_n$ by $\mathbb{Q}_n$ and $B_{n-1,i}$ by $B_{n-1,i} - A_{n-1,i}\Gamma$, the proof follows that of Theorem 3.3.1 almost verbatim. □

It follows from the size of the modified Jacobi matrices that $\mathbb{Q}_n$ can have at most $\dim \Pi_{n-1}$ zeros. Our next theorem characterizes when $\mathbb{Q}_n$ can have a maximal number of zeros.

Theorem 4.1.3. *For $n = 1, 2, \ldots$, $\mathbb{Q}_n = \mathbb{P}_n + \Gamma\mathbb{P}_{n-1}$ has $N = \dim \Pi_{n-1}^d$ common zeros if and only if the matrix Γ satisfies, for $1 \le i \le d$,*

$$A_{n-1,i}\Gamma = \Gamma^{\mathsf{T}} A_{n-1,i}^{\mathsf{T}}, \tag{4.1}$$

and, for $1 \le i < j \le d$,

$$\begin{aligned} &\Gamma^{\mathsf{T}}(A_{n-1,i}^{\mathsf{T}}A_{n-1,j} - A_{n-1,j}^{\mathsf{T}}A_{n-1,i})\Gamma - (A_{n-1,i}A_{n-1,j}^{\mathsf{T}} - A_{n-1,j}A_{n-1,i}^{\mathsf{T}}) \\ &= (B_{n-1,i}A_{n-1,j} - B_{n-1,j}A_{n-1,i})\Gamma - \Gamma^{\mathsf{T}}(B_{n-1,i}A_{n-1,j} - B_{n-1,j}A_{n-1,i})^{\mathsf{T}}. \end{aligned} \tag{4.2}$$

Proof As in the proof of Theorem 3.3.3, $\mathbb{Q}_n$ has N zeros if and only if the matrices $J_{n,i}(\Gamma)$ commute. Thus, it is sufficient to show that $J_{n,i}(\Gamma)J_{n,j}(\Gamma) = J_{n,j}(\Gamma)J_{n,i}(\Gamma)$ for $1 \le i, j \le n$ if and only if (4.1) and (4.2) hold.

Now, by the commutativity relations (3.15), a quick verification shows that $J_{n,i}J_{n,j} = J_{n,j}J_{n,i}$ holds if and only if

$$A_{n-1,i}\Gamma A_{n-2,j}^{\mathsf{T}} = A_{n-1,j}\Gamma A_{n-2,i}^{\mathsf{T}} \tag{4.3}$$

and

$$\begin{aligned} &A_{n-2,i}^{\mathsf{T}}A_{n-2,j} + (B_{n-1,i} - A_{n-1,j}\Gamma)(B_{n-1,i} - A_{n-1,j}\Gamma) \\ &= A_{n-2,j}^{\mathsf{T}}A_{n-2,i} + (B_{n-1,j} - A_{n-1,i}\Gamma)(B_{n-1,j} - A_{n-1,i}\Gamma). \end{aligned}$$

Using (3.15) again and that $B_{n,i}$ are symmetric, the last identity can be shown to be equivalent to (4.2) provides that (4.1) holds. Thus, it is sufficient to prove that (4.3) is equivalent to (4.1). One direction is easy: if (4.1) holds, then $A_{n-1,i}\Gamma$ is symmetric, so that the first identity in (3.15) implies

$$A_{n-1,i}\Gamma A_{n-2,j}^{\mathsf{T}} = \Gamma^{\mathsf{T}}A_{n-1,i}^{\mathsf{T}}A_{n-2,j} = \Gamma^{\mathsf{T}}A_{n-1,j}^{\mathsf{T}}A_{n-2,i} = A_{n-1,j}\Gamma A_{n-2,i}^{\mathsf{T}}.$$

In the other direction, let $L_{n,i}$ be matrices defined at (3.8) and G_n be the leading coefficient matrix of $\mathbb{P}_n$, then (4.3) is equivalent to

$$L_{n,i}H_nL_{n-1,j}^{\mathsf{T}} = L_{n,j}H_nL_{n-1,i}^{\mathsf{T}}, \quad 1 \le i,j \le d,$$

where $H_n = G_n^{-1}\Gamma G_{n-1}^{-1}$, which implies that H_n is a Hankel matrix of the form $H_n = (h_{\alpha+\beta})_{|\alpha|=n,|\beta|=n-1}$, proved in the Lemma 3.5.1. The special form of H_n allows us to define a linear functional, $\mathcal{L}$, on Π^d by setting $\mathcal{L}(x^\alpha) = h_\alpha$. It follows that we can write $H_n = \mathcal{L}(\mathbf{x}^n(\mathbf{x}^{n-1})^{\mathsf{T}})$. Thus, $L_{n,i}H_n = \mathcal{L}(x_i\mathbf{x}^{n-1}(\mathbf{x}^{n-1})^{\mathsf{T}})$, which is evidently symmetric. By the definition of H_n, this means

$$L_{n,i}G_n^{-1}\Gamma(G_{n-1}^{-1})^{\mathsf{T}} = G_{n-1}^{-1}\Gamma^{\mathsf{T}}(G_n^{-1})^{\mathsf{T}}L_{n-1,i}^{\mathsf{T}},$$

which, by (3.7), implies that $A_{n-1,i}\Gamma$ is symmetric. This completes the proof. □

When $\mathbb{Q}_n$ has the maximal number $N = \dim\Pi_{n-1}^d$ zeros, the zeros satisfy further properties.

Theorem 4.1.4. *Assume the quasi-orthogonal polynomials of $\mathbb{Q}_n$ have $N = \dim\Pi_{n-1}^d$ zeros. Then*

(1) *All zeros of $\mathbb{Q}_n$ are real, distinct and simple. Moreover, polynomials $\mathbb{Q}_n$ and $\mathbb{P}_{n-1}$ do not have common zeros.*
(2) *Let $\mathbf{x}_1,\ldots,\mathbf{x}_N$ be zeros of $\mathbb{Q}_n$. Then the Lagrange interpolation polynomial $L_n(f)$ of degree $n-1$ based on these zeros is given by*

$$L_n(f,\mathbf{x}) = \sum_{k=1}^{N} f(\mathbf{x}_k)\frac{\mathbf{K}_{n-1}(\mathbf{x},\mathbf{x}_k)}{\mathbf{K}_{n-1}(\mathbf{x}_k,\mathbf{x}_k)}. \tag{4.4}$$

(3) *The Gauss cubature rule of degree $2n-2$ with nodes $\mathbf{x}_1,\ldots,\mathbf{x}_N$ is positive and its weights are given by $\Lambda_k = [\mathbf{K}_{n-1}(\mathbf{x}_k,\mathbf{x}_k)]^{-1} > 0$, $1 \le k \le N$.*

Proof If $\mathbb{Q}$ has $\dim\Pi_{n-1}^d$ zeros, then the matrix $J_{n,i}$ is symmetric by (4.1) and that $B_{n,i}$ is symmetric. Hence, as joint eigenvalues of symmetric matrices, all zeros of $\mathbb{Q}_n$ are real. Adding and subtracting $\mathbb{Q}_n^{\mathsf{T}}(\mathbf{x})A_{n-1,i}^{\mathsf{T}}\mathbb{P}_{n-1}(\mathbf{x})$ in (3.9) we obtain

$$\mathbf{K}_{n-1}(\mathbf{x},\mathbf{x}) = \mathbb{P}_{n-1}^{\mathsf{T}}(\mathbf{x})A_{n-1,i}^{\mathsf{T}}\frac{\mathbb{Q}_n(\mathbf{x})-\mathbb{Q}(\mathbf{y})}{x_i-y_i} - \mathbb{Q}_n^{\mathsf{T}}(\mathbf{x})A_{n-1,i}^{\mathsf{T}}\frac{\mathbb{P}_{n-1}(\mathbf{x})-\mathbb{P}_{n-1}(\mathbf{y})}{x_i-y_i},$$

and, taking the limit $y_i \to x_i$, we further deduce

$$\mathbf{K}_{n-1}(\mathbf{x},\mathbf{x}) = \mathbb{P}_{n-1}^{\mathsf{T}}(\mathbf{x})A_{n-1,i}^{\mathsf{T}}\partial_i\mathbb{Q}_n(\mathbf{x}) - \mathbb{Q}_n^{\mathsf{T}}(\mathbf{x})A_{n-1,i}^{\mathsf{T}}\partial_i\mathbb{P}_{n-1}(\mathbf{x}),$$

where $\partial_i = \partial/\partial x_i$. If λ is a zero of $\mathbb{Q}_n$, then neither $\mathbb{P}_{n-1}(\lambda)$ nor $\partial_i\mathbb{Q}_n(\lambda)$ can be zero as $\mathbf{K}_{n-1}(\lambda,\lambda) \ge \mathbf{K}_0(\lambda,\lambda) > 0$. Thus, the zeros of $\mathbb{Q}_n$ have the stated properties in (1). Using the two displayed formulas of $\mathbf{K}_{n-1}$ given above, we see that the polynomial $\ell_k(\mathbf{x}) = \mathbf{K}_{n-1}(\mathbf{x},\mathbf{x}_k)/\mathbf{K}_{n-1}(\mathbf{x}_k,\mathbf{x}_k)$ satisfies $\ell_k(\mathbf{x}_j) = \delta_{k,j}$, $1 \le k, j \le N$. This proves (2). Moreover, by Lemma 2.2.6, the points $\mathbf{x}_k$ are not on an algebraic hypersurface of degree $n-1$ so that $L_n(f)$ is the unique Lagrange interpolation polynomial of degree $n-1$. Thus, integrating $L_n(f)$ gives the Gauss cubature rule with $\Lambda_k = \mathcal{I}(L_{n-1}(\ell_k)) = [\mathbf{K}_{n-1}(\mathbf{x}_k,\mathbf{x}_k)]^{-1}$ by orthogonality. □

We end this section with the following characterization of Gauss cubature rules in terms of structural matrices of orthogonal polynomials, derived from the characterization in Theorem 4.1.1 and combined with Theorem 3.3.3 and Theorem 4.1.3.

Theorem 4.1.5. *Let $\mathcal{I}$ be a positive definite linear functional. Let $A_{n,i}$ and $B_{n,i}$ be coefficient matrices of the three-term relation* (3.5). *For $n \ge 2$,*

(i) *$\mathcal{I}$ admits a Gauss cubature rule of degree $2n-1$ if and only if* (3.17) *holds.*

(ii) *$\mathcal{I}$ admits a Gauss cubature rule of degree $2n-2$ if and only if there is a matrix $\Gamma : r_n^d \times r_{n-1}^d$ that satisfies* (4.1) *and* (4.2).

Remark 4.1.6. *Since* (4.1) *is equivalent to* (3.53), *which shows that $H_n = G_n^{-1}\Gamma(G_{n-1}^{-1})^{\mathsf{T}}$ is a Hankel matrix (h_{i+j}); that is,*

$$\Gamma = G_n H_n G_{n-1}^{\mathsf{T}} \quad \textit{with} \quad H_n = [h_{\alpha+\beta}]_{|\alpha|=n,|\beta|=n-1}. \tag{4.5}$$

The existence of a Gauss cubature rule of degree $2n-2$ amounts to finding a Hankel matrix H_n so that Γ in (4.5) *solves* (4.2). *The matrix Γ has r_{2n-1}^d parameters, which is the cardinality of $\{\alpha+\beta : |\alpha| = n, |\beta| = n-1\} = \{\gamma \in \mathbb{N}_0^d : |\gamma| = 2n-1\}$, and the system of equation* (4.2) *consists of $\frac{1}{2}(r_n^d)^2 - r_n^d$ distinct equations, since the matrices on both sides of the system are anti-symmetric, by* (4.1). *Thus, solving* (4.2) *amounts to solving a system of $\frac{1}{2}\left[\binom{n+d-1}{d-1}^2 - \binom{n+d-1}{d-1}\right]$ quadratic equations with $\binom{2n+d-1}{d-1}$ variables. In particular, for $d = 2$, $H_n = [h_{i+j}]_{i=0}^{n}{}_{j=0}^{n-1}$ has $2n+1$ parameters, and solving* (4.2) *amounts to solving $\binom{n+1}{2}$ quadratic equations of $2n+1$ variables, which consist of a severely overdetermined nonlinear system of equations. This indicates that, in most cases, the system may not have a solution.*

The above discussion provides another indication that the Gauss cubature rules of degree $2n-2$ exist rarely. Nevertheless, the presence of Γ suggests they may exist when the Gaussian cubature rules do not. An example in Subsection 4.3 shows that this can indeed be the case.

4.2 Examples of Gauss Cubature Rules

As we discussed in Section 1.2, Gauss cubature rules exist rarely. They do not exist when both the weight function and the domain of an integral are symmetric to the origin, as we shall show in the next section. Nevertheless, they do exist in some cases. We give a family of examples in this section.

4.2.1 Example on Curved Domain: $d = 2$

Let Ω be the curved domain (3.40) bounded in part by a parabola, including the bounded domain (3.41) depicted in Figure 3.1 of Subsection 3.4.5. For a weight function w on $\mathbb{R}$, let $W_{\pm\frac{1}{2}}$ be defined by (3.42) with $\gamma = -\frac{1}{2}$ or $\frac{1}{2}$,

$$W_{\pm\frac{1}{2}}(u_1, u_2) = w(x_1)w(x_2)(u_1^2 - 4u_2)^{\pm\frac{1}{2}}, \qquad (u_1, u_2) \in \Omega,$$

where $u_1 = x_1 + x_2$ and $u_2 = x_1 x_2$, which includes Examples (3.4.12)–(3.4.14) as special cases.

We now show that the Gauss cubature rules exist for $W_{\pm\frac{1}{2}}$ by identifying the common zeros of orthogonal polynomials. Let $P_k^{n,\pm\frac{1}{2}}$ be the orthogonal polynomials for $W_{\pm\frac{1}{2}}$ given in Proposition 3.4.11. Let $\mathbb{P}_n^{\pm\frac{1}{2}} = (P_0^{n,\pm\frac{1}{2}}, \ldots P_n^{n,\pm\frac{1}{2}})$. Let $t_{k,n}$, $1 \le k \le n$, be the zeros of $p_n(w)$ and define

$$u_{j,k}^{(n)} = t_{j,n} + t_{k,n} \quad \text{and} \quad v_{j,k}^{(n)} = t_{j,n} t_{k,n}, \qquad 1 \le k, j \le n.$$

Proposition 4.2.1. *The set $\mathbb{P}_n^{-\frac{1}{2}}$ has $\binom{n+1}{2}$ common zeros given by*

$$Z_n^{-\frac{1}{2}} := \left\{ \left(u_{j,k}^{(n)}, v_{j,k}^{(n)} \right) : 1 \le j \le k \le n \right\}, \tag{4.6}$$

and the set $\mathbb{P}_n^{\frac{1}{2}}$ has $\binom{n+1}{2}$ common zeros given by

$$Z_n^{\frac{1}{2}} := \left\{ \left(u_{j,k}^{(n+1)}, v_{j,k}^{(n+1)} \right) : 1 \le j < k \le n+1 \right\}. \tag{4.7}$$

Proof For $\gamma = -\frac{1}{2}$, it is evident that the right-hand side of (3.44) vanishes at $(t_{k,n}, t_{j,n})$ for $1 \le k, j \le n$. Half of these zeros satisfy $t_{k,n} \le t_{j,n}$, so that they are in the domain $\triangle$ that is mapped to Ω by (3.39). Thus, every point in $Z_n^{-\frac{1}{2}}$ is a common zero of $\mathbb{P}_n^{-\frac{1}{2}}$. There are $\binom{n+1}{2}$ such zeros, which is the maximal number of common zeros for $\mathbb{P}_n^{-\frac{1}{2}}$, so that $Z_n^{-\frac{1}{2}}$ is uniquely determined. For $\gamma = \frac{1}{2}$, the right-hand side of (3.44) vanishes at $(t_{k,n+1}, t_{j,n+1})$ whenever $k \ne j$

and $1 \le k, j \le n+1$. The same argument shows that $Z_n^{\frac{1}{2}}$ is the set of common zeros of $\mathbb{P}_n^{\frac{1}{2}}$. □

By Theorem 4.1.1, the proposition shows that the Gauss cubature rules exist for $W_{\pm\frac{1}{2}}$ on the domain Ω. Moreover, a moment of reflection shows that these cubature rules can be given explicitly. Let the Gauss quadrature rule for w be given by

$$\int_{\mathbb{R}} f(t)w(t)\mathrm{d}t = \sum_{k=1}^{n} \lambda_{k,n} f(t_{k,n}), \qquad \forall f \in \Pi_m,$$

where $m = 2n-1$ or $m = 2n-2$.

Theorem 4.2.2. *For $\mathcal{W}_{-\frac{1}{2}}$ on Ω, the Gauss cubature rule of degree $2n-1$ or $2n-2$ is given by*

$$\int_{\Omega} f(\mathbf{u})\mathcal{W}_{-\frac{1}{2}}(\mathbf{u})\mathrm{d}\mathbf{u} = 2\sum_{k=1}^{n}\sum_{j=1}^{k}{}' \lambda_{k,n}\lambda_{j,n} f\left(u_{j,k}^{(n)}, v_{j,k}^{(n)}\right), \tag{4.8}$$

where $\sum'$ means that the term for $j = k$ is divided by 2. For $\mathcal{W}_{\frac{1}{2}}$ on Ω, the Gauss cubature rule of degree $2n-1$ or $2n-2$ is given by

$$\int_{\Omega} f(\mathbf{u})\mathcal{W}_{\frac{1}{2}}(\mathbf{u})\mathrm{d}\mathbf{u} = 2\sum_{k=2}^{n+1}\sum_{j=1}^{k-1} \lambda_{j,k}^{(n)} f\left(u_{j,k}^{(n+1)}, v_{j,k}^{(n+1)}\right), \tag{4.9}$$

where $\lambda_{j,k}^{(n)} = \lambda_{j,n+1}\lambda_{k,n+1}(t_{j,n+1} - t_{k,n+1})^2$.

Proof We use the product Gauss cubature rule of degree m given by

$$\int_{\mathbb{R}^2} g(x_1,x_2)w(x_1)w(x_2)\mathrm{d}x_1\mathrm{d}x_2 = \sum_{k=1}^{n}\sum_{j=1}^{n} \lambda_{k,n}\lambda_{j,n} g(t_{k,n}, t_{j,n}), \tag{4.10}$$

which is exact for $g \in \Pi_m \times \Pi_m$ with $m = 2n-1$ or $m = 2n-2$ and apply it to the integral in the right-hand side of (3.43) with $\gamma = -\frac{1}{2}$. This gives a cubature rule of degree m for the integral in the left-hand side, since if $f \in \Pi_m^2$ then $g(x_1,x_2) = f(x_1+x_2, x_1x_2)$ belongs to $\Pi_m \times \Pi_m$. Moreover, since $f(x_1+x_2, x_1x_2)$ is symmetric, we can write the sum (4.10) as twice the sum over $g(t_{k,n}, t_{j,n})$ with $j \le k$. This proves (4.8), and its number of nodes is $(n^2-n)/2 + n = n(n+1)/2$, so that it is a Gauss cubature of degree m. The cubature rule in (4.9) is established similarly by applying the product cubature (4.10) of degree $2n+1$ or $2n$, taking into account of the factor $(x_1-x_2)^2$ in the right-hand side of (3.43). □

In the case that $w = w_{\alpha,\beta}$ is the Jacobi weight, the weight function $W_{\pm\frac{1}{2}} =$

$W_{\alpha,\beta,\pm\frac{1}{2}}$ is given in Example (3.4.12). In particular, for $\alpha = \beta = -\frac{1}{2}$, the weight function is

$$W_{-\frac{1}{2},-\frac{1}{2},-\frac{1}{2}}(u_1,u_2) = \left((1+u_2)^2 - u_1^2\right)^{-\frac{1}{2}} (u_1^2 - u_2)^{-\frac{1}{2}}.$$

The nodes of the Gauss cubature rule of degree $2n-1$ with $n = 20$ are depicted in Figure 4.1.

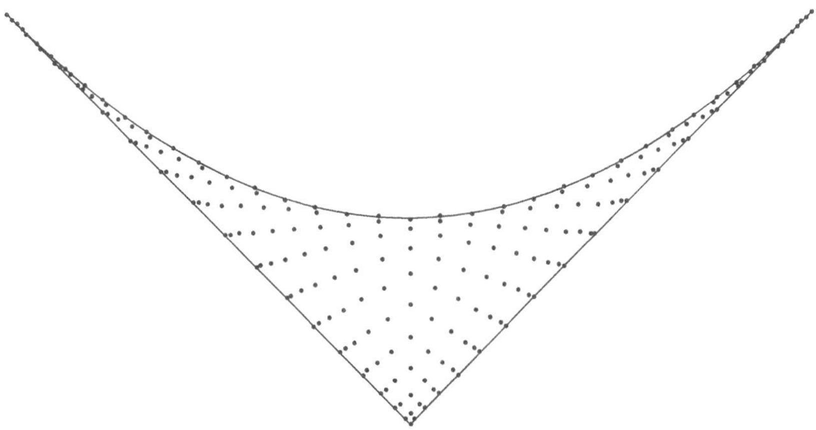

Figure 4.1 The nodes of Gauss cubature of degree 39 for $W_{-\frac{1}{2},-\frac{1}{2},-\frac{1}{2}}$.

To illustrate the characterization in Theorem 4.1.5, we state the coefficient matrices for the three-term relations

$$u_i \mathbb{P}_n^{(\pm\frac{1}{2})} = A_{n,i}^{(\pm\frac{1}{2})} \mathbb{P}_{n+1}^{(\pm\frac{1}{2})} + B_{n,i}^{(\pm\frac{1}{2})} \mathbb{P}_n^{(\pm\frac{1}{2})} + A_{n-1,i}^{(\pm\frac{1}{2})} \mathbb{P}_{n-1}^{(\pm\frac{1}{2})}, \qquad i = 1,2,$$

for the polynomials $P_k^{n,-\frac{1}{2}}$ defined in (3.44) and the polynomials $P_k^{n,\frac{1}{2}}$ defined in (3.45). Let $\{a_n\}$ and $\{b_n\}$ be the coefficients of the three-term relation for $p_n(w)$. We define three auxiliary matrices,

$$\mathsf{A}_{n,1}(\rho) = \begin{bmatrix} 1 & & & 0 \\ & \ddots & & \vdots \\ & & 1 & 0 \\ & & & \rho & 0 \end{bmatrix}, \quad \mathsf{A}_{n,2}(\rho) = \begin{bmatrix} b_0 & a_0 & & & \\ a_0 & b_1 & a_1 & & \\ & \ddots & \ddots & \ddots & \\ & & a_{n-2} & b_{n-1} & \rho a_{n-1} \\ & & & \rho a_{n-1} & \rho b_n \end{bmatrix}.$$

where $\mathsf{A}_{n,i}(\rho)$ is of size $(n+1)\times(n+2)$ and $\mathsf{A}_{n,2}(\rho)$ is of size $(n+1)\times(n+1)$, and

$$\mathsf{B}_n(\rho)=\begin{bmatrix} b_0 & a_0 & & & \\ a_0 & b_1 & a_1 & & \\ & \ddots & \ddots & \ddots & \\ & & a_{n-2} & b_{n-1} & \rho a_{n-1} \\ & & & \rho a_{n-1} & b_n \end{bmatrix}$$

of size $(n+1)\times(n+1)$. Then, the coefficient matrices of the three-term relations for $\gamma=-\frac{1}{2}$ are given by

$$A_{n,1}^{(-\frac{1}{2})}=a_n\mathsf{A}_{n,1}(\rho) \quad \text{and} \quad B_{n,1}^{(-\frac{1}{2})}=\mathsf{B}_n(\rho)+b_nI_{n+1}, \quad \rho=\sqrt{2},$$

$$A_{n,2}^{(-\frac{1}{2})}=a_n\left[\mathsf{A}_{n,2}(\rho)\middle|\begin{matrix}0\\a_n\end{matrix}\right] \quad \text{and} \quad B_{n,2}^{(-\frac{1}{2})}=\mathsf{B}_n(\rho)+a_{n-1}^2\begin{bmatrix}\bigcirc & & \\ & 1 & 0\\ & 0 & 0\end{bmatrix}, \quad \rho=\sqrt{2};$$

and the coefficient matrices of the three-term relations for $\gamma=\frac{1}{2}$ are given by

$$A_{n,1}^{(\frac{1}{2})}=a_{n+1}\mathsf{A}_{n,1}(\rho) \quad \text{and} \quad B_{n,1}^{(\frac{1}{2})}=\mathsf{B}_n(\rho)+b_{n+1}I_{n+1}, \qquad \rho=1,$$

$$A_{n,2}^{(\frac{1}{2})}=a_{n+1}\left[\mathsf{A}_{n,2}(\rho)\middle|\begin{matrix}0\\a_n\end{matrix}\right] \quad \text{and} \quad B_{n,2}^{(\frac{1}{2})}=\mathsf{B}_n(\rho)-a_n^2\begin{bmatrix}\bigcirc & \\ & 1\end{bmatrix}, \qquad \rho=1,$$

where I_{n+1} is the identity matrix of size $n+1$.

These coefficient matrices are derived from the explicit formulas of $P_k^{n,\pm\frac{1}{2}}$ in (3.44) and (3.45) by applying the three-term relation for $p_n(w)$. The computation is straightforward but tedious. For example, multiplying $P_k^{n,-\frac{1}{2}}(u)$ by u_2 means multiplying the right-hand side of (3.44) by x_1x_2, which can be carried out by applying the three-term relation of $p_n(w)$ twice, one with $x_1p_n(w;x_1)$ and the other with $x_2p_n(w;x_2)$.

Using the explicit formulas of these matrices, a straightforward computation shows that (3.17) holds, which is the necessary and sufficient condition for $\mathbb{P}_n^{(\pm\frac{1}{2})}$ to have the maximal number of common zeros. Furthermore, if we define

$$\Gamma^{(-\frac{1}{2})}=c\begin{bmatrix}1 & & & \\ & \ddots & & \\ & & 1 & \\ & & & \sqrt{2}\\ 0 & \dots & 0 & -c\end{bmatrix} \quad \text{and} \quad \Gamma^{(\frac{1}{2})}=ca_n\begin{bmatrix}1 & & \\ & \ddots & \\ & & 1\\ 0 & \dots & 0\end{bmatrix},$$

then one can verify that $\Gamma^{(\pm\frac{1}{2})}$ satisfies the equations (4.1) and (4.2).

Consequently, by Theorem 4.1.3, the quasi-orthogonal polynomials

$$\mathbb{P}_n^{(\pm\frac{1}{2})} + \Gamma^{(\pm\frac{1}{2})}\mathbb{P}_{n-1}^{(\pm\frac{1}{2})}$$

has $\dim \Pi_{n-1}^2$ zeros, which are nodes of the Gauss cubature rule of degree m by Theorem 4.1.5, where $m = 2n-2$ if $c \neq 0$ and $m = 2n-1$ if $c = 0$.

4.2.2 Example via Symmetric Polynomials: $d > 2$

We extend the Gauss cubature rules in Subsection 4.2.1 beyond two dimensions. Let $W_{\pm\frac{1}{2}}$ be the weight function defined at (3.51). We show that these weight functions admit the Gauss cubature rule of degree $2n-1$ on the domain Ω, which is the image of $\triangle^d$ under the mapping (3.50). For $d = 3$, the domain Ω is depicted in Figure 3.2.

For $1 \le k \le n$, let $t_{k,n}$ be the zeros of $p_n(w)$ ordered by $t_{1,n} < t_{2,n} < \cdots < t_{n,n}$ and let $\lambda_{k,n}$ be the weight associated with $t_{k,n}$ in the Gauss quadrature rule.

Theorem 4.2.3. *The weight function $W_{\pm\frac{1}{2}}$ admits Gauss cubature rules of degree $2n-1$ for $n = 1,2,\ldots$. More precisely,*

$$\int_\Omega f(\mathbf{u})W_{-\frac{1}{2}}(\mathbf{u})d\mathbf{u} = \sum_{\gamma_1=1}^{n}\sum_{\gamma_2=1}^{\gamma_1}\cdots\sum_{\gamma_d=1}^{\gamma_{d-1}} \Lambda_{\gamma,n}^{(-\frac{1}{2})} f(\mathbf{u}_{\gamma,n}), \tag{4.11}$$

where $\mathbf{u}_{\gamma,n}$ is the image of $\mathbf{x}_{\gamma,n} = (t_{\gamma_1,n}, t_{\gamma_2,n}, \ldots, t_{\gamma_d,n})$ under the mapping (3.50), and

$$\int_\Omega f(\mathbf{u})W_{\frac{1}{2}}(\mathbf{u})d\mathbf{u} = \sum_{\gamma_1=1}^{n+d-1}\sum_{\gamma_2=1}^{\gamma_1-1}\cdots\sum_{\gamma_d=1}^{\gamma_{d-1}-1} \Lambda_{\gamma,n}^{(\frac{1}{2})} f(\mathbf{u}_{\gamma,n+d-1}), \tag{4.12}$$

where $\mathbf{u}_{\gamma,n+d-1}$ is the image of $\mathbf{x}_{\gamma,n+d-1} = (t_{\gamma_1,n+d-1}, t_{\gamma_2,n+d-1}, \ldots, t_{\gamma_d,n+d-1})$ under the mapping (3.50) and, furthermore,

$$\Lambda_{\gamma,n}^{(-\frac{1}{2})} = \frac{\lambda_{\hat\gamma_1,n}^{m_1}\cdots\lambda_{\hat\gamma_{d'},n}^{m_{d'}}}{m_1!\cdots m_{d'}!} \quad \textit{and} \quad \Lambda_{\gamma,n}^{(\frac{1}{2})} = J(\mathbf{x}_{\gamma,n})^2\lambda_{\gamma_1,n}\cdots\lambda_{\gamma_d,n},$$

where $\hat\gamma_1\ldots,\hat\gamma_{d'}$ denote the distinct elements in γ and $m_1,\ldots,m_{d'}$ denote their respective multiplicities, and $J(\mathbf{x}) = \prod_{i<j}(x_i - x_j)$.

Proof For $n = \alpha_1 \ge \cdots \ge \alpha_d$, it is easy to see, by (3.47), that the polynomial $Q_\alpha^{-\frac{1}{2}}$ vanishes on $\mathbf{x}_{\gamma,n} = (t_{\gamma_1,n}, t_{\gamma_2,n}, \ldots, t_{\gamma_d,n})$ whenever $1 \le \gamma_i \le d$. Consequently, under the mapping (3.50), $\mathbb{P}_n^{-\frac{1}{2}}$ vanishes on the set $\{\mathbf{u}_{\gamma,n} : 1 \le \gamma_1 \le \gamma_2 \le \cdots \le n\}$, which has the cardinality $\binom{n+d-1}{n}$. Similarly, by (3.49), the polynomial $Q_\alpha^{\frac{1}{2}}$ vanishes on $\mathbf{x}_{\gamma,n+d-1}$ whenever $1 \le \gamma_1 < \gamma_2 < \cdots\gamma_d \le n+d-1$, so that $\mathbb{P}_n^{\frac{1}{2}}$

vanishes on the set $\{\mathbf{u}_{\gamma,n+d-1} : 1 \le \gamma_1 < \gamma_2 < \cdots \gamma_d \le n+d-1\}$, which also has the cardinality $\binom{n+d-1}{n}$. Thus, the Gauss cubature rules exist by Theorem 4.1.1. We only need to determine the values of $\Lambda_{\gamma,n}^{(\pm\frac{1}{2})}$.

Under the mapping (3.50) and using (3.51) and (3.52), we see that (4.11) is equivalent to

$$\frac{1}{d!}\int_{\mathbb{R}^d} f(\mathbf{x})\prod_{i=1}^{d} w(x_i)\mathrm{d}\mathbf{x} = \sum_{\alpha_1=1}^{n}\sum_{\alpha_2=1}^{\gamma_1}\cdots\sum_{\alpha_d=1}^{\alpha_{d-1}} \Lambda_{\alpha,n}^{(-\frac{1}{2})} f(\mathbf{x}_{\alpha,n}), \tag{4.13}$$

and it is exact for symmetric polynomials of degree $2n-1$ in each variable. Let $\ell_k(t)$ be the unique polynomial of degree $n-1$ that satisfies $\ell_k(t_j) = \delta_{k,j}$. For $\gamma \in \mathbb{N}_0^d$, we define

$$f_\gamma(\mathbf{x}) = \sum_{\beta\in\mathcal{S}_d} \ell_{\gamma_1}(x_{\beta_1})\cdots\ell_{\gamma_d}(x_{\beta_d}),$$

which is a symmetric polynomial that has degree $n-1$ in each of its variables, and the sum contains $d!$ terms. Applying the cubature (4.13) on f_γ, the left-hand side becomes $\lambda_{\gamma_1,n}\cdots\lambda_{\gamma_d,n} = \lambda_{\hat{\gamma}_1,n}^{m_1}\cdots\lambda_{\hat{\gamma}_{d'},n}^{m_{d'}}$, whereas the right-hand side becomes $m_1!\ldots m_{d'}!\Lambda_{\gamma,n}^{(-\frac{1}{2})}$ since only those terms with $\alpha = \gamma$ in the sum are nonzero by the property of $\ell_{\alpha_k}(\mathbf{x})$. This establishes the formula of $\Lambda_{\gamma,n}^{(-\frac{1}{2})}$.

Likewise, under the mapping (3.50), the cubature rule (4.12) is equivalent to,

$$\int_{\mathbb{R}^d} f(\mathbf{x})[J(\mathbf{x})]^2\prod_{i=1}^{d} w(x_i)\mathrm{d}\mathbf{x} = \sum_{\gamma_1=1}^{n+d-1}\sum_{\gamma_2=1}^{\gamma_1-1}\cdots\sum_{\gamma_d=1}^{\gamma_{d-1}-1} \Lambda_{\gamma,n}^{(\frac{1}{2})} f(\mathbf{x}_{\gamma,n+d-1}), \tag{4.14}$$

which is exact for symmetric polynomials of degree $2n-1$. Let $\ell_k(t)$ be the unique polynomial of degree $n-1$ that satisfies $\ell_k(t_j) = \delta_{k,j}$. For $\gamma \in \mathbb{N}_0^d$, we define

$$f_\gamma(\mathbf{x}) = \frac{[\det(\ell_{\gamma_j}(\mathbf{x}_k))]_{j,k=1}^d}{[J(\mathbf{x})]^2} = \frac{1}{[J(\mathbf{x})]^2}\sum_{\beta\in\mathcal{S}_d}\operatorname{sign}(\beta)\ell_{\gamma_1}(x_{\beta_1})\cdots\ell_{\gamma_d}(x_{\beta_d}),$$

where the second equality follows from the Lagrange expansion of a determinant, which is a symmetric polynomial of degree $n-1$. Applying the cubature (4.14) on f_γ gives the formula of $\Lambda_{\gamma,n}^{(\frac{1}{2})}$. □

Two comments are in order. First, the Gauss cubature rule of degree $2n-2$ also exists for $W_{\pm\frac{1}{2}}$. Indeed, if $\operatorname{supp}(w) = [a,b]$, we let $\{t_{k,n} \mid 1 \le k \le n\}$ be the zeros of quasi-orthogonal polynomials $q_n = p_n(w) + \eta_n p_{n-1}(w)$, where $\eta_n \ne 0$ is chosen so that all $t_{k,n}$ are inside $[a,b]$, which holds if $q_n(b) > 0$ and $(-1)^n q_n(a) > 0$. As shown in Theorem 1.3.2, these zeros are the nodes of a

Gauss quadrature rule of degree $2n-2$ for w. Let $\lambda_{k,n}$ be the weight associated with $t_{k,n}$ in the quadrature. Then, the proof of Theorem 4.2.3 can be modified slightly to show the existence of the Gauss cubature rule of degree $2n-2$.

Second, the cubature rule on Ω is equivalent to a cubature rule for symmetric polynomials. In particular, let Sym_n^d be the space of symmetric polynomials of degree at most n in each variable; that is,

$$\mathrm{Sym}_n^d = \mathrm{span}\left\{m_\alpha : n \ge \alpha_1 \ge \cdots \ge \alpha_d,\, \alpha \in \Lambda^d\right\},$$

where $m_\alpha(x) = \sum_{\sigma\in\mathcal{S}} x^{\alpha\sigma}$ denotes symmetrical monomial of x^α. We state this as the following theorem.

Theorem 4.2.4. *Let w be a weight function on $\mathbb{R}$ and adopt the same notation of Theorem 4.2.3. Then, both* (4.13) *and* (4.14) *are cubature rules for the space Sym_{2n-1}^d of symmetric polynomials.*

While the cubature (4.13) can be deduced immediately from the product Gauss cubature rule of degree $2n-1$ restricted on Sym_{2n-1}^d, the deduction of the cubature (4.14) from the product Gauss cubature rule of degree $2n+1$ restricted on Sym_{2n+1}^d, using $[J(x)]^2$ being a symmetric polynomial of degree 2, relies essentially on the symmetric orthogonal polynomials $Q_\alpha^{\frac{1}{2}}$ defined in (3.48).

4.3 Gauss Cubature for Product Chebyshev Weight

There are no Gauss cubature rules of degree $2n-1$ for the product Gegenbauer weight functions $W_\lambda(\mathbf{x}) = (1-x_1^2)^{\lambda-\frac{1}{2}}(1-x_2^2)^{\lambda-\frac{1}{2}}$ on the square $[-1,1]^2$ since the weight is centrally symmetric. However, the Gauss cubature rules of degree $2n-2$ could still exist because of Γ in $\mathbb{P}_n + \Gamma\mathbb{P}_{n-1}$. We show this is the case when $\lambda = 1$.

The weight W_1 is the product Chebyshev weight functions of the second kind,

$$W_1(\mathbf{x}) = \sqrt{1-x_1^2}\sqrt{1-x_2^2}, \quad \mathbf{x} = (x_1, x_2) \in [-1,1]^2.$$

Since the product Chebyshev polynomials $U_k(x_1)U_{n-k}(x_2)$, $0 < k \le n$, are orthonormal with respect to the normalized weight function $\frac{4}{\pi^2}W_1$ on $[-1,1]^2$, the three-term relations satisfied by

$$\mathbb{U}_n(\mathbf{x}) = [U_n(x_1), U_{n-1}(x_1)U_1(x_2), \ldots, U_1(x_1)U_{n-1}(x_2), U_n(x_2)]^{\mathsf{T}}$$

have coefficient matrices $A_{n-1,1} = \frac{1}{2}(I_n|0)$ and $A_{n-1,2} = \frac{1}{2}(I_n|0)$, as well as $B_{n,i}=0$.

Among the family of W_λ, the three-term relations have the simplest structure for W_1.

Theorem 4.3.1. *The Gauss cubature rules of degree $2n-2$ exist for the product Chebyshev weight $W_1(\mathbf{x}) = \sqrt{1-x_1^2}\sqrt{1-x_2^2}$ for all $n \ge 2$. More precisely,*

$$\frac{4}{\pi^2}\int_{[-1,1]^2} f(\mathbf{x})W_1(\mathbf{x})d\mathbf{x} = \sum_{\mathbf{u}\in X_n} \lambda_n(\mathbf{u})f(\mathbf{u}), \qquad \forall f \in \Pi_{2n-2},$$

where setting $x_k = \cos\frac{k\pi}{n+2}$ and $y_j = \cos\frac{j\pi}{n+1}$,

$$X_n = \left\{\left(x_{n+3-2k}, y_{n+2-2j}\right), 1 \le k \le \lfloor\tfrac{n+2}{2}\rfloor, 1 \le j \le \lfloor\tfrac{n+1}{2}\rfloor\right\}$$
$$\bigcup\left\{\left(x_{n+2-2k}, y_{n+1-2j}\right), 1 \le k \le \lfloor\tfrac{n+1}{2}\rfloor, 1 \le j \le \lfloor\tfrac{n}{2}\rfloor\right\},$$

and $\lambda_n(\mathbf{u}) = 1/\mathbf{K}_{n-1}(\mathbf{u},\mathbf{u})$ is given explicitly by the formula

$$\lambda_n\left(x_k, y_j\right) = \frac{8(1-x_k^2)(1-y_j^2)}{(n+2)(n+1)}.$$

Moreover, X_n consists of common zeros of quasi-orthogonal polynomials

$$U_{n-k}(x_1)U_k(x_2) + U_k(x_1)U_{n-1-k}(x_2), \quad 0 \le k \le n-1, \quad U_n(x_2). \tag{4.15}$$

Proof By Theorem 4.1.5, we need to choose a matrix Γ to satisfy (4.1) and (4.2). Since $U_k(t) = 2^k t^k + \cdots$, it follows readily that $\mathbb{U}_n = 2^n I_n \mathbf{x}^n + \cdots$, where $\mathbf{x}^n = (x_1^n, x_1^{n-1}x_2, \ldots, x_1 x_2^{n-1}, x_2^n)^\mathsf{T}$, so that the leading coefficient matrix G_n of $\mathbb{U}_n$ is $2^n I_n$. Consequently, by (4.5), the matrix Γ is a Hankel matrix $\Gamma = (h_{i+j})$. We choose the matrix Γ as

$$\Gamma_n = \begin{bmatrix} \bigcirc & & 1 \\ & \iddots & \\ 1 & & \bigcirc \\ 0 & \cdots & 0 \end{bmatrix},$$

and this choice solves (4.1), as noted in Remark 4.1.6. A straightforward calculation shows that this choice also satisfies (4.2). Thus, the Gauss quadrature exists. Moreover, the quasi-orthogonal polynomials $\mathbb{U}_n + \Gamma\mathbb{U}$, given in (4.15), have $\dim \Pi_{n-1}^d$ common zeros, which are the nodes of the Gauss cubature rules of degree $2n-2$. The nodes can be determined by the first and the last two polynomials, $U_n(x_1) + U_{n-1}(x_2)$ and $U_n(x_2)$ by using $U_n(\cos\theta) = \sin(n+1)\theta/\sin\theta$. Indeed, the zeros of $U_n(x_2)$ are y_j, and it is easy to see that $U_{n-1}(y_j) = (-1)^{j+1}$, so that $U_n(x_1) + U_{n-1}(y_j) = 0$ becomes $U_n(x_1) = (-1)^{j+1}$. Solving the last equation gives the point set X_n.

By Theorem 4.1.4, the cubature weight $\lambda_n(u,v) = 1/\mathbf{K}_{n-1}(\mathbf{u},\mathbf{u})$, for $\mathbf{u} \in X_n$, where $\mathbf{K}_{n-1}(\cdot,\cdot)$ is the reproducing kernel of Π^2_{n-1}. Since $U_j(x_1)U_{k-j}(x_2)$ is orthonormal, writing U_n in terms of $x_1 = \cos\theta$ and $x_2 = \cos\phi$,

$$\mathbf{K}_{n-1}(\mathbf{x},\mathbf{x}) = \sum_{k=0}^{n-1}\sum_{j=0}^{k} \frac{\sin^2(j+1)\theta}{\sin^2\theta}\frac{\sin^2(k-j+1)\phi}{\sin^2\phi}$$
$$= \frac{1}{\sin^2\theta\sin^2\phi}\sum_{k=0}^{n-1}\sum_{j=0}^{k}(1-2\sin 2(j+1)\theta)(1-2\sin 2(k-j+1)\phi).$$

Multiplying out, we write the last sum as four sums and find close formulas using elementary trigonometric identities. A tedious calculation leads to the closed formula

$$\mathbf{K}_{n-1}(\mathbf{x},\mathbf{x}) = \frac{1}{8(1-x_1^2)(1-x_2^2)}\Bigg[2\binom{n+2}{2} - U_n^2(x_1) - U_n^2(x_2) \\ + \frac{x_1T_{n+1}(x_1)U_n(x_1) - x_2T_{n+1}(x_2)U_n(x_2)}{8(x_2^2-x_1^2)} \\ - \frac{x_2^2T_{n+2}(x_1)U_n(x_1) - x_1^2T_{n+2}(x_2)U_n(x_2)}{8(x_2^2-x_1^2)}\Bigg]. \tag{4.16}$$

Evaluating at $\mathbf{x} = (x_k, y_j)$ shows that the terms inside the square bracket become $2\binom{n+2}{2}$, which completes the proof. □

The nodes of the cubature rule when $n = 50$ are depicted in Figure 4.2.

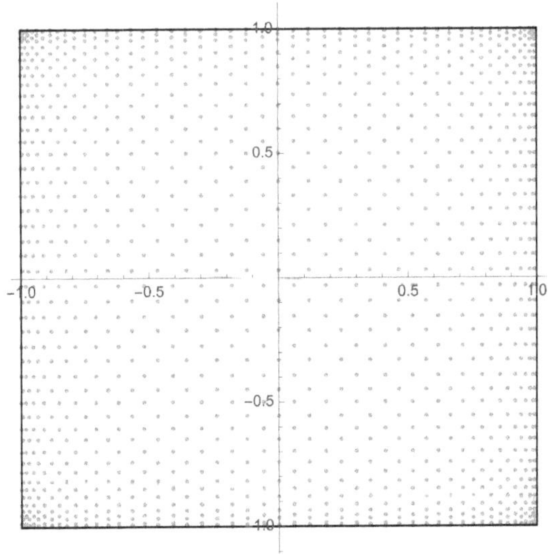

Figure 4.2 The set of nodes with $n = 50$.

An immediate question is if the Gauss cubature rule of degree $2n-2$ for W_1 also holds when $d \ge 3$. In the case of $d = 3$, the answer is positive for $n = 2$ and $n = 3$ but unknown for $n > 3$. Let us first state the two known cases. For W_1, the coefficient matrices $A_{n,i}$ are of the simplest form

$$A_{n,i} = \frac{1}{2} L_{n,i}, \qquad i = 1,2,3,$$

where $L_{n,i}$ are matrices of 0 and 1 defined by (3.8).

$d = 3$, *Gauss cubature of degree* 2. The matrix $\Gamma : 6 \times 3$ given by

$$\Gamma = \begin{bmatrix} 0 & 0 & 0 \\ 0 & 0 & 1 \\ 0 & 1 & 0 \\ 0 & 0 & 0 \\ 1 & 0 & 0 \\ 0 & 0 & 0 \end{bmatrix}$$

solves the system of equations (4.1) and (4.2). The six quasi-orthogonal polynomials that are components of $\mathbb{P}_2 + \Gamma \mathbb{P}_1$ are

$$U_2(x_1), U_1(x_1)U_1(x_2) + U_1(x_3), U_1(x_1)U_1(x_3) + U_1(x_2),$$
$$U_2(x_2), U_1(x_2)U_1(x_3) + U_1(x_1), U_2(x_3),$$

and they have $\dim \Pi_1^3 = 4$ common zeros. This gives a Gauss cubature rule of degree 2 for W_1

$$\frac{8}{\pi^3} \int_{[-1,1]^3} f(\mathbf{x}) W_1(\mathbf{x}) d\mathbf{x}$$
$$= \frac{1}{4} \left[f(-\tfrac{1}{2}, -\tfrac{1}{2}, -\tfrac{1}{2}) + f(-\tfrac{1}{2}, \tfrac{1}{2}, \tfrac{1}{2}) + f(\tfrac{1}{2}, -\tfrac{1}{2}, \tfrac{1}{2}) + f(\tfrac{1}{2}, \tfrac{1}{2}, -\tfrac{1}{2}) \right],$$

$d = 3$, *Gauss cubature of degree* 4. One solution of the system of equations (4.1) and (4.2) is given by of the matrix $\Gamma : 10 \times 6$ defined by

$$\Gamma = \frac{1}{\sqrt{2}} \begin{bmatrix} 0 & -1 & 0 & 0 & 0 & 0 \\ -1 & 0 & 0 & -1 & 0 & 1 \\ 0 & 0 & 0 & 0 & 1 & 0 \\ 0 & -1 & 0 & 0 & 0 & 0 \\ 0 & 0 & 1 & 0 & 0 & 0 \\ 0 & 1 & 0 & 0 & 0 & 0 \\ -1 & 0 & 0 & 1 & 0 & 1 \\ 0 & 0 & 0 & 0 & -1 & 0 \\ 1 & 0 & 0 & -1 & 0 & -1 \\ 0 & 0 & 0 & 0 & -1 & 0 \end{bmatrix}.$$

The 10 quasi-orthogonal polynomials in $\mathbb{P}_3 + \Gamma\mathbb{P}_2$ are given by

$$
\begin{aligned}
&2x_1(-2+4x_1^2+\sqrt{2}x_2), \quad -2x_2+2\sqrt{2}x_2^2+2x_1^2(\sqrt{2}+4x_2)-(1+4x_3^2)/\sqrt{2},\\
&2(-1+4x_1^2-\sqrt{2}x_2)x_3, \quad 2x_1(-1+\sqrt{2}x_2+4x_2^2),\\
&-2x_1(\sqrt{2}-4x_2)x_3, \quad -2x_1(1+\sqrt{2}x_2-4x_3^2),\\
&2\sqrt{2}x_1^2-4x_2-2\sqrt{2}x_2^2+8x_2^3+(-1+4x_3^2)/\sqrt{2}, \quad 2(-1+\sqrt{2}x_2+4x_2^2)x_3,\\
&-2\sqrt{2}x_1^2+2\sqrt{2}x_2^2+(-1+4x_3^2)/\sqrt{2}+x_2(-2+8x_3^2), \quad 2x_3(-2+\sqrt{2}x_2+4x_3^2).
\end{aligned}
$$

They have 10 common zeros, which are nodes of the Gauss cubature rule of degree 4 for W_1,

$$
\begin{aligned}
&\frac{8}{\pi^3}\int_{[-1,1]^3} f(\mathbf{x})W_1(\mathbf{x})\mathrm{d}\mathbf{x}\\
&=\frac{1}{8}\Big[f\Big(0,-\tfrac{1}{\sqrt{2}},-\tfrac{\sqrt{3}}{2}\Big)+f\Big(0,-\tfrac{1}{\sqrt{2}},\tfrac{\sqrt{3}}{2}\Big)+f\Big(0,-\tfrac{1}{\sqrt{2}},-\tfrac{\sqrt{3}}{2}\Big)+f\Big(0,\tfrac{1}{\sqrt{2}},-\tfrac{\sqrt{3}}{2}\Big)\Big]\\
&+\frac{1}{9}\Big[f\Big(-\tfrac{1}{2}\sqrt{\tfrac{3}{2}},\tfrac{\sqrt{2}}{4},-\tfrac{1}{2}\sqrt{\tfrac{3}{2}}\Big)+f\Big(-\tfrac{1}{2}\sqrt{\tfrac{3}{2}},\tfrac{\sqrt{2}}{4},\tfrac{1}{2}\sqrt{\tfrac{3}{2}}\Big)\\
&\qquad\qquad +f\Big(\tfrac{1}{2}\sqrt{\tfrac{3}{2}},\tfrac{\sqrt{2}}{4},-\tfrac{1}{2}\sqrt{\tfrac{3}{2}}\Big)+f\Big(\tfrac{1}{2}\sqrt{\tfrac{3}{2}},\tfrac{\sqrt{2}}{4},\tfrac{1}{2}\sqrt{\tfrac{3}{2}}\Big)\Big]\\
&+\frac{\sqrt{3}}{8}(1+\sqrt{3})\Big[f\Big(0,\tfrac{\sqrt{2}}{4}(1-\sqrt{3}),0\Big)+f\Big(0,\tfrac{\sqrt{2}}{4}(1+\sqrt{3}),0\Big)\Big].
\end{aligned}
$$

This rule is found by solving (4.2) with $\Gamma = (\gamma_{\alpha+\beta})$ given in (4.5). The nonlinear system of equations (4.2) is solved with the help of *Mathematica* and by setting a subset of parameters, judiciously chosen, as zero. As noted in Remark 4.1.6, $\Gamma = (\gamma_{\alpha+\beta})_{|\alpha|=3,|\beta|=2}$ has $r_{3+2}^3 = 21$ independent parameters and the system (4.2) consists of $\binom{r_2^3}{2} = 45$ equations, which consist of an overdetermined system of quadratic equations, and the existence of this solution is far from trivial. We do not know if the system of (4.1) and (4.2) has a solution for $n > 3$.

For $d = 2$, the existence of the Gauss cubature rule of even degree for W_1 on $[-1,1]^2$ raises the question of whether the same holds for other product weight functions on the square. Again, what we need is to solve (4.2) with Γ given in (4.5). This is explored with the help of *Mathematica*. We report the results in two cases.

Chebyshev weight of the first kind. With $W_0(x_1,x_2) = 1/(\sqrt{1-x_1^2}\sqrt{1-x_2^2})$, the coefficient matrices $A_{n,i}$ are given in (6.25). It turns out that the Gauss cubature rules of degree $2n-2$ exist if $n = 2,\ldots,6$, but do not exist for $n > 6$. Moreover, for $n \geq 4$, the cubature rule has at least one node lying outside the domain $[-1,1]^2$, which is undesirable.

Constant weight. Here, $W_{\frac{1}{2}}(\mathbf{x}) = 1$, and the orthogonal polynomials are the product Legendre polynomials. In this case, the Gauss cubature rules of degree $2n-2$ exist if $n = 2,\dots,5$, but do not exist for $n > 5$. Moreover, the cubature has node(s) lying outside the domain $[-1,1]^2$ for $n = 4$ and 5.

4.4 Nonexistence of Gauss Cubature Rules

In this section, we show that the conditions (3.17) do not hold, which implies that the Gauss cubature rules of degree $2n-1$ do not exist, for two classes of weight functions.

4.4.1 Centrally Symmetric Weight Functions

By Theorem 3.2.5, W is centrally symmetric if and only if $B_{n,i} = 0$ for all i. More generally, we call W *quasi-centrally symmetric* if

$$B_{n,i}B_{n,j} = B_{n,j}B_{n,i}, \qquad 1 \le i,j \le d.$$

We also extend this definition to positive linear functionals. If W in centrally symmetric, then for any invertible $d \times d$ matrix T and $\boldsymbol{\alpha} \in \mathbb{R}^d$, the weight function $W(Tx+\boldsymbol{\alpha})$ is quasi-centrally symmetric. This can be verified by changing variables in the integral $B_{n,i} = \mathcal{I}(x_i\mathbb{P}_n\mathbb{P}_n^{\mathsf{T}})$, which leads to $B_{n,i} = a_i I$.

For our purpose, we need to group the matrices $A_{n-1,i}A_{n-1,j}^{\mathsf{T}} - A_{n-1,j}A_{n-1,i}^{\mathsf{T}}$, $1 \le i < j \le d$ as one matrix. For this, we introduce the following notation: Let $C_1,\dots,C_d$ be matrices of size $p \times q$. For $1 \le j \le d-1$, we denote by $\Xi_{j,C}$ the block matrix of size $(d-j)p \times dq$,

$$\Xi_{j,C} = \begin{bmatrix} 0 & \cdots & 0 & C_{j+1} & -C_j & & & \bigcirc \\ 0 & \cdots & 0 & C_{j+2} & & -C_j & & \\ \vdots & \cdots & \vdots & \vdots & & & \ddots & \\ 0 & \cdots & 0 & C_d & \bigcirc & & & -C_j \end{bmatrix},$$

where the first nonzero block column is the jth one. Furthermore, we denote by C the block column matrix that has $C_1,\dots C_d$ as its elements and define Ξ_C likewise using $\Xi_{1,C},\dots\Xi_{d,C}$; that is,

$$C = (C_1^{\mathsf{T}},\dots,C_d^{\mathsf{T}})^{\mathsf{T}} \quad \text{and} \quad \Xi_C = (\Xi_{1,C}^{\mathsf{T}},\dots,\Xi_{d,C}^{\mathsf{T}})^{\mathsf{T}}. \tag{4.17}$$

Then, C is a matrix of $dp \times q$, and Ξ_C is a matrix of $\binom{d}{2}p \times dq$. In particular, A_n denotes the column matrix of $A_{n,1},\dots,A_{n,d}$. We further denote by $(A^{\mathsf{T}})_n$ the column matrix of $A_{n,1}^{\mathsf{T}}, \ldots, A_{n,d}^{\mathsf{T}}$. With these notations, it follows readily that

$$\Xi_{A_{n-1}}(A^{\mathsf{T}})_{n-1} = 0 \iff A_{n-1,i}A_{n-1,j}^{\mathsf{T}} = A_{n-1,j}A_{n-1,i}^{\mathsf{T}}, \quad 1 \le i, j \le d, \tag{4.18}$$

where the matrix on the left is of the size $\binom{d}{2} r_{n-1}^d \times r_{n-1}^d$.

Theorem 4.4.1. *Let W be a quasi-centrally symmetric weight function. Then,*

$$\operatorname{rank}\left[\Xi_{A_{n-1}}(A^{\mathsf{T}})_{n-1}\right] = \begin{cases} r_{n-1}^d, & n \text{ even}, \\ r_{n-1}^d - 1, & n \text{ odd}. \end{cases} \tag{4.19}$$

In particular, for $d = 2$, the identity (4.19) *becomes*

$$\operatorname{rank}\left(A_{n-1,1}A_{n-1,2}^{\mathsf{T}} - A_{n-1,2}A_{n-1,1}^{\mathsf{T}}\right) = 2\left\lfloor \frac{n}{2} \right\rfloor. \tag{4.20}$$

Proof Let $\mathcal{A}_n = \Xi_{A_n}(A^{\mathsf{T}})_n$. Since $\mathcal{A}_n$ is of size $\binom{d}{2} r_n^d \times r_n^d$ and its rank is equal to $r_n^d - \rho_n$, where ρ_n is the dimension of the null space $\operatorname{null}(\mathcal{A}_n)$ of $\mathcal{A}_n$, it follows that (4.19) is equivalent to $\rho_{n-1} = 0$ if n is even and $\rho_{n-1} = 1$ if n is odd. Let $X_n \in \mathbb{R}^{r_n^d}$ be a vector that satisfies $\mathcal{A}_n X_n = 0$. Then, since $\mathcal{A}_n = \Xi_{A_{n-1}^{\mathsf{T}}} A_{n-1}$ by the third commuting condition in (3.15), $A_{n-1}X_n$ belongs to the null space $\operatorname{null}(\Xi_{A_{n-1}^{\mathsf{T}}})$. Furthermore, by the first commuting condition in (3.15), $\Xi_{A_{n-1}^{\mathsf{T}}}(A^{\mathsf{T}})_{n-2} = 0$ and the rank of $(A^{\mathsf{T}})_{n-2}$ is r_{n-2}^d by (3.6), the columns of $(A^{\mathsf{T}})_{n-2}$ is a base of $\operatorname{null}(\Xi_{A_{n-1}^{\mathsf{T}}})$. Consequently, there is a vector $X_{n-1} \in \mathbb{R}^{r_{n-1}^d}$ such that

$$A_{n-1}X_n = (A^{\mathsf{T}})_{n-2}X_{n-1}. \tag{4.21}$$

By (3.12), it follows that $X_n = D_{n-1}^{\mathsf{T}}(A^{\mathsf{T}})_{n-1}X_{n-1}$, so that X_n depends on X_{n-1}. Furthermore, by the first commuting condition in (3.15), $\Xi_{A_{n-2}}A_{n-1} = 0$, so that (4.21) also leads to $\Xi_{A_{n-2}}(A^{\mathsf{T}})_{n-2}X_{n-1} = 0$, which is $\mathcal{A}_{n-2}X_{n-1} = 0$. This shows that $\rho_n = \rho_{n-2}$. Since $\mathcal{A}_0$ is a 1×1 skew symmetric matrix, $\mathcal{A}_0 = 0$, there is one vector in $\operatorname{null}(\mathcal{A}_0)$. Hence, $\rho_n = \rho_0 = 1$ if n is even. Moreover, for $n = 1$, $\mathcal{A}_1 = \Xi_{A_0^{\mathsf{T}}} A_0$ by the third commuting condition in (3.15) and A_0 is a $d \times d$ matrix and invertible by (3.12), it follows that $\operatorname{rank}\mathcal{A}_1 = \operatorname{rank}\Xi_{A_0^{\mathsf{T}}} = d$. Consequently, $\rho_n = \rho_1 = 1$ when n is odd. □

Corollary 4.4.2. *The Gauss cubature rules of degree $2n-1$ do not exist for any quasi centrally symmetric linear functional.*

By (4.19) and the identity (4.18), the relations $A_{n-1,i}A_{n-1,j}^{\mathsf{T}} = A_{n-1,j}A_{n-1,i}^{\mathsf{T}}$ failed to hold in the quasi-centrally symmetric setting. Hence, the Gauss cubature rules of degree $2n-1$ do not exist by Theorem 4.1.5.

4.4.2 Wrapt Product Weight Functions in Two Variables

We consider the weight function

$$W(x_1,x_2) = w_1(x_1)w_2\left(\frac{x_2}{\rho(x_1)}\right),$$

defined in (3.21), where ρ is either a linear polynomial or the square root of a quadratic polynomial as in (3.22). Let $A_{n,i}$ be the coefficient matrices of the three-term relations for the orthogonal polynomials with respect to W, which are given explicitly in Subsection 3.4.2.

Proposition 4.4.3. *Let W be defined in* (3.21) *with ρ given in* (3.22). *Assume $\rho_0 < 0$ in Case II. Then,*

$$\operatorname{rank}\left[A_{n-1,1}A_{n-1,2}^{\mathsf{T}} - A_{n-1,2}A_{n-1,1}^{\mathsf{T}}\right] = 2\left\lfloor\frac{n}{2}\right\rfloor.$$

In particular, this identity holds for the Jacobi polynomials on the disk, triangle, and parabolic domain.

Proof Using the formula for $A_{n,i}$ given in Subsection 3.4.2, we deduce immediately that

$$\mathcal{A}_{n-1} := A_{n-1,1}A_{n-1,2}^{\mathsf{T}} - A_{n-1,2}A_{n-1,1}^{\mathsf{T}} = \begin{bmatrix} 0 & h_0 & & & \bigcirc \\ -h_0 & 0 & h_1 & & \\ & \ddots & \ddots & \ddots & \\ & & -h_{n-2} & 0 & h_{n-1} \\ \bigcirc & & & -h_{n-1} & 0 \end{bmatrix},$$

where

$$h_k := h_{k,n-1} = a_{n-k,k-1}c_{k,n-1} - a_{n-k-1,k}d_{k-1,n-1}, \qquad 1 \le k \le n-1,$$

in terms of $c_{k,n}$ and $d_{k,n}$ given in (3.25) and the coefficient $a_{n,k}$ in the three-term relation of the orthogonal polynomial $\{q_{n,k}\}_{n=0}^{\infty}$. Since $\mathcal{A}_{n-1}$ is a $n\times n$ skew-symmetric matrix, its rank must be even. If all h_k are nonzero, then $\operatorname{rank}\mathcal{A}_{n-1} = n-1$ for n odd since the elements on the subdiagonal of $\mathcal{A}_{n-1}$ are all nonzero, and $\operatorname{rank}\mathcal{A}_{n-1} = n$ if n is even since

$$\det \mathcal{A}_{n-1} = h_0^2 h_2^2 \ldots h_{[n/2]}^2 > 0.$$

Thus, we only need to show $h_k \neq 0$ for $1 \le k \le n-1$. By (3.25),

$$h_k = a_{k-1}(w_2)\left(\tau a_{n-k,k-1}\frac{\gamma_{n-k-1,k}}{\gamma_{n-k+1,k-1}} - a_{n-k-1,k}\frac{\gamma_{n-k,k-1}}{\gamma_{n-k,k}}\right).$$

Furthermore, let $D_{n,k}$ be the determinant of the moment matrix for $\rho(t)^{2k+1}$ $w(t)$; since $a_{n,k} = \gamma_{n,k}/\gamma_{n+1,k}$, we deduce using (1.4) that

$$h_k = a_{k-1}(w_2)\gamma_{n-k-1,k}\gamma_{n-k,k-1}\left(\tau\frac{D_{n-k+2,k-1}}{D_{n-k+1,k-1}} - \frac{D_{n-k+1,k}}{D_{n-k,k}}\right).$$

Because $\gamma_{m,k}$ are positive and $\tau = \rho_0 \le 0$ in Case II by our assumption, we conclude that $h_\kappa < 0$ for $1 \le k \le n$ in Case II. Next, we consider Case I, for which $\rho(t) = \rho_0 t + \rho_1$. If $\rho_0 = 0$, then $\rho(t) = \rho_1$ is a positive constant and $\tau = \rho_0^2 = 0$. It follows readily that $h_k < 0$ for all $1 \le k \le n$. If $\rho_0 > 0$, then by a translation and a dilation of w_2 if necessary, we can assume $\rho(t) = t$ without loss of generality. For $\rho(t) = t$, the moments $\mu_{i,k}$ of $\rho(t)^{2k+1}w_1(t)$ can be given in terms of the moments μ_k of $w_1(t)$,

$$\mu_{j,k} = \int_a^b t^j(\rho(t))^{2k+1}w_1(t)\mathrm{d}t = \int_a^b t^{j+2k+1}w_1(t)\mathrm{d}t = \mu_{j+2k+1}.$$

For $r = 1,2,\ldots$, let $D_m^{(r)} = \det[\mu_{i+j+r}]_{i,j=0}^n$. It follows that

$$D_{m,k} = \det\left[\mu_{i+j,k}\right]_{i,j=0}^m = \det\left[\mu_{i+j+2k+1}\right]_{i,j=0}^m = D_m^{(2k+1)}.$$

Consequently, we obtain

$$\begin{aligned} h_k &= \frac{a_{k-1}(w_2)\gamma_{n-k-1,k}\gamma_{n-k,k-1}}{D_{n-k}^{(2k+1)}D_{n-k+1}^{(2k-1)}}(D_{n-k+2}^{(2k-1)}D_{n-k}^{(2k+1)} - D_{n-k+1}^{(2k+1)}D_{n-k+1}^{(2k-1)}) \\ &= -\frac{a_{k-1}(w_2)\gamma_{n-k-1,k}\gamma_{n-k,k-1}}{D_{n-k}^{(2k+1)}D_{n-k+1}^{(2k-1)}}\left(D_{n-k+1}^{(2k)}\right)^2, \end{aligned}$$

where we have used $D_m^{(r)}D_m^{(r+2)} - D_{m+1}^{(r)}D_{m-1}^{(r)} = [D_m^{(r+1)}]^2$, see [128, p. 96, Prob. 19], for example. Now, the matrix $(\mu_{i+j})_{i,j=0}^n$ is positive definite, so are its principal minors. In particular, $D_m^{(r)} > 0$. Thus, $h_k < 0$. This completes the proof. □

Corollary 4.4.4. *The Gauss cubature rules of degree* $2n-1$ *do not exist for the wrapt product weight function W in* (3.21). *In particular, they do not exist for the Jacobi weight on the disk, triangle, and parabolic domain.*

4.5 Notes and Further Results

The characterization of Gauss cubature rules in Theorem 4.1.1 is classical [115]. In terms of the structural matrices, the characterization of Theorem 4.1.5 was established in [164] for cubature rules of degree $2n-1$ and in [165] for cubature rules of degree $2n-2$. Another characterization is given in [84], which shows in particular that the Gauss cubature rule of degree $2n-1$ exists for the integral $\mathcal{L}$ if and only if there is a polynomial $Q \in \mathcal{V}_{2n}^d$ of degree $2n$ such that $\mathcal{I}\left(\mathbb{P}_n\mathbb{P}_n^{\mathsf{T}}Q\right) = I_{r_{2n}^d}$, where $\mathbb{P}_n$ consists of orthonormal polynomials for $\mathcal{I}$. Polynomial interpolation on nodes of the Gauss cubature rule is discussed in [160].

The existence of Gauss cubature rules of degree $2n-1$ was known only for small n until the example in the Subsection 4.11 appeared in [140], followed by the example in Subsection 4.2.2 appeared in [12]. Another family of examples will be given in Section 8.4. These, however, are the only Gauss cubature rules known for all $n \in \mathbb{N}$ up to now.

For $d = 2$, the Gauss cubature rules of $2n-2$ in Subsection 4.3 are first established in [110], but the weights $\lambda(u,v)$ are not explicitly given. The product Chebyshev weight of the second kind is the only centrally symmetric weight function for which the Gauss cubature rules of degree $2n-2$ are known for all n. For $d = 3$, the Gauss cubature of degree 4 is new; it would be interesting to find out if the same can be established for larger n. Could it hold for all n?

Radon [130] observed that the Gauss cubature rule of degree 5 does not exist for the integral on a square, triangle, or disk. A quantitative result of this nonexistence is the profound result of Möller's lower bound, to be discussed in the next chapter, for the number of nodes of a cubature rule in the centrally symmetric setting, for which the rank of the matrices computed in Subsection 4.4 will be needed. The rank identity for the quasi-central symmetric case appeared in [167] and for the wrapt product weight in [166]. A necessary condition for the existence of Gauss cubature rule of degree $2n-1$ imposes a quantitative condition on the rank of the matrix $B_{n,i}B_{n,j} - B_{n,j}B_{n,i}$ and it shows, in particular, (3.19), which also implies that the Gauss cubature rules do not exist for quasi-centrally symmetric integrals.

As it is shown in the proof of Theorem 4.2.3, the Gauss cubature rule (4.12) is equivalent to the exact cubature rule (4.14) for symmetric polynomials, and the nodes of the latter are zeros of generalized Schur functions associated with the orthogonal polynomials $\{p_n\}_{n\ge 0}$. This symmetric cubature rule is rediscovered and used in [43], and the construction is generalized from Schur to Hall–Littlewood polynomials in [44], which, however, does not lift to a cubature rule of high precision for algebraic polynomials and, as a result, does not lead to further examples of Gauss cubature rules.

Throughout this chapter, we assume that the integral is square positive. We could consider the Gauss cubature rules for integrals that are not necessarily positive definite. In [134], this is discussed in the context of extending the relation between the Gauss quadrature and continued fractions (cf. [133]) to more than one variable.

5
Lower Bounds for the Number of Nodes

When the Gauss cubature rule does not exist for an integral, we are interested in cubature rules having as few nodes as possible. This suggests the following definition.

Definition 5.0.1. A cubature rule $\mathsf{C}_m(f)$ is called minimal if its number of nodes is the smallest among all cubature rules of the same degree for the same integral.

The existence of the minimal cubature rule is a tautology. Verifying that a cubature rule is minimal requires knowing a lower bound for the number of nodes. The Gauss cubature rules, when they exist, are minimal since their number of nodes attains the lower bound (2.2) for the number of nodes in Theorem 2.1.1. The nonexistence of the Gauss cubature rule in Chapter 4 implies that the lower bound (2.2) is not sharp for integrals that are centrally symmetric or with wrapt product weight functions. In this chapter, we give several improved lower bounds for subclasses of integrals.

A generic lower bound, given in terms of the rank of the moment matrix of the positive cubature rule, is stated in Section 5.1. For cubature rules of degree $2n-1$, a lower bound due to Möller is given for centrally symmetric weight functions by a quantitative estimate on the rank of the moment matrix of degree n in Section 5.2. A second lower bound is established via a different method in Section 5.3, which agrees with the first one when $d = 2$ but is stronger for $d > 2$. A different lower bound is established in Section 5.4, using the linear programming method, for the surface measure on the unit sphere and the Chebyshev weight functions on the unit ball and on the simplex, which shows that the improved lower bounds in the previous two subsections are not sharp for these weight functions. Finally, in Section 5.5, we show that Möller's lower bound is sharp for one rotationally symmetric integral, which, however, is an integral with the weight function depending on n.

5.1 A Generic Lower Bound

Let $\mathcal{I}(f)$ be a square positive linear functional. Let $\mathsf{C}_m(f)$ be a positive cubature rule of degree m for $\mathcal{I}(f)$ with N nodes,

$$\mathsf{C}_m(f) = \sum_{k=1}^{N} \lambda_k f(\mathbf{x}_k) = \mathcal{I}(f), \qquad \forall f \in \Pi_m^d,$$

where $\mathbf{x}_k \in \mathbb{R}^d$ are pairwise distinct points and $\lambda_k > 0$. Associate with the nodes of $\mathsf{C}_m(f)$, we define a matrix M_k of size $\dim \Pi_k^d \times N$ by

$$\mathsf{M}_k = \left[\mathbf{x}_j^\alpha\right]_{|\alpha| \le k, 1 \le j \le N}, \quad k = 0, 1, 2, \ldots, m,$$

where the rows of M_k are ordered according to the graded lexicographical order of monomials in Π_k^d. For $m = 2n - 1$ and $m = 2n - 2$, we define

$$\sigma_k = \operatorname{rank} \mathsf{M}_{n+k} - \operatorname{rank} \mathsf{M}_{n+k-1}, \qquad k = 0, 1, 2, \ldots, m - n. \tag{5.1}$$

Theorem 5.1.1. *Let $m = 2n - 1$ or $m = 2n - 2$. Then, the number of nodes of the positive cubature rule $\mathsf{C}_m(f)$ satisfies*

$$N \ge \operatorname{rank} \mathsf{M}_m = \dim \Pi_{n-1}^d + \sigma_0 + \cdots + \sigma_{m-n}. \tag{5.2}$$

Moreover, the inequality becomes equality if $\mathsf{C}_m(f)$ is minimal.

Proof If $\mathsf{C}_m(f)$ is minimal, then its number of nodes is $N = \operatorname{rank} \mathsf{M}_m$. Indeed, if $N > \operatorname{rank} \mathsf{M}_m$, then there is an $\mathbf{a} \in \mathbb{R}^N$ such that $\mathsf{M}_m^\mathsf{T} \mathbf{a} = 0$, which shows $\sum_{k=1}^N a_k \mathbf{x}_k^\alpha = 0$ for $|\alpha| \le m$. In particular, this shows $\sum_{k=1}^N a_k P(\mathbf{x}_k) = 0$ for all polynomials in Π_m^d. Since the cubature weights λ_k are positive, a_k cannot be all of the same sign. Let $\mu = \min\{a_i^{-1}\lambda_i : \alpha_i > 0\}$. Then,

$$\sum_{k=1}^{N} (\lambda_k - \mu a_k) P(\mathbf{x}_k) = \sum_{k=1}^{N} \lambda_k P(\mathbf{x}_k) = \mathcal{I}(f), \qquad \forall P \in \Pi_m^d.$$

Since $\lambda_k - \mu a_k \ge 0$ and $= 0$ for at least one index, the sum on the left is a cubature rule of degree m with at most $N - 1$ nodes. This contradicts the assumption that $\mathsf{C}_m(f)$ is minimal.

Let N be the number of nodes of any positive cubature $\mathsf{C}_m(f)$. Then, N is greater or equal to the number of minimal cubature rules of the same degree. Hence, $N \ge \operatorname{rank} \mathsf{M}_m$. By the definition of σ_k,

$$N \ge \operatorname{rank} \mathsf{M}_m = \sigma_{m-n} + \cdots + \sigma_1 + \sigma_0 + \operatorname{rank} \mathsf{M}_{n-1}.$$

Thus, it suffices to show that $\operatorname{rank} \mathsf{M}_{n-1} = \dim \Pi_{n-1}^d$. If $\xi = (\xi_\alpha)_{|\alpha|=n-1}$ satisfies $\mathsf{M}_{n-1}\xi = 0$, then the polynomial $p(\xi) = \sum_{|\alpha|\le n-1}^N \xi_\alpha \mathbf{x}^\alpha$ satisfies $p(\mathbf{x}_k) = 0$ for $1 \le k \le N$. Since $0 = \mathsf{C}_m(p^2) = \mathcal{I}(p^2)$, p much be 0. This shows that $\xi = 0$ and M_{n-1} has the full rank, which is $\dim \Pi_{n-1}^d$. This completes the proof. □

We state the lower bound (5.2) as a sum of σ_i instead of $\operatorname{rank} \mathsf{M}_m$ because the nodes of a positive cubature rule of degree $m = 2n-1$ are zeros of certain polynomials of degree up to m, among which the number of independent polynomials of degree exactly $n+k$ is determined via σ_k. The first case is σ_0, for which the corresponding polynomials are of degree n and vanish on the nodes of C_{2n-1}.

Proposition 5.1.2. *Let $m = 2n-1$. Then, there are $r_n^d - \sigma_0$ linearly independent orthogonal polynomials of degree n that vanish on the nodes of $\mathsf{C}_{2n-1}(f)$.*

Proof Each solution of $\mathsf{M}_n\xi = 0$ leads to a polynomial p_n of degree n that vanishes on the nodes of $\mathsf{C}_{2n-1}(f)$, and it is an orthogonal polynomial since $\mathcal{I}(p_n q) = \mathsf{C}_{2n-1}(p_n q) = 0$ for $q \in \Pi_{n-1}^d$. Such polynomials consist of a subspace with dimension $\dim \Pi_n^d - \operatorname{rank} \mathsf{M}_n = \dim \Pi_{n-1}^d + r_n^d - \operatorname{rank} \mathsf{M}_n$, which is equal to $r_n^d - \sigma_0$ since $\sigma_0 = \operatorname{rank} \mathsf{M}_n - \operatorname{rank} \mathsf{M}_{n-1}$ and $\operatorname{rank} \mathsf{M}_{n-1} = \dim \Pi_{n-1}^d$. □

For σ_k with $k > 0$, we can state an analog result, but we need to extend the orthogonality of polynomials to the notion of m-orthogonality.

Definition 5.1.3. Let $\mathcal{I}$ be a positive definite linear functional. A polynomial p is called m-orthogonal with respect to $\mathcal{I}$ if $\mathcal{I}(pq) = 0$ for all polynomials q such that $\deg p + \deg q \le m$.

Ordinary orthogonal polynomials of degree n are $(2n-1)$-orthogonal. There are other $(2n-1)$-orthogonal polynomials, necessarily of higher degrees. Let $\mathbb{P}_n$ be a basis of orthogonal polynomials for $\mathcal{I}$, which we again regard as a column vector. Then, the components of

$$\mathbb{P}_{n+1} + \Gamma_1 \mathbb{P}_n + \Gamma_2 \mathbb{P}_{n-1},$$

where $\Gamma_1 : r_{n+1}^d \times r_n^d$ and $\Gamma_2 : r_{n+1}^d \times r_{n-1}^d$ are matrices, are $(2n-1)$-orthogonal polynomials of degree $n+1$, since they are orthogonal to polynomials in Π_{n-2}^d, by the orthogonality of $\mathbb{P}_{n-1}$, and $n+1+(n-2) = 2n-1$. We sum up the properties of m-orthogonal polynomials in the following lemma.

Lemma 5.1.4. *Let $m = 2n-1$ or $m = 2n-2$. Let $\Gamma_j : r_{n+1}^d \times r_{n+\ell-j}^d$ be matrices.*

1. *For each* $0 \le \ell \le n - \ell - \delta_m$, *the components of*

$$\mathbb{P}_{n+\ell} + \Gamma_1 \mathbb{P}_{n+\ell-1} + \cdots + \Gamma_{2\ell-\delta_m} \mathbb{P}_{n-\ell-\delta_m}, \qquad \delta_m = \begin{cases} 0 & m = 2n-1 \\ 1 & m = 2n-2 \end{cases},$$

are m-orthogonal polynomials of degree $n + \ell$.

2. *If P is an m-orthogonal polynomial of degree k and h is a polynomial of degree j with* $j \le m - k$, *then hP is an m-orthogonal polynomial of degree* $k + j$.

Proof Part 1 follows immediately from $n + \ell + (n - \ell - \delta_m) - 1 = m$ and the orthogonality of $\mathbb{P}_k$. Part 2 follows from the three-term relation (3.5), which shows immediately that $\mathbf{x}_i P$ is an m-orthogonal polynomial of degree $k+1$ for $1 \le i \le d$, so that part 2 holds for h being a linear polynomial. Applying the three-term relation repeatedly completes the proof. □

We note that the definition includes the extreme case of $\ell = n - 1 - \delta_m$, which shows that every polynomial of degree m that does not have the constant term is m-orthogonal. What is of interest are polynomials with small ℓ. For example, the quasi-orthogonal polynomials of $\mathbb{P}_n + \Gamma \mathbb{P}_{n-1}$ are $(2n-2)$-orthogonal and $\mathbf{x}_i \mathbb{P}_n$ consist of $(2n-1)$-orthogonal polynomials of degree $n+1$.

We can now establish an analog of Proposition 5.1.2 for σ_k with $k > 0$.

Proposition 5.1.5. *For* $0 \le k \le m - n$, *there are* $r^d_{n+k} - \sigma_k$ *linearly independent m-orthogonal polynomials of degree n that vanish on the nodes of* $\mathsf{C}_m(f)$.

Proof As in the proof of the Proposition 5.1.2, each nontrivial solution ξ of $\mathsf{M}_{n+k}\xi = 0$ corresponds to a polynomial P_ξ of degree $n+k$ that vanishes on the nodes of the cubature rule, which is m-orthogonal, and such polynomials consist of a space of dimension $r^d_{n+k} - \tau_k$, where $\tau_k = \dim \Pi^d_{n-1} + \sigma_0 + \cdots + \sigma_k$. Moreover, the space contains $\dim \Pi^d_{n+k-1} - \tau_{k-1}$ many linearly independent polynomials of degree at most $n+k-1$, which come from the null space of M_{n+k-1}. Therefore, there are exactly

$$\dim \Pi^d_{n+k} - \tau_k - \left(\dim \Pi^d_{n+k-1} - \tau_{k-1}\right) = r^d_{n+k} - \sigma_k$$

linearly independent polynomials of degree $n+k$ that vanish on the nodes of the cubature. □

5.2 Möller's Lower Bound

We consider cubature rules of degree $m = 2n-1$. By Theorem 5.1.1, a lower bound for the number of nodes of the cubature rule can be deduced from a lower bound of σ_0. Let $\mathbb{P}_n$ be an orthonormal basis of $\mathcal{V}_n^d$. We write

$$r_n = r_n^d = \dim \mathcal{V}_n^d = \binom{n+d-1}{n}.$$

By Proposition 5.1.2, a cubature rule $\mathsf{C}_{2n-1}(f)$ of degree $2n-1$ has $r_n - \sigma_0$ linearly independent orthogonal polynomials that vanish on its nodes. Written these polynomials as $\mathbf{u}_j^\mathsf{T}\mathbb{P}_n$, $\mathbf{u}_j \in \mathbb{R}^{r_n}$ and $1 \le j \le r_n - \sigma_0$, and denote by U the matrix $U = [\mathbf{u}_1, \ldots, \mathbf{u}_{r_n-\sigma_0}]$, then the columns of U are linearly independent as the polynomials are linearly independent. Thus, the orthogonal polynomials that vanish on the nodes of $\mathsf{C}_{2n-1}(f)$ are the components of

$$U^\mathsf{T}\mathbb{P}_n \quad \text{with} \quad U : r_n \times (r_n - \sigma_0), \qquad \operatorname{rank} U = r_n - \sigma_0.$$

Let $A_{n,i}$ be the coefficient matrices of the three-term relation (3.5). We use the notation $\Xi_{A_{n-1}}(A^\mathsf{T})_{n-1}$ as in (4.4.1).

Theorem 5.2.1. *For $d \ge 2$, the number of nodes of a cubature rule of degree $2n-1$ satisfies*

$$N \ge \dim \Pi_{n-1}^d + \frac{1}{d}\operatorname{rank}\left[\Xi_{A_{n-1}}(A^\mathsf{T})_{n-1}\right]. \tag{5.3}$$

In particular, if $d = 2$, then

$$N \ge \dim \Pi_{n-1}^2 + \frac{1}{2}\operatorname{rank}\left[A_{n-1,1}A_{n-1,2}^\mathsf{T} - A_{n-1,2}A_{n-1,1}^\mathsf{T}\right]. \tag{5.4}$$

Proof We assume that the polynomials of degree n that vanish on the nodes of the cubature rule are given by $U^\mathsf{T}\mathbb{P}_n$ and give a lower bound for σ_0. For this purpose, we consider the space $\mathcal{W} = \operatorname{span}\{U^\mathsf{T}\mathbb{P}_n, x_i U^\mathsf{T}\mathbb{P}_n : 1 \le i \le d\}$. Let γ denote the number of linearly dependent polynomials among $\{U^\mathsf{T}\mathbb{P}_n, x_i U^\mathsf{T}\mathbb{P}_n, 1 \le i \le d\}$. Then, since $x_i\mathbb{P}_n$ are polynomials of degree $n+1$, we obtain

$$(d+1)(r_n - \sigma_0) - \gamma \le \dim \mathcal{W} \le r_n - \sigma_0 + r_{n+1}, \tag{5.5}$$

which implies that

$$\sigma_0 \ge r_n - \frac{1}{d}(r_{n+1} + \gamma). \tag{5.6}$$

The number γ is equal to the linearly independent solutions of the system

$$\sum_{i=1}^d \xi_i^\mathsf{T} x_i U^\mathsf{T}\mathbb{P}_n = \eta^\mathsf{T} U^\mathsf{T}\mathbb{P}_n, \qquad \xi_i, \eta \in \mathbb{R}^{r_n - \sigma_0}.$$

Applying the three-term relation (3.5) on the left-hand side and comparing the coefficient of $\mathbb{P}_n$, we see that a solution of the system must satisfy

$\sum_{i=1}^d \xi_i^{\mathsf{T}} U^{\mathsf{T}} B_{n,i} = \eta^{\mathsf{T}} U^{\mathsf{T}}$, which determines the value of η since $U^{\mathsf{T}} U$ is an invertible matrix; moreover, the coefficients of $\mathbb{P}_{n+1}$ and $\mathbb{P}_{n-1}$ must be zero, so that

$$\sum_{i=1}^{d} \zeta_i^{\mathsf{T}} A_{n,i} = 0 \quad \text{and} \quad \sum_{i=1}^{d} \zeta_i^{\mathsf{T}} A_{n-1,i}^{\mathsf{T}} = 0, \tag{5.7}$$

where $\zeta_i = U\xi_i$. These two equations can be written as $\mathcal{A}\zeta = 0$, where the matrix $\mathcal{A}$ is defined by

$$\mathcal{A} = \begin{pmatrix} A_{n,1}^{\mathsf{T}} & \cdots & A_{n,d}^{\mathsf{T}} \\ A_{n-1,1} & \cdots & A_{n-1,d} \end{pmatrix},$$

and $\mathcal{A}$ is of size $(r_{n+1} + r_{n-1}) \times dr_n$. Hence, $\gamma = dr_n - \operatorname{rank}\mathcal{A}$. To compute the rank of $\mathcal{A}$, we consider the linear system $\mathcal{A}^{\mathsf{T}}\xi = 0$, which is equivalent to

$$A_{n,i}\xi_1 + A_{n-1,i}^{\mathsf{T}}\xi_2 = 0, \quad 1 \le i \le d, \qquad \xi_1 \in \mathbb{R}^{r_{n+1}^d}, \quad \xi_2 \in \mathbb{R}^{r_{n-1}^d}.$$

By (3.12), $\xi_1 = \sum_{i=1}^d D^{\mathsf{T}} A_{n-1,i}^{\mathsf{T}} \xi_2$, so that ξ_1 depends on ξ_2. Furthermore, by the first commuting relation in (3.15), we further deduce that

$$\left(A_{n-1,i}A_{n-1,j}^{\mathsf{T}} - A_{n-1,j}A_{n-1,i}^{\mathsf{T}}\right)\xi_2 = \left(A_{n-1,i}A_{n,j} - A_{n-1,j}A_{n-1,i}\right)\xi_1 = 0$$

for $1 \le i, j \le d$. This system is equivalent to $\Xi_{A_{n-1}}(\mathcal{A}^{\mathsf{T}})_{n-1}\xi_2 = 0$ and the matrix $\Xi_{A_{n-1}}(\mathcal{A}^{\mathsf{T}})_{n-1}$ is of size $\binom{d}{2} r_{n-1} \times r_{n-1}$, so that it has

$$\rho_n = r_{n-1} - \operatorname{rank}[\Xi_{A_{n-1}}(\mathcal{A}^{\mathsf{T}})_{n-1}]$$

linearly independent solutions. Consequently,

$$\operatorname{rank}\mathcal{A} = \operatorname{rank}\mathcal{A}^{\mathsf{T}} = r_{n+1} + r_{n-1} - \rho_n = r_{n+1} + \operatorname{rank}\left[\Xi_{A_{n-1}}(\mathcal{A}^{\mathsf{T}})_{n-1}\right].$$

Thus, $\gamma = dr_n - r_{n+1} - \operatorname{rank}\left[\Xi_{A_{n-1}}(\mathcal{A}^{\mathsf{T}})_{n-1}\right]$ and, by (5.6), we conclude that $\sigma_0 \ge \frac{1}{d}\operatorname{rank}\left[\Xi_{A_{n-1}}(\mathcal{A}^{\mathsf{T}})_{n-1}\right]$. Thus, (5.3) follows from $N \ge \dim \Pi_{n-1}^d + \sigma_0$. □

If Gauss cubature rules exist, then $A_{n-1,i}A_{n-1,j}^{\mathsf{T}} - A_{n-1,j}A_{n-1,i}^{\mathsf{T}} = 0$ by Theorem 4.1.5, so that $\Xi_{A_{n-1}}(\mathcal{A}^{\mathsf{T}})_{n-1} = 0$ and the lower bound (5.3) agrees with the classical lower bound (2.2). In the centrally symmetric setting, the rank of $\Xi_{A_{n-1}}(\mathcal{A}^{\mathsf{T}})_{n-1}$ is determined in Theorem 4.4.1, and we obtain an improved lower bound.

Theorem 5.2.2. *Let $\mathcal{L}$ be a quasi-centrally symmetric linear functional. Then, the number of nodes of a cubature rule of degree $2n-1$ satisfies*

$$N \ge \dim \Pi_{n-1}^d + \frac{1}{d}(r_{n-1} - \delta_n) = \binom{n+d-1}{d} + \frac{1}{d}\left(\binom{n+d-2}{d-1} - \delta_n\right), \tag{5.8}$$

where $\delta_n = 0$ if n is even and $\delta_n = 1$ if n is odd.

The most useful cases of this lower bound are when $d = 2$, for which the same lower bound also holds for the case of wrapt product weight functions by Proposition 4.4.3. This is stated in the following theorem, and its last statement follows from Proposition 5.1.2.

Theorem 5.2.3. *Let $d = 2$ and let W be a quasi-centrally symmetric weight function or the wrapt product weight function defined in* (3.21). *Then the number of nodes of a cubature rule of degree $2n-1$ satisfies*

$$N \ge \dim \Pi_{n-1}^2 + \left\lfloor \frac{n}{2} \right\rfloor = \frac{1}{2}n(n+1) + \left\lfloor \frac{n}{2} \right\rfloor. \tag{5.9}$$

Moreover, the nodes of the cubature rule are common zeros of $n+1-\lfloor \frac{n}{2} \rfloor$ linearly independent orthogonal polynomials of degree n.

We shall call (5.9) Möller's lower bound since it was first established by H. Möller for the centrally symmetric weight function in [107].

The lower bound (5.9) holds for the wrapt product weight function discussed in (3.21), which includes product weight on the square and the Jacobi weight functions on the unit disk and the triangle, and they are not necessarily centrally symmetric. For later reference, we state it as a corollary.

Corollary 5.2.4. *For the wrapt product weight function defined in* (3.21), *the number of nodes of a cubature rule of degree $2n-1$ satisfies* (5.9).

5.3 A Second Lower Bound for Centrally Symmetric Weight

We state a second lower bound for cubature rules of degree $2n-1$ in the centrally symmetric setting, which is given in terms of even polynomials and odd polynomials. We denote by $\mathscr{P}_{2k}^d$ the space of even polynomials of degree at most $2k$ and $\mathscr{P}_{2k+1}^d$ the space of odd polynomials of degree at most $2k+1$. Then,

$$\mathscr{P}_{2k}^d = \Pi_0^d \cup \Pi_2^d \cup \cdots \cup \Pi_{2k}^d \quad \text{and} \quad \mathscr{P}_{2k+1}^d = \Pi_1^d \cup \Pi_3^d \cup \cdots \cup \Pi_{2k+1}^d.$$

Lemma 5.3.1. *The dimension of the space $\mathscr{P}_n^d$ satisfies*

$$2\dim \mathscr{P}_n^d = \dim \Pi_n^d + \dim \mathscr{P}_n^{d-1} = \dim \Pi_n^d + \sum_{\ell=1}^{d-1} 2^{\ell-d} \dim \Pi_n^\ell. \tag{5.10}$$

In particular,

$$2\dim \mathscr{P}_n^d = \binom{n+d}{d} + \frac{1}{2}\binom{n+d-1}{d-1} + \cdots + \frac{1}{2^{d-1}}(n+1) = \frac{n^d}{d!} + \mathcal{O}(n^{d-1}).$$

Proof For $n = 2k-1$, using $r_k^d = \binom{k+d-1}{d-1}$ and the identity for the binomial coefficients $\binom{p}{q} = \binom{p-1}{q} + \binom{p-1}{q-1}$, we obtain

$$\dim \mathscr{P}_{2k-1}^d - \dim \mathscr{P}_{2k-1}^{d-1} = \sum_{j=1}^{k} \left(r_{2j-1}^d - r_{2j-1}^{d-1} \right) = \sum_{j=1}^{k} \left[\binom{2j+d-2}{d-1} - \binom{2j+d-3}{d-2} \right]$$
$$= \sum_{j=1}^{k} \binom{2j+d-3}{d-1} = \sum_{j=1}^{k} r_{2j-2}^d = \dim \mathscr{P}_{2k-2}^d.$$

The same proof shows that this identity holds if $2k-1$ is replaced by $2k-2$. Thus, we have shown that

$$\dim \mathscr{P}_n^d - \dim \mathscr{P}_n^{d-1} = \dim \mathscr{P}_{n-1}^d.$$

Moreover, by definition, $\mathscr{P}_n^d \cup \mathscr{P}_{n-1}^d = \Pi_n^d$ so that

$$\dim \mathscr{P}_n^d + \dim \mathscr{P}_{n-1}^d = \dim \Pi_n^d.$$

Combining the two identities proves the first equality of (5.10), which iterates to the second identity. □

Theorem 5.3.2. *Let $\mathcal{L}$ be a centrally symmetric positive linear functional. Then the number of nodes of a cubature rule of degree $2n-1$ for $\mathcal{L}$ satisfies*

$$N \ge N_{n,d} \quad \text{with} \quad N_{n,d} = 2 \dim \mathscr{P}_{n-1}^d - \delta_n, \tag{5.11}$$

where $\delta_n = 1$ if n is odd and $\delta_n = 0$ if n is even. In particular,

$$N_{n,d} = \dim \Pi_{n-1}^d + \dim \mathscr{P}_{n-1}^{d-1} - \delta_n. \tag{5.12}$$

Proof Let C be a cubature rule of degree $2n-1$. We need $\mathsf{M}_k = \left[\mathbf{x}_j^\alpha\right]_{1\le j\le N, |\alpha|\le k}$ defined via the nodes of C. By Theorem 5.1.1, $N \ge \operatorname{rank} \mathsf{M}_m \ge \operatorname{rank} \mathsf{M}_n$. We need a lower bound for the rank of M_n. The rows of M_n are ordered as monomials in Π_n^d. We define an analog of M_n with $\mathbf{x}^\alpha \in \Pi_n^d$ replaced by $\mathbf{x}^\alpha \in \mathscr{P}_n^d$,

$$\mathscr{M}_n = \left[\mathbf{x}_j^\alpha\right]_{1\le j\le N, \alpha\in \mathscr{I}(n)}, \quad \text{with} \quad \mathscr{I}(n) = \{\alpha \in \mathbb{N}_0^d : \mathbf{x}^\alpha \in \mathscr{P}_n^d\}.$$

Likewise, for a vector $\xi = \{\xi_\alpha : |\alpha| \le n\}$, we denote $\xi(n) = \{\xi_\alpha : \alpha \in \mathscr{I}(n)\}$. The rows of $\mathscr{M}_n$ and $\mathscr{M}_{n-1}$ are disjoint, and their union consists of the rows of M_n. The same holds for ξ, $\xi(n)$ and $\xi(n-1)$, treating as column vectors.

Let $I(\mathsf{C})$ be the space of polynomials that vanish on the nodes of C. The matrix $\mathscr{M}_{n-1}$ must have full rank since a nontrivial solution of $\mathscr{M}_{n-1}^\mathsf{T}\xi = 0$ would lead to a polynomial degree $n-1$ in $I(\mathsf{C})$, but $\Pi_{n-1}^d \cap I(\mathsf{C}) = \varnothing$. Now, each nontrivial solution ξ of $\mathsf{M}_n^\mathsf{T}\xi = 0$ gives a polynomial P_ξ of degree n in $I(\mathsf{C})$, which is an orthogonal polynomial as C is of degree $2n-1$. Moreover, since $\mathcal{I}$ is centrally symmetric, P_ξ belongs to $\mathscr{P}_n$. Hence, $\xi(n-1)$ must be zero and $\xi(n)$

satisfies $\mathcal{M}_n^{\mathsf{T}} \xi(n) = 0$. It follows that there is a one-to-one correspondence between the null space of $\mathsf{M}_n^{\mathsf{T}}$ and the null space of $\mathcal{M}_n^{\mathsf{T}}$. In particular, they have the same dimension,

$$\dim \Pi_n^d - \operatorname{rank} \mathsf{M}_n^{\mathsf{T}} = \dim \mathscr{P}_n^d - \operatorname{rank} \mathcal{M}_n^{\mathsf{T}}.$$

Consequently, by $\Pi_n^d = \mathscr{P}_n^d \cup \mathscr{P}_{n-1}^d$, we conclude that

$$\operatorname{rank} \mathsf{M}_n = \operatorname{rank} \mathcal{M}_n + \dim \mathscr{P}_{n-1}^d. \tag{5.13}$$

Next, we derive a lower bound for the rank of $\mathcal{M}_n$. First, let n be an even integer. We choose a hyperplane h that passes through the origin but contains none of the nonzero nodes; that is, $h(0) = 0$ but $h(\mathbf{x}_k) \neq 0$ for all $\mathbf{x}_k \neq 0$. If $P \in \mathscr{P}_{n-1}^d$ then $P \notin I(\mathsf{C})$ and $P(0) = 0$ since $n-1$ is odd. Hence, P is not zero at some nonzero node. But then $hP \notin I(\mathsf{C})$ and, since $h(\mathbf{x}) = \sum_i c_i x_i$, $hP \in \mathscr{P}_n^d$. A polynomial in $\mathscr{P}_n^d \cap I(\mathsf{C})$ is of the form $\xi^{\mathsf{T}} \mathbb{P}_n$ with ξ satisfying $\mathcal{M}_n^{\mathsf{T}} \xi = 0$, so that $hP \notin I(\mathsf{C})$ means that it does not correspond to a vector in the null space of $\mathcal{M}_n$ for any $P \in \mathscr{P}_{n-1}$. Consequently,

$$\dim \mathscr{P}_n^d - \operatorname{rank} \mathcal{M}_n^{\mathsf{T}} = \dim \ker \mathcal{M}_n^{\mathsf{T}} \leq \dim \mathscr{P}_n^d - \dim \mathscr{P}_{n-1}^d,$$

or $\operatorname{rank} \mathcal{M}_n \geq \dim \mathscr{P}_{n-1}^d$. Together with (5.13), this established the lower bound $\operatorname{rank} \mathsf{M}_n \geq 2 \dim \mathscr{P}_{n-1}^d$ and, as a consequence, the lower bound (5.11) for even n.

Second, let n be odd. In this case, all orthogonal polynomials of degree n vanish on the origin, so $\mathbf{x} = 0$ must be a node of the cubature rule. Let $\mathscr{O}_{n-1}$ be the subspace of $\mathscr{P}_{n-1}^d$ that consists of polynomials Q, $Q(0) \neq 0$ but $Q(\mathbf{x}_k) = 0$ for all $\mathbf{x}_k \neq 0$. Defining h as in the case of n even. Then, for $P \in \mathscr{P}_{n-1}^d \setminus \mathscr{O}_{n-1}$, $hP \notin I(\mathsf{C})$ and $hP \in \mathscr{P}_n$. Hence,

$$\dim \mathscr{P}_n^d - \operatorname{rank} \mathcal{M}_n^{\mathsf{T}} \leq \dim \mathscr{P}_n^d - (\dim \mathscr{P}_{n-1}^d - \dim \mathscr{O}_{n-1}),$$

so that $\operatorname{rank} \mathcal{M}_n \geq \dim \mathscr{P}_{n-1}^d - \dim \mathscr{O}_{n-1}$. If Q_1 and Q_2 are two polynomials in $\mathscr{O}_{n-1}$, then $Q_1(0) Q_2(\mathbf{x}) - Q_2(0) Q_1(\mathbf{x})$ belongs to $I(\mathsf{C}) \cap \Pi_{n-1}^d$ so that Q_2 is a constant multiple of Q_1. Thus, $\dim \mathscr{O}_{n-1} \leq 1$. Hence, $\operatorname{rank} \mathcal{M}_n \geq \dim \mathscr{P}_{n-1}^d - 1$. Together with (5.13), we have the desired lower bound for $\operatorname{rank} \mathsf{M}_n$, hence N, for n odd.

Finally, (5.12) follows from (5.11) by (5.10) with n replaced by $n-1$. □

Remark 5.3.3. *First, for $d = 2$, one can easily verify that the lower bound* (5.11) *agrees with* (5.9). *For $d > 2$, this lower bound is of the form*

$$N \geq N_{n,d} = \dim \Pi_{n-1}^d + \frac{1}{2} \dim \Pi_{n-1}^{d-1} + \cdots = \binom{n+d-1}{d} + \frac{1}{2}\binom{n+d-2}{d-1} + \cdots,$$

which is stronger than that of (5.8) *if* $n \geq 3$. *Second, by* (5.10) *and* (5.12),

$$N_{n,d} = \dim \Pi^d_{n-1} + \dim \mathscr{P}^{d-1}_{n-1} - \delta_n < \dim \Pi^d_{n-1} + \dim \Pi^{d-1}_n,$$

so that, using $\dim \Pi^{d-1}_n = \dim \mathcal{P}^d_n = r^d_n$, *we obtain*

$$N_{n,d} < \dim \Pi^d_{n-1} + r^d_n = \dim \Pi^d_n. \tag{5.14}$$

Corollary 5.3.4. *If a cubature rule of degree* $2n-1$ *attains the lower bound* $N \geq N_{n,d}$ *in* (5.11), *then its nodes are zeros of* $r^d_n + \dim \Pi^d_{n-1} - N_{n,d}$ *linearly independent orthogonal polynomials of degree n.*

Proof The lower bound is derived from $N \geq \operatorname{rank} \mathsf{M}_n$. If a cubature attains the lower bound, then we must have $N = \operatorname{rank} \mathsf{M}_n = N_{n,d}$. By Theorem 5.1.1 and $\sigma_\ell = \operatorname{rank} \mathsf{M}_{n+\ell}$ for $\ell = 0, 1, \ldots$, we see that $\sigma_1 = \ldots = \sigma_{n-1} = 0$ and $\sigma_0 = N_{n,d} - \dim \Pi^d_{n-1}$. Consequently, the nodes are zeros of $U^\mathsf{T} \mathbb{P}_n$, where $\operatorname{rank} U = r^d_n - \sigma_0 = r^d_n + \dim \Pi^d_{n-1} - N_0$. □

In particular, for $d = 2$, there are $n + 1 - \lfloor \frac{n}{2} \rfloor$ linearly independent orthogonal polynomials that vanish on the cubature, which agrees with Theorem 5.2.3.

5.4 A Linear Programming Lower Bound

Minimal cubature rules that attain Möller's lower bounds in Theorem 5.2.3 will be characterized in the next chapter, where several examples will be given. In the present section, we provide another lower bound for the Gegenbauer weight functions W_μ on the unit ball $\mathbb{B}^d$ and the Jacobi weight function W_κ on the simplex when the parameters are half-integers. The lower bound is asymptotically superior to Möller's lower bounds in Theorem 5.2.3 and Theorem 5.3.2 for cubature rules of high degree, showing that Möller's lower bounds are far from sharp for these two families of weight functions.

This lower bound is derived by considering the unit sphere first, where such a bound has been studied extensively for the spherical designs, which are cubature rules with equal weights, and then applying the connection between the cubature rules on the sphere, the ball, and the simplex. We establish the lower bound on the sphere in Subsection 5.4.1 and the lower bound for W_μ in Subsection 5.4.2, and extend it to the Jacobi weight W_κ on the simplex in Subsection 5.4.3.

5.4.1 Lower Bound on the Unit Sphere

Let $\mathbb{S}^{d-1}$ denote the unit sphere in $\mathbb{R}^d$ and $\mathrm{d}\sigma$ be the surface measure on $\mathbb{S}^{d-1}$. Define

$$\mathcal{I}_{\mathbb{S}}(f) = \frac{1}{\omega_d} \int_{\mathbb{S}^{d-1}} f(\xi) \mathrm{d}\sigma(\xi),$$

where $\omega_d = 2\pi^{\frac{d}{2}}/\Gamma(\frac{d}{2})$ is the surface measure of $\mathbb{S}^{d-1}$. A cubature rule C_m of degree m with respect to $\mathrm{d}\sigma$ on the sphere is defined by

$$\mathsf{C}_m(f) := \sum_{k=0}^{N} \lambda_\xi f(\xi_k) \quad \text{so that} \quad \mathcal{I}_{\mathbb{S}}(f) = \mathsf{C}_m(f), \quad \forall f \in \Pi_m(\mathbb{S}^{d-1}),$$

where $\Pi_m(\mathbb{S}^{d-1})$ denotes the space of polynomial degree at most m restricted on $\mathbb{S}^{d-1}$, and there is at least one $f^* \in \Pi_{m+1}(\mathbb{S}^{d-1})$ so that $\mathcal{I}_{\mathbb{S}}(f^*) \neq \mathsf{C}_m(f^*)$.

The orthogonal polynomials with respect to the inner product $\langle f, g\rangle = \mathcal{I}_{\mathbb{S}}(fg)$ are spherical harmonics, by Theorem 3.4.3, which are homogeneous harmonic polynomials restricted on the sphere. Recall that $\mathcal{H}_n^d$ denotes the space of spherical harmonics of degree n, and its dimension is given in (3.32). Moreover, the space of polynomials of degree n restricted on $\mathbb{S}^{d-1}$ satisfies $\Pi_n(\mathbb{S}^{d-1}) = \bigoplus_{k=0}^{n} \mathcal{H}_k^d$ and, as shown in (2.19),

$$\dim \Pi\left(\mathbb{S}^{d-1}\right) = \binom{n+d-1}{n} + \binom{n+d-2}{n-1}.$$

One of the most important properties of spherical harmonics is the addition formula. It states that if $\{Y_k : 1 \le k \le \dim \mathcal{H}_n^d\}$, is an orthonormal basis of $\mathcal{H}_n^d$, then

$$\sum_{k=1}^{\dim \mathcal{H}_n^d} Y_k(\xi) Y_k(\eta) = Z_n(\langle \xi, \eta\rangle), \qquad \xi, \eta \in \mathbb{S}^{d-1}, \tag{5.15}$$

where Z_n is a constant multiple of the Gegenbauer polynomial of one variable,

$$Z_n(t) := \frac{n+\lambda}{\lambda} C_n^\lambda(t), \qquad \lambda = \frac{d-2}{2}.$$

The generic structure of cubature rules remains to hold on the sphere. In particular, if $m = 2n-1$, then the nodes of the cubature rules are zeros of orthogonal polynomials of degree n. Of particular interest are cubature rules that have equal weights; that is, $\lambda_k = 1/N$ for all k. The set of nodes of an equal-weight cubature rule of degree m is called a *spherical design* or m-design, a name originated from combinatorics. Spherical designs have generated substantial attention, especially those that have the smallest number of points. A lower bound for the cardinality of m-designs often holds for cubature rules of degree

m without the requirement of equal weights. Several such lower bounds are based on Proposition 5.4.1.

Let $w_\lambda(t) = (1-t^2)^{\lambda-\frac{1}{2}}$. For $F \in L^2(w_\lambda;[-1,1])$, define its Fourier–Gegenbauer series by

$$F(t) = \sum_{k=0}^{\infty} \hat{F}_k^\lambda C_k^\lambda(t) \quad \text{with} \quad \hat{F}_k^\lambda = \frac{1}{h_k^\lambda}\int_{-1}^{1} F(s) C_k^\lambda(s) w_\lambda(s) \mathrm{d}s,$$

where $h_k^\lambda = \int_{-1}^{1} [C_k^\lambda(s)]^2 w_\lambda(s) \mathrm{d}s$.

Proposition 5.4.1. *For $\lambda = \frac{d-2}{2}$, $d \geq 3$, let F be a nonnegative function defined on $[-1,1]$ such that, for a positive integer m, $\hat{F}_0^\lambda > 0$ and $\hat{F}_k^\lambda \leq 0$ for $k > m$. Then the number of nodes, N, of a positive cubature formula of degree m for the integral $\mathcal{I}_{\mathbb{S}}$ on $\mathbb{S}^{d-1}$ satisfies*

$$N \geq \frac{F(1)}{\hat{F}_0^\lambda}. \tag{5.16}$$

Proof Let C_m be a cubature formula of degree m for $\mathcal{I}_{\mathbb{S}}$. Then $\sum_{k=1}^{N} \lambda_k = 1$ by setting $f(\xi) = 1$, since $\mathcal{I}_{\mathbb{S}}(1) = 1$. By the addition formula (5.15), for $k \geq 1$,

$$\sum_{i=1}^{N}\sum_{j=1}^{N} \lambda_i \lambda_j Z_k(x_i,x_j) = \sum_{\ell=1}^{\dim \mathcal{H}_k^d} \left| \sum_{i=1}^{N} \lambda_i Y_\ell(x_i) \right|^2 \geq 0,$$

where $\{Y_\ell : 1 \leq \ell \leq \dim \mathcal{H}_k^d\}$ is an orthonormal basis of $\mathcal{H}_k^d$. Moreover, the sum on the left-hand side is equal to zero for $1 \leq k \leq m$ since the inner sum on the right-hand side is equal to $\mathcal{I}_{\mathbb{S}}(Y_\ell) = 0$ by the cubature rule. Hence, using $\hat{F}_k \leq 0$ for $k > m$, we obtain

$$\begin{aligned} I := \sum_{i=1}^{N}\sum_{j=1}^{N} \lambda_i \lambda_j F(\langle x_i, x_j \rangle) &= \sum_{k=0}^{\infty} \hat{F}_k^\lambda \sum_{i=1}^{N}\sum_{j=1}^{N} \lambda_i \lambda_j \frac{\lambda}{k+\lambda} Z_k(x_i,x_j) \\ &\leq \hat{F}_0^\lambda \sum_{i=1}^{N}\sum_{j=1}^{N} \lambda_i \lambda_j = \hat{F}_0^\lambda. \end{aligned}$$

Now, since F is nonnegative and $\lambda_j \geq 0$, by the Cauchy–Schwarz inequality,

$$\begin{aligned} I \geq \sum_{j=1}^{N} \lambda_j^2 F(\langle x_j, x_j \rangle) &= F(1) \sum_{j=1}^{N} \lambda_j^2 \\ &\geq F(1) \frac{1}{N} \left(\sum_{j=1}^{N} \lambda_j \right)^2 = \frac{F(1)}{N}. \end{aligned}$$

Together, the two inequalities establish the desired lower bound for N. □

For cubature rules of degree $2n-2$, the lower bound $N \ge \dim \Pi_{n-1}(\mathbb{S}^{d-1})$ holds as in Theorem 2.1.1. While the latter holds for both $m = 2n-2$ and $m = 2n-1$ for cubature rules on a solid domain, a stronger lower bound holds when $m = 2n-1$ for cubature rules on the unit sphere.

Theorem 5.4.2. *The number of roots of a cubature rule of degree m for $\mathcal{I}_{\mathbb{S}}$ satisfies*

$$N \ge \begin{cases} 2\binom{n+d-2}{n-1}, & m = 2n-1, \\ \binom{n+d-2}{n-1} + \binom{n+d-3}{n-2}, & m = 2n-2. \end{cases} \tag{5.17}$$

Proof For $m = 2n-2$, this lower bound is $N \ge \dim \Pi_{n-1}(\mathbb{S}^{d-1})$, proved exactly as in Theorem 2.1.1. For $m = 2n-1$, we use Proposition 5.4.1 with F defined by

$$F(t) = (1+t)H(t)^2 \quad \text{with} \quad H(t) = \sum_{0 \le 2k \le n-1} Z_{n-1-2k}(t).$$

Since F is a nonnegative polynomial of degree m, it is evident that $\hat{F}_0^\lambda > 0$ and $\hat{F}_k^\lambda = 0$ for $k > m$. Integrating the addition formula with $\eta = \xi$ over $\mathbb{S}^{d-1}$ shows $Z_k(1) = \dim \mathcal{H}_k^d$, which leads to $H(1) = \dim \mathcal{P}_{n-1}^d$, where $\mathcal{P}_{n-1}^d$ is the space of homogeneous polynomial of degree $n-1$, and $F(1) = 2(\dim \mathcal{P}_{n-1}^d)^2$. Since H^2 is an even function, a change of variable shows

$$\hat{F}_0^\lambda = \frac{1}{h_0^\lambda} \int_{-1}^{1} H^2(t) w_\lambda(t) \mathrm{d}t = \frac{1}{\omega_d} \int_{\mathbb{S}^{d-1}} H^2(\langle \xi, \eta \rangle) \mathrm{d}\sigma(\xi),$$

which implies by the addition formula that $\hat{F}_0^\lambda = \dim \mathcal{P}_{n-1}^d$. □

The lower bound (5.17) is classical, and the cubature rules attaining this bound are akin to the Gauss cubature rules, and they are known to exist only for a few cases of moderate degrees and dimensions. The bound (5.17) is not sharp for cubature rules of higher degrees, as can be seen in another lower bound given below, for which we need to construct a specific function F and apply Proposition 5.4.1. The function is defined in terms of the Gegenbauer polynomials as follows.

Let $\gamma_m = \gamma_m^\lambda$ be the largest zero of the Gegenbauer polynomial $C_m^{\lambda+1}$. Let functions f and g be defined by

$$f(t) = \begin{cases} C_{m+1}^\lambda(t) - C_{m+1}^\lambda(\gamma_m), & t \ge \gamma_m, \\ 0, & t < \gamma_m, \end{cases} \quad \text{and} \quad g(t) = \begin{cases} 1, & t \ge \gamma_m, \\ 0, & t < \gamma_m, \end{cases} \tag{5.18}$$

where $\lambda > -\frac{1}{2}$. For $t \in [-1,1]$, define

$$F(t) = \int_{-1}^{1} g(u) c_\lambda \int_{-1}^{1} f\left(ut + v\sqrt{1-u^2}\sqrt{1-t^2}\right)(1-v^2)^{\lambda-1} \mathrm{d}v\, w_\lambda(u) \mathrm{d}u, \tag{5.19}$$

where c_λ is the normalized constant given in (1.17).

We will show that this F meets the requirement specified in Proposition 5.4.1. First, however, we need information on the Fourier–Gegenbauer coefficients of f and F.

Lemma 5.4.3. *For f in* (5.18), *if $k \neq m+1$ and $k \geq 0$, then*

$$\hat{f}_k^\lambda = \frac{(m+1)(m+2\lambda+1)}{h_k^\lambda (k-m-1)(k+m+1+2\lambda)} C_{m+1}^\lambda(\gamma_m) \int_{\gamma_m}^{1} C_k^\lambda(u) w_\lambda(u) \mathrm{d}u.$$

Proof The Gegenbauer polynomials satisfy a differential equation [148, p. 80]

$$\frac{\mathrm{d}}{\mathrm{d}t}\left[(1-t^2)^{\lambda+1/2} \frac{\mathrm{d}}{\mathrm{d}t} C_n^\lambda(t)\right] = -\mu_n (1-t^2)^{\lambda-1/2} C_n^\lambda(t), \quad \mu_n = n(n+2\lambda).$$

Using this equation with $n = k$ and $n = m+1$, it follows that

$$\begin{aligned}
&(\mu_k - \mu_{m+1}) C_k^\lambda(t) C_{m+1}^\lambda(t) w_\lambda(t) \\
&= \frac{\mathrm{d}}{\mathrm{d}t}\left[(1-t^2)^{\lambda+1/2} C_k^\lambda(t)(t) \frac{\mathrm{d}}{\mathrm{d}t} C_{m+1}^\lambda(t)\right] - \frac{\mathrm{d}}{\mathrm{d}t}\left[(1-t^2)^{\lambda+1/2} C_{m+1}^\lambda(t) \frac{\mathrm{d}}{\mathrm{d}t} C_k^\lambda(t)\right].
\end{aligned}$$

Furthermore, integrating the differential equation with $n = k$ gives

$$\mu_k \int_t^1 C_k^\lambda(u)(1-u^2)^{\lambda-\frac{1}{2}} \mathrm{d}u = (1-t^2)^{\lambda+\frac{1}{2}} \frac{\mathrm{d}}{\mathrm{d}t} C_k^\lambda(t). \tag{5.20}$$

Since $\frac{\mathrm{d}}{\mathrm{d}t} C_{m+1}^\lambda(t) = 2\lambda C_m^{\lambda+1}(t)$ vanishes at $t = \gamma_m$, these two identities lead to

$$\begin{aligned}
(\mu_k - \mu_{m+1}) \int_{\gamma_m}^1 C_k^\lambda(t) C_{m+1}^\lambda(t) w_\lambda(t) \mathrm{d}t &= -\left[(1-t^2)^{\lambda+1/2} C_{m+1}^\lambda(t) \frac{\mathrm{d}}{\mathrm{d}t} C_k^\lambda(t)\right]_{t=\gamma_m}^1 \\
&= \mu_k C_{m+1}^\lambda(\gamma_m) \int_{\gamma_m}^1 C_k^\lambda(u) w_\lambda(u) \mathrm{d}u.
\end{aligned}$$

Consequently, for $k \neq m+1$, we deduce

$$\begin{aligned}
\hat{f}_k^\lambda &= \frac{1}{h_k^\lambda} \int_{\gamma_m}^1 \left(C_{m+1}^\lambda(u) - C_{m+1}^\lambda(\gamma_m)\right) C_k^\lambda(u) w_\lambda(u) \mathrm{d}u \\
&= \frac{\mu_{m+1}}{\mu_k - \mu_{m+1}} C_{m+1}^\lambda(\gamma_m) \frac{1}{h_k^\lambda} \int_{\gamma_m}^1 C_k^\lambda(u) w_\lambda(u) \mathrm{d}u.
\end{aligned}$$

This is the stated formula since $\mu_k - \mu_{m+1} = (k-m-1)(k+m+1+2\lambda)$. □

Lemma 5.4.4. *Let f and g in (5.18). The function F in (5.19) is continuous and nonnegative on $[-1,1]$; its Fourier coefficients $\hat{F}_k^\lambda$ satisfy*

$$\hat{F}_k^\lambda = \hat{f}_k^\lambda \hat{g}_k^\lambda \frac{h_k^\lambda}{C_k^\lambda(1)}, \qquad k \ge 0,$$

and, moreover, $\hat{F}_{m+1}^\lambda = 0$ and $\hat{F}_k^\lambda < 0$ for $k > m+1$.

Proof Since f is evidently continuous, F is a continuous function on $[-1,1]$. Moreover, $f(t) \ge 0$ on $[-1,1]$ since λ_m is the largest zero of the derivative of C_{m+1}^λ, which implies that F is nonnegative on $[-1,1]$. Moreover, since $C_k^\lambda(t) = \mathcal{O}(k^{\lambda-1})$ for $-1 < t < 1$, and $\int_0^1 |C_k^\lambda(t)|(1-t^2)^{\lambda-1/2}\mathrm{d}t = \mathcal{O}(k^{\lambda-1})$ ([148, (7.34.1), p. 173]), the formula of $\hat{f}_k^\lambda$ shows $|\hat{f}_k^\lambda| = \mathcal{O}(k^{-\lambda-1})$. Hence, the expansion of f converges uniformly on every compact set inside $[-1,1]$. The Gegenbauer polynomials satisfy a product formula

$$\frac{C_k^\lambda(s)C_k^\lambda(t)}{C_k^\lambda(1)} = c_\lambda \int_{-1}^1 C_k^\lambda(st + u\sqrt{1-s^2}\sqrt{1-t^2})(1-u^2)^{\lambda-1}\mathrm{d}t,$$

which implies, using the expansion of f in the Gegenbauer polynomials, that

$$\begin{aligned} F(t) &= \int_{-1}^1 g(u) \sum_{k=0}^\infty \hat{f}_k^\lambda \frac{1}{C_k^\lambda(1)} C_k^\lambda(u) C_k^\lambda(t) w_\lambda(u)\mathrm{d}u \\ &= \sum_{k=0}^\infty \hat{f}_k^\lambda \hat{g}_k^\lambda \frac{h_k^\lambda}{C_k^\lambda(1)} C_k^\lambda(t). \end{aligned}$$

This gives the stated formula for $\hat{F}_k$, and it shows that the sign of $\hat{F}_k$ is the same as that of $\hat{f}_k^\lambda \hat{g}_k^\lambda$. By the definition of $g(t)$,

$$\hat{g}_k^\lambda = h_k^\lambda \int_{\gamma_m}^1 C_k^\lambda(u) w_\lambda(u)\mathrm{d}u,$$

which shows, in particular, that $\hat{g}_{m+1}^\lambda = 0$ by (5.20), so that $\hat{F}_{m+1}^\lambda = 0$. Moreover, together with the expression of $\hat{f}_k^\lambda$ in Lemma 5.4.3, we see that the sign of $\hat{F}_k^\lambda$ for $k > m+1$ is the same as that of $C_{m+1}^\lambda(\gamma_m)$. Since the zeros of C_m^λ and $C_m^{\lambda+1}$ interlace and C_m^λ maximizes at $t = 1$ on $[-1,1]$, $\frac{\mathrm{d}}{\mathrm{d}t}C_m^\lambda$ must be positive at γ_m. Consequently, by the Christoffel–Darboux formula, we obtain $0 < K_n(t,t) = -d_m C_{m+1}^\lambda(\gamma_m)\frac{\mathrm{d}}{\mathrm{d}t}C_m^\lambda(\gamma_m)$, where d_m is a positive number, we conclude that $C_{m+1}^\lambda(\gamma_m) < 0$. Hence, $\hat{F}_k < 0$ if $k > m+1$. This completes the proof. □

We are now ready to establish the new lower bound for the number of nodes of the cubature rules on the unit sphere. To obtain a sharp lower bound, we need further information on the largest zero of the Gegenbauer polynomial, for which we state one more lemma.

Lemma 5.4.5. *Let $\gamma_m(\lambda)$ be the largest zero of C_m^λ. For $m \ge 4$ and $\lambda > -\frac{1}{2}$,*

$$1 - \gamma_m(\lambda) < \frac{(2\lambda+1)(2\lambda+3)(2\lambda+7)}{(10\lambda+17)[m(m+2\lambda)+\frac{1}{8}(2\lambda+1)^2]}. \tag{5.21}$$

This lemma is [117, Theorem 3], and we shall not give its proof here. There are several other relevant estimates for $\gamma_m(\lambda)$, which will be briefly discussed in Section 5.6.

Theorem 5.4.6. *Let $N(m,d)$ denote the number of nodes of a positive cubature rule of degree m on the unit sphere $\mathbb{S}^{d-1}$. Then,*

$$N(m,d) \ge 2\frac{\int_0^1 (1-u^2)^{\frac{d-3}{2}} \mathrm{d}u}{\int_{\gamma_m}^1 (1-u^2)^{\frac{d-3}{2}} \mathrm{d}u}. \tag{5.22}$$

where $\gamma_m = \gamma_m(\frac{d}{2})$ is the largest zero of $C_m^{d/2}(t)$. In particular, for $d = 3$, the cubature rule of degree m for $\mathbb{S}^2$ satisfies

$$N(m,2) > \frac{4}{15}[m(m+3)+2], \qquad m \ge 4, \tag{5.23}$$

which is stronger than (5.17) *if $m \ge 11$.*

Proof The lower bound (5.22) follows from Proposition 5.4.1 by using the function F defined in (5.19), with $\lambda = \frac{d-2}{2}$, which is warranted by Lemma 5.4.4. Since

$$F(1) = \int_{\gamma_m}^1 f(u) w_\lambda(u) \mathrm{d}u = \hat{f}_0^\lambda h_0^\lambda \quad \text{and} \quad \hat{F}_0^\lambda = \hat{f}_0^\lambda \hat{g}_0^\lambda h_0^\lambda,$$

so that the lower bound (5.16) becomes $N \ge 1/\hat{g}_0^\lambda$. By the definition of g in (5.18) and write $h_0^\lambda = 2\int_0^1 (1-t^2)^\lambda (t)\mathrm{d}t$, we see that $1/\hat{g}_0^\lambda$ is precisely the right-hand side of (5.22).

In the case $d = 3$, $\gamma_m = \gamma_m(\frac{3}{2})$, the lower bound and the inequality (5.21) with $\lambda = \frac{3}{2}$ become

$$N(m,2) \ge \frac{2}{1-\gamma_m} \quad \text{and} \quad 1-\gamma_m < \frac{15}{2[m(m+3)+2]}.$$

Together, they lead to the lower bound (5.23) on $\mathbb{S}^2$. For the unit sphere $\mathbb{S}^2$, the right-hand side of the lower bound (5.23) is stronger than the lower bound (5.17), which gives, for $d = 3$,

$$N(m,2) \ge N_c = \begin{cases} \frac{1}{4}(m+1)(m+3), & m = 2n-1, \\ \frac{1}{4}(m+2)^2, & m = 2n-1. \end{cases}$$

A simple calculation shows that the bound (5.23) is stronger if $m \ge 11$. □

The new lower bound (5.23) is not only stronger than the first lower bound (5.17), it is stronger in the order of magnitude, for m large, since it gives

$$N(m,2) \ge \frac{4}{15}m^2 + \mathcal{O}(m), \quad \text{whereas} \quad N_c = \frac{1}{4}m^2 + \mathcal{O}(m), \tag{5.24}$$

and $\frac{4}{15} > \frac{1}{4}$. The factor $4/15$ is likely not sharp.

For an analog of (5.23) for $d > 3$, we need a stronger estimate of the largest zero of the Gegenbauer polynomial; see the discussion in Section 5.6.

5.4.2 Lower Bound on the Unit Ball

We consider cubature rules for the Jacobi weight $W_\mu(\mathbf{x}) = (1 - \|\mathbf{x}\|^2)^{\mu - \frac{1}{2}}$, $\mu \ge 0$, on the unit ball $\mathbb{B}^d$. The weight function W_0 is the Chebyshev weight on the ball. The cubature rules on $\mathbb{B}^d$ are closely related to those on the unit sphere, as shown in Theorem 2.5.4.

Theorem 5.4.7. *If a positive cubature rule of degree m for W_0 has all nodes on $\mathbb{B}^d$, then its number of nodes satisfies*

$$N \ge \frac{\int_0^1 (1-u^2)^{\frac{d-2}{2}} du}{\int_{\gamma_m}^1 (1-u^2)^{\frac{d-2}{2}} du}. \tag{5.25}$$

where γ_m is the largest zero of $C_m^{\frac{d+1}{2}}(t)$. In particular, for the cubature rules of degree $m \ge 4$ for W_0 that has all nodes on the unit disk $\mathbb{B}^2$,

$$N > \frac{2}{15}m(m+3)+4] = \frac{8}{15}\begin{cases} n(n+\frac{1}{2})+\frac{1}{2}, & m = 2n-1, \\ n(n-\frac{1}{2})+\frac{1}{2}, & m = 2n-2, \end{cases} \tag{5.26}$$

which shows that Möller's lower bound is not sharp if $n \ge 21$.

Proof If the cubature rule (2.28) for W_0, with $H(\mathbf{y}) = 1$, on the unit ball is of degree m and has N nodes all in $\mathbb{B}^d$, then the corresponding cubature rule (2.29) on the sphere $\mathbb{S}^d$ has at most $2N$ nodes, and it has exactly $2N$ nodes if all nodes of (2.28) are in the interior of $\mathbb{B}^d$. Hence, if N were less than the lower bound in (5.25), then the corresponding cubature rule on $\mathbb{S}^d$ would have less nodes than the lower bound (5.22) with $d-1$ replaced by d, which is not possible.

For $d = 2$, the cubature rule for W_0 on $\mathbb{B}^2$ corresponds to a cubature rule on $\mathbb{S}^2$, so that (5.26) follows as a consequence of (5.23). A quick calculation shows that the new lower bound with $m = 2n-1$ is stronger than Möller's lower bound when $n \ge 21$. □

The lower bound in (5.26) shows that Möller's lower bound (5.9) is not sharp for a cubature rule for W_0 that has all its nodes on the unit disk. There is a possibility, however, that some minimal cubature rules on the unit ball could have some nodes outside the ball. Indeed, this is the case for W_0 on the disk $\mathbb{B}^2$ when $m = 9$, for which the minimal cubature rule for W_0 has 18 nodes, two are outside the disk (cf. [30]).

The lower bound (5.26) shows that (5.9) is far from sharp for n large, since it gives

$$N \geq \frac{8}{15}n^2 + \mathcal{O}(n) = \left(\frac{1}{2} + \frac{1}{30}\right)n^2 + \mathcal{O}(n),$$

for cubature rules of degree $2n-1$ on the disk, whereas Möller's lower bound is of the order $\frac{1}{2}n^2 + \mathcal{O}(n)$. By considering polynomials $(1-\|\mathbf{x}\|^2)^\ell f(\mathbf{x})$, the asymptotic of this lower bound for n large also applies to the weight function $W_\ell(\mathbf{x}) = (1-\|\mathbf{x}\|^2)^{\ell-\frac{1}{2}}$ if ℓ is a positive integer. Indeed, for $f \in \Pi^2_{2n-1}$, applying the cubature rule (2.28) of degree $2n+2\ell-1$ for W_0 gives

$$\int_{\mathbb{B}^2} f(\mathbf{x})W_\ell(\mathbf{x})\mathrm{d}\mathbf{x} = \int_{\mathbb{B}^2} f(x)(1-\|\mathbf{x}\|^2)^\ell W_0(\mathbf{x})\mathrm{d}\mathbf{x} = \sum_{k=1}^{N} \lambda_k \hat{\lambda}_k f(\mathbf{x}_k),$$

where $\hat{\lambda}_k = (1-\mathbf{x}_k^2)^\ell \lambda_k$ and N satisfies the lower bound (5.26) with n replaced by $n+\ell$, which does not change the main term in the asymptotic of the lower bound. We state this as a corollary.

Corollary 5.4.8. *Let ℓ be a nonnegative integer. For the weight function W_ℓ on the unit disk, Möller's lower bound* (5.9) *is not sharp for cubature rules of degree $2n-1$ that have all nodes on $\mathbb{B}^2$ if n is sufficiently large.*

The above result does not apply to the weight function $W_{\frac{1}{2}}(\mathbf{x}) = 1$, which is the integral against the Lebesgue measure on the disk. However, since the coefficient matrices in the three-term relations for W_μ have similar structures, any variance between the lower bounds for W_μ with different values of μ is likely small.

5.4.3 Lower Bound on the Simplex

The argument in Subsection 5.4.2 also applies to cubature rules on the simplex Δ^d with the classical Jacobi weight $W_\kappa(\mathbf{x}) = \prod_{i=1}^d x_i^{\kappa_i}(1-|\mathbf{x}|)^{\kappa_{d+1}}$, $\kappa_i \geq -\frac{1}{2}$. With $\mathbf{1} = (1,\ldots,1)$, $W_{-\mathbf{1}/2}$ is the Chebyshev weight

$$W_{-\mathbf{1}/2}(\mathbf{x}) = \frac{1}{\sqrt{x_1\cdots x_d(1-|\mathbf{x}|)}}, \qquad \mathbf{x} \in \Delta^d.$$

The cubature rules on Δ^d are closely related to those on the unit sphere, as shown in Theorem 2.5.4.

Theorem 5.4.9. *If a positive cubature rule of degree m for $W_{-1/2}$ has all nodes on the simplex Δ^d, then its number of nodes satisfies*

$$N \geq \frac{\int_0^1 (1-u^2)^{\frac{d-2}{2}} \mathrm{d}u}{2^d \int_{\gamma_{2m+1}}^1 (1-u^2)^{\frac{d-2}{2}} \mathrm{d}u}. \tag{5.27}$$

where γ_{2m+1} is the largest zero of $C_{2m+1}^{\frac{d+1}{2}}(t)$. In particular, for the cubature rules of degree m for $W_{-1/2}$ that have all nodes on the triangle Δ^2,

$$N \geq \frac{2}{15}\left[(m+\tfrac{1}{2})(m+2)+1\right], \tag{5.28}$$

which shows that Möller's lower bound for cubature rules of degree $m = 2n-1$ is not sharp if $n \geq 25$.

Proof If a cubature rule of degree m for $W_{-1/2}$ has N nodes, which is of the form (2.23) with $W(\mathbf{x}) = \frac{1}{\sqrt{1-|\mathbf{x}|}}$, then the corresponding cubature rule of degree $2m+1$ for the Chebyshev weight on the unit ball (2.24) has degree at most $2^d N$. Hence, if N were less than the lower bound in (5.27), then the number of nodes $2^d N$ of the cubature rule for W_0 on the ball would be less than the lower bound on the right-hand side of (5.25), which is not possible.

For $d = 2$, the lower bound (5.28) is deduce from

$$N > \frac{1}{4} \times \frac{2}{15}[2m+1)(2m+4)+4] = \frac{2}{15}\left[(m+\tfrac{1}{2})(m+2)+1\right],$$

by using (5.28) with m replaced by $2m+1$. This lower bound is stronger than Möller's lower bound when $m = 2n-1$ if $n \geq 25$, as a quick calculation shows. □

Like the case of the unit ball, the lower bound (5.28) gives the asymptotic order for large n

$$N \geq \frac{8}{15} n^2 + \mathcal{O}(n),$$

for cubature rules of degree $2n-1$ on the triangle, which surpasses Möller's lower bound $\frac{n^2}{2} + \mathcal{O}(n)$. Moreover, the lower bound with this asymptotic order holds for $W_{\boldsymbol{\kappa}}$ if $\kappa_i + \frac{1}{2}$ are nonnegative integers for $1 \leq i < d+1$.

The lower bounds (5.26) on the unit disk and (5.28) on the simplex are likely not sharp. What makes them important is that they demonstrate that Möller's lower bound is far from sharp for these domains.

5.5 Minimal Cubature Rule for a Rotational Invariant Weight Function

For cubature rules of degree $2n-1$, the first lower bound for the number of nodes is $N \ge \dim \Pi_{n-1}^d$ established in Theorem 2.1.1. The bound is attained by the Gauss cubature rules, which exist rarely but do exist for some weight functions. Any improved lower bound has to exclude these weight functions and, as a result, holds necessarily only for some restricted class of integrals or linear functionals. For example, the new bounds in Theorem 5.2.3 and Theorem 5.3.2 are established for the class of centrally symmetric integrals. One may ask if a stronger lower bound is possible when the weight function possesses stronger symmetry. The lower bound in Section 5.4 gives one for the Chebyshev weight function on the unit ball, which is invariant under rotation. Thus, one may speculate that there is a stronger lower bound for the class of rotationally symmetric weight functions. However, this is false, as shown by the example in Theorem 5.5.2 of this section, which shows that the lower bound (5.9) is sharp for at least one rotationally symmetric weight function.

We consider a rotationally symmetric weight function on $\mathbb{R}^2$. Let $W(\mathbf{x}) = w(\|\mathbf{x}\|)$, where w is a weight function on $[0,R)$, where R could be $+\infty$. Let

$$\mathcal{I}_{\mathrm{rot}} f = \int_\Omega f(\mathbf{x}) w(\|\mathbf{x}\|) \mathrm{d}\mathbf{x}, \qquad \Omega = \{\mathbf{x} \in \mathbb{R}^d : 0 \le \|\mathbf{x}\| < R\}, \tag{5.29}$$

and normalize w so that $\mathcal{I}_{\mathrm{rot}} 1 = 1$. Then, $\mathcal{I}_{\mathrm{rot}} f$ is invariant under the orthogonal group. We first give sufficient conditions for a simple-minded construction of cubature rules that may attain Möller's lower bound for rotationally invariant weight.

Proposition 5.5.1. *Let $\mathcal{I}_{\mathrm{rot}}$ be defined as above. For $\ell = 2,3,\ldots$, define*

$$\mathsf{C}_{2\ell}(f) = \frac{1}{\ell+1} \sum_{i=1}^{\ell} \lambda_i \sum_{j=0}^{2\ell+1} f\left(r_i \cos \tfrac{(2j+\tau_i)\pi}{2\ell+2}, r_i \sin \tfrac{(2j+\tau_i)\pi}{2\ell+2}\right), \tag{5.30}$$

$$\mathsf{C}_{2\ell+1}(f) = \lambda_0 f(0,0) + \frac{1}{\ell+2} \sum_{i=1}^{\ell} \lambda_i \sum_{j=0}^{2\ell+3} f\left(r_i \cos \tfrac{2j+\tau_i}{2\ell+4}, r_i \sin \tfrac{2j+\tau_i}{2\ell+4}\right), \tag{5.31}$$

where $\tau_i = 0$ if $\ell + i$ is even and $\tau_i = 1$ if $\ell + i$ is odd. Then C_n is a minimal cubature rule of degree $2n-1$ for $\mathcal{I}_{\mathrm{rot}}$ and it attains Möller's lower bound (5.9) *provided* λ_i *and* r_i *satisfy*

$$\sum_{i=1}^{\lfloor \frac{n}{2} \rfloor} \lambda_i r_i^{2k} = \frac{1}{2\pi} \int_0^{\infty} r^{2k+1} w(r) \mathrm{d}r, \quad \delta \le k \le n-1, \tag{5.32}$$

$$\sum_{i=1}^{\lfloor \frac{n}{2} \rfloor} (-1)^i \lambda_i r_i^{2k} = 0, \quad \lfloor \tfrac{n+3}{2} \rfloor \le k \le n-1, \tag{5.33}$$

where $\delta = 0$ *if* $n = 2\ell$ *and* $\delta = 1$ *if* $n = 2\ell + 1$.

Proof We write $f(x) = f(r\cos\theta, r\sin\theta)$ in the polar coordinates $x_1 = r\cos\theta$ and $x_2 = r\cos\theta$. Let D_q denote the dihedral group of order q, the group of symmetries of a regular polygon of q sides. If f is invariant under D_{2q}, then it can be written as a polynomial of r^2 and $r^{2q}\cos(2q\theta)$. Let $n = 2\ell$. Then the cubature rule (5.30) is invariant under $D_{2\ell+2}$ and, moreover, $\mathrm{C}_{2\ell}(f) = \mathcal{I}_{\mathrm{rot}}(f)$ becomes (5.32) if $f = r^{2k}$, $0 \le k \le n-1$, and (5.33) if $f = r^{2k}\cos(n+2)\theta$, $\frac{n+2}{2} \le k \le n-1$, so that (5.32) and (5.33) imply that (5.30) holds for all polynomials of degree $2n-1$ that are invariant under $D_{2\ell+2}$. Hence, it is of degree $2n-1$ by Theorem 2.5.2. The case $n = 2\ell+1$ is proved similarly by using $D_{2\ell+4}$, where $\delta = 1$ since invariant polynomials, except $f(x) = 1$, vanish on $(0,0)$, whereas the cubature rule for $f(x) = 1$ determines $\lambda_0 = 1 - \lambda_1 - \cdots - \lambda_\ell$. □

The system of equations (5.30) and (5.31) depends on the weight function w; it does not, however, have solutions for most weight functions. Later, we give one solution to the system. For a cubature rule of degree m to exist, we only need the moment functional $\mathcal{I}_{\mathrm{rot}}$ to be defined for polynomials up to degree m.

Theorem 5.5.2. *Let n be a positive integer. Define w_n on $\mathbb{R}_+$ by*

$$w_n(r) = \frac{\sqrt{r^2-1}}{r^{2n+4}} \quad \textit{if} \quad r \ge 1, \quad \textit{and} \quad w_n(r) = 0 \quad \textit{if} \quad 0 \le r < 1.$$

Then, the weight function $w_n(\|x\|)$ admits a cubature formula of degree $2n-1$ that attains Möller's lower bound (5.9).

Proof The linear functional $\mathcal{I}_{\mathrm{rot}}$ is well defined on the space Π_{2n+1}^d. We show that the system (5.32) and (5.33) with $w = w_n$ is solvable for $n = 2\ell+1$. Changing variable $r \mapsto \sqrt{2/(t+1)}$, we have

$$\int_0^{\infty} r^{2j+1} w_n(r) \mathrm{d}r = \int_1^{\infty} r^{2j+1} \frac{\sqrt{r^2-1}}{r^{4\ell+6}} \mathrm{d}r = \frac{1}{8} \int_{-1}^{1} \left(\frac{1+t}{2}\right)^{2\ell-j} \sqrt{1-t^2} \mathrm{d}t.$$

The Gauss quadrature rule for the Chebyshev weight function $\sqrt{1-t^2}$ gives

$$\frac{2}{\pi} \int_{-1}^{1} g(t)\sqrt{1-t^2} \mathrm{d}t = \sum_{i=1}^{\ell} \sin^2 \frac{i\pi}{\ell+1} g\left(\cos \frac{i\pi}{\ell+1}\right), \quad g \in \Pi_{2\ell-1}.$$

Applying this identity for $g(t) = (\frac{1+t}{2})^{2\ell-j} = (\cos\frac{t}{2})^{2(2\ell-j)}$ with $t = \cos\theta$, we obtain

$$\int_0^\infty r^{2j+1} w_n(r)\mathrm{d}r = \frac{\pi}{8(\ell+1)} \sum_{i=1}^{\ell} \left(\cos\tfrac{i\pi}{2\ell+2}\right)^{2(2\ell-j)} \sin^2\tfrac{i\pi}{\ell+1},$$

and it holds for $0 \le 2\ell - j \le 2\ell - 1$ or $1 \le j \le 2\ell$. This provides the solution for (5.32) with

$$r_i = \left[\cos\frac{i\pi}{2\ell+2}\right]^{-1} \quad \text{and} \quad \lambda_i = \frac{1}{2(\ell+1)}\left(\cos\frac{i\pi}{2\ell+2}\right)^{4\ell} \sin^2\frac{i\pi}{\ell+1}.$$

We still need to verify (5.33). With the given r_i and λ_i and using the double angle formula for the sine function,

$$\sum_{i=1}^{m} (-1)^i \lambda_i r_i^{2j} = \frac{1}{2(\ell+1)} \sum_{i=1}^{\ell} (-1)^i \left(\cos\frac{i\pi}{2\ell+2}\right)^{2(2\ell-j)+2} \sin^2\frac{i\pi}{2\ell+2},$$

where $\ell+2 \le j \le 2\ell$, which implies $1 \le 2\ell - j + 1 \le \ell - 1$. Thus, we need to show

$$J_\ell(j) := \sum_{i=1}^{\ell} (-1)^i \left(\cos\frac{i\pi}{2\ell+2}\right)^{2j} \sin^2\frac{i\pi}{2\ell+2} = 0, \qquad 1 \le j \le \ell - 1.$$

Changing summation index $i \mapsto 2\ell + 2 - i$ in $\sum_{i=\ell+2}^{2\ell+1}$, it follows that

$$2J_m(j) = \sum_{i=1}^{2\ell+1} (-1)^i \left(\cos\frac{i\pi}{2\ell+2}\right)^{2j} \sin^2\frac{i\pi}{2\ell+2},$$

where we have also used the fact that the term for $j = \ell + 1$ is zero. Now, splitting the last sum into two sums according to i even and odd and applying the Gauss quadrature rules with respect to $1/\sqrt{1-t^2}$ and $\sqrt{1-t^2}$, respectively, we conclude

$$\begin{aligned} 2J_\ell(j) &= \sum_{i=1}^{\ell} \left(\cos\frac{i\pi}{\ell+1}\right)^{2j} \sin^2\frac{i\pi}{\ell+1} - \sum_{i=0}^{\ell} \left(\cos\frac{(2i+1)\pi}{2\ell+2}\right)^{2j} \sin^2\frac{(2i+1)\pi}{2\ell+2} \\ &= \frac{\ell+1}{\pi}\int_{-1}^{1} t^{2j}\sqrt{1-t^2}\mathrm{d}t - \frac{\ell+1}{\pi}\int_{-1}^{1} t^{2j}(1-t^2)\frac{\mathrm{d}t}{\sqrt{1-t^2}} = 0 \end{aligned}$$

for $0 \le j \le \ell - 1$. This completes the proof for the case of degree $n = 2\ell + 1$. The proof for the case of degree $n = 2\ell$ follows along the same line with little modification. □

The weight function $w_n(\|\mathbf{x}\|)$ depends on n, but it defines a moment functional on the space of polynomials of degree $2n + 1$, which could be extended to a moment functional that is well-defined for all polynomials and remains positive and invariant under the rotation group. Thus, the theorem shows that

Möller's lower bound (5.9) is sharp in the class of rotationally invariant moment functionals so that, since rotationally invariant is stronger than invariant under any subgroup, restricting to a class of weight functions with stronger symmetry is unlikely to yield a better lower bound for cubature rules of degree $2n-1$. The weight function $w_n(\|x\|)$ is supported on an unbounded domain. Thus, it does not preclude the possibility of an improved lower bound for a class of invariant weight functions supported on the unit ball, although it does indicate that such a class of functions will likely be fairly restricted and beyond the assumption of symmetry under a larger subgroup of the orthogonal group.

5.6 Notes and Further Results

Radon initiated the study of minimal cubature rules in [130], where he determined that the minimal number of nodes for a cubature of degree 5 for the integration over $[-1,1]^2$ is 7.

While the lower bound (2.2) is classical, the first nontrivial lower bound (5.9) for centrally symmetric integrals was established by H. N. Möller in his doctoral thesis [106] in 1973 and published in [107]. The lower bounds in Theorem 5.2.1, stated in terms of the coefficient matrices of the three-term relations, were formulated in [167]. That the lower bound holds for weight functions beyond the centrally symmetric setting was first established on the triangle in [11, 13] and then in [166] for all wrapt product weight functions. The second lower bound for the centrally symmetric weight functions in Theorem 5.3.2 was also given by Möller and appeared in [108].

As spherical designs, cubature rules with equal weights have been studied extensively [5, 25]. The linear programming lower bound in Proposition 5.4.1 was first proved in [38] and used to establish the bound (5.17). The lower bound in (5.22) appeared in [183], derived by choosing a more refined function F. Although these bounds are stated for cubature rules with equal weights, they remain valid for all positive cubature rules.

In [183], the lower bound (5.22) is shown to imply the lower bound

$$N(m,d) \ge 2\frac{\int_{\gamma_m}^1 t(1-t^2)^{\frac{d-3}{2}}\,\mathrm{d}t}{(1-\gamma_m)(1-\gamma_m^2)^{\frac{d-3}{2}}} \ge \frac{(1-\gamma_m^2)^{\frac{d-1}{2}}}{d-1},$$

which is then used to give the asymptotic lower bound [183, Corollary]

$$N \ge \frac{2^{d-1}}{d^d} m^{d-1}\left(1+\mathcal{O}\left(m^{-\frac{2}{3}}\right)\right), \tag{5.34}$$

based on an estimate for the largest zero γ_m of $C_n^{(\frac{d}{2})}$ given in [91] by

$$\gamma_m \ge \frac{\sqrt{4(m-1)(d+m-1}}{d+2(m-1)} - \frac{(4\pi)^{2/3}}{(m-1)^{1/6}(d+m-1)^{1/2}}.$$

However, the argument is flawed since this bound of γ_m is not strong enough to yield (5.34). The sharp estimate on the largest zero of the Gegenbauer polynomial C_n^λ turns out to be a fairly delicate problem and has been studied by several authors; see, for example, [53, 82, 116, 117] and their references. Lemma 5.4.5 is a fairly recent result in [117]. It seems likely that a sharper upper bound for $1-\gamma_m^2$ could be proved eventually and that it would yield the lower bound (5.34). What is needed is an upper bound that matches the lower bound [53, Lemma 3]

$$1-\gamma_m^2 > \frac{(2\lambda+1)^2}{4(n+\lambda)^2-1}, \qquad \lambda > -\frac{1}{2}, \quad n \ge 1.$$

Since the lower bound (5.17) gives, by the Stirling formula, an asymptotic order

$$N \ge \left(\frac{e}{2d}\right)^{d-1} m^{d-1}\left(1+O(m^{-1})\right),$$

the bound in (5.34) would be stronger by a factor $2(\frac{e}{4})^d$ if it were true.

The asymptotic for $N(m,2)$ in (5.24) has the main term $\frac{4}{15}m^2 = 0.2666\bar{6}m^2$, which however is not sharp. Indeed, the largest zero γ_m for C_n^λ satisfies an asymptotic formula [1, (22.16.2)]

$$\gamma_m = 1 - \frac{j_{\lambda-\frac{1}{2}}}{2m^2}\left[1 - \frac{2\lambda-1}{m} + \mathcal{O}\left(m^{-2}\right)\right],$$

where $j_{\lambda-\frac{1}{2}}$ is the first positive zero of the Bessel function $J_{\lambda-\frac{1}{2}}(x)$. In particular, it follows that

$$N(m,2) \ge \frac{2}{1-\gamma_m} = \frac{4m^2}{j_1^2}\left(1+\mathcal{O}(m^{-1}\right),$$

where j_1 is the first positive zero of the Bessel function $J_1(x)$, which can be computed numerically and leads to the asymptotic lower bound

$$N(m,2) \ge 0.27244 \times m^2\left(1+\mathcal{O}(m^{-1}\right).$$

This last comment is not important quantitatively since none of the lower bounds in Section 5.4 are likely to be sharp. They are significant since they tell how far off the lower bounds could be for classical weight functions on the unit ball and the simplex. For the existence of cubature rules on the sphere, it is conjectured in [58], based on numerical evidence, that there exists a spherical design of degree m with $m^2/2 + \mathcal{O}(m)$ nodes for $d = 3$.

The lower bound on the unit ball in (5.25) first appeared in [176], whereas the bound for the simplex in (5.27) has not appeared elsewhere. These lower bounds show, in particular, that Möller's lower bound is far from sharp for at least a subclass of the Jacobi weight functions on these two domains. An earlier study [155] in this respect analyzed when the lower bound is sharp, and it leads to a result in [32] that shows a cubature rule of degree $4n+1$ for the weight function $(x_1^2 + x_2^2)^{\alpha}(1 - x_1^2 - x_2^2)^{\beta}$, $\alpha, \beta > -1$, on the unit disk needs at least $2(n+1)^2$ nodes, one more than Möller's lower bound for $n > 1$. For the unit ball $\mathbb{B}^d$ with the constant weight, it was shown in [120] that Möller's lower bound (5.11) is sharp for cubature rules of degree 5 if $d = 2, 3, 7$ and 23 and possibly for $d = (2m+1)^2 - 2$ with $m \geq 5$, which was later confirmed in [68], but not sharp for all other dimensions. The latter shows that cubature rules of degree 5 with one more than the lower bound (5.11) in those other dimensions, which holds, for example, if $d = 4, 5, 6, 7$ (cf. [146]). The work in this direction is further extended in [67].

The product type of cubature rules of degree m for the sphere $\mathbb{S}^d$, the unit ball $\mathbb{B}^d$, and the simplex Δ^d all have nodes in the order of $(m/2)^d$. Using the lower bounds for the sphere and central symmetric and wrapt product weight functions, we see that the order m^d in the lower bound is sharp. That is, a minimal cubature rule of degree m for these weight functions and domains must have nodes in the order of $\mathcal{O}(m^d)$. For the spherical design, the requirement of equal weight adds complexity, making the proof of existence more difficult. It was conjectured in [83] that the number should remain in the order of $\mathcal{O}(m^d)$, and the conjecture was settled positively in [15].

The construction in Theorem 5.5.2 was given in [173], where Proposition 5.5.1 was proved by working out the closed form of sums by trigonometric identities, and the more succinct proof was given in [121].

6
First Minimal Cubature Rules

In this chapter, we study the positive cubature rules of degree $m = 2n - 1$ for an integral on $\mathbb{R}^d$ with

$$N = \dim \Pi_{n-1}^d + \sigma_0, \qquad 1 \le \sigma_0 \le r_n^d,$$

number of nodes, which means $\sigma_1 = \cdots = \sigma_{m-n} = 0$ in Proposition 5.1.1. The polynomials that vanish on the nodes of the cubature rules consist of, by Proposition 5.1.2, $r_n^d - \sigma_0$ linearly independent orthogonal polynomials of degree n and, by Proposition 5.1.5 with $\sigma_1 = 0$, r_{n+1}^d linearly independent $(2n-1)$-orthogonal polynomials of degree $n+1$. These polynomials form a ring generated by

$$\mathcal{Q}_n = \left\{ U^{\mathsf{T}} \mathbb{P}_n, \mathbb{Q}_{n+1} \right\} \quad \text{with} \quad \mathbb{Q}_{n+1} = \mathbb{P}_{n+1} + \Gamma_1 \mathbb{P}_n + \Gamma_2 \mathbb{P}_{n-1}, \tag{6.1}$$

where $U : r_n^d \times (r_n^d - \sigma_0)$ and $\Gamma_i : r_{n+1}^d \times r_{n+1-i}^d$ are real matrices and U has the full rank, $\operatorname{rank} U = r_n^d - \sigma_0$. This class of cubature rules includes minimal cubature rules that attain Möller's lower bound.

In Section 6.1, we show that the cubature rule of degree $2n-1$ with $\dim \Pi_{n-1}^d + \sigma_0$ nodes exists if the nodes are common zeros of $\mathcal{Q}_n$. Section 6.2 provides a characterization of positive cubature rules based on the common zeros of $\mathcal{Q}_n$, and it shows that the matrices U and Γ_i, which determine $\mathcal{Q}_n$, need to satisfy a nonlinear system of matrix equations given in terms of coefficient matrices in the three-term relations. In Section 6.3, we restrict the characterization to those cubature rules that attain Möller's lower bound, especially in the centrally symmetric setting. Based on the characterization, examples of minimal cubature rules are given for the Chebyshev weight on the square in Section 6.4. A set of related nodes, called Padua points, that permits unique interpolation by polynomials of total degree and are nodes of cubature rules of degree $2n - 1$ are discussed in Section 6.5. Finally, the approach for the

Padua points is extended to polynomial interpolation for the chessboard data in Section 6.6.

6.1 Cubature Rules Based on Zeros of $\mathcal{Q}_n$

We start with a simple result on the existence of cubature rules based on interpolation and orthogonality. Let $U^{\mathsf{T}}\mathbb{P}_n$ and $\mathbb{Q}_{n+1}$ be defined as in (6.1). Let $V : r_n^d \times \sigma_0$ be a matrix satisfying

$$U^{\mathsf{T}}V = 0 \quad \text{and} \quad \operatorname{rank} V = \sigma_0.$$

Furthermore, let $V^+ : \sigma_0 \times r_n^d$ be a generalized inverse of V, so that $\operatorname{rank} V^+ = \sigma_0$. We then define

$$\Pi_n^d(V) = \Pi_n^d \setminus \operatorname{span} U^{\mathsf{T}}\mathbb{P}_n = \Pi_{n-1}^d \bigoplus \operatorname{span} V^+\mathbb{P}_n, \tag{6.2}$$

which contains polynomials of degree up to n that do not vanish on the zeros of $\mathcal{Q}_n$. Then, $\Pi_{n-1}^d \subset \Pi_n^d(V) \subset \Pi_n^d$ and

$$\dim \Pi_n^d(V) = \dim \Pi_{n-1}^d + \sigma_0. \tag{6.3}$$

Let $U^+ : (r_n^{\mathrm{d}} - \sigma_0) \times r_n^d$ be the generalized inverses of U. Assume both U^+ and V^+ are defined as in (3.11), so that

$$U^+U = I_{r_n^d - \sigma_0} \qquad \text{and} \qquad V^+V = I_{\sigma_0}.$$

Since U and V both have full rank, $U^+ = (U^{\mathsf{T}}U)^{-1}U^{\mathsf{T}}$ and $V^+ = (V^{\mathsf{T}}V)^{-1}V^{\mathsf{T}}$. Thus, $U^{\mathsf{T}}V = 0$ also implies $V^+U^{+\mathsf{T}} = 0$. Consequently, it follows that

$$\begin{bmatrix} U^{\mathsf{T}} \\ V^+ \end{bmatrix} \begin{bmatrix} U^{+\mathsf{T}} & V \end{bmatrix} = I_{r_n^d} \quad \text{and} \quad U^{+\mathsf{T}}U^{\mathsf{T}} + VV^+ = I_{r_n^d}, \tag{6.4}$$

where the second identity follows by reversing the multiplication order on the left-hand side of the first identity.

Proposition 6.1.1. *If the polynomials $U^{\mathsf{T}}\mathbb{P}_n$ and $\mathbb{Q}_{n+1}$ have $N = \dim \Pi_{n-1}^d + \sigma_0$ simple, real common zeros, denoted by $\mathbf{x}_k$, $1 \le k \le N$, then there is a unique polynomial $L_n f \in \Pi_n(V)^d$ that interpolates a generic function f,*

$$L_n f(\mathbf{x}_k) = f(\mathbf{x}_k), \qquad 1 \le k \le N.$$

Proof Let $\{\mathbf{x}_k : 1 \le k \le \dim \Pi_{n-1}^d + \sigma_0\}$ be the set of zeros of $U^{\mathsf{T}}\mathbb{P}_n$ and $\mathbb{Q}_{n+1}$. Let Y be the column vector defined by

$$Y(\mathbf{x}) = \left[\mathbb{P}_0^{\mathsf{T}}(\mathbf{x}), \mathbb{P}_1^{\mathsf{T}}(\mathbf{x}), \ldots, \mathbb{P}_{n-1}^{\mathsf{T}}(\mathbf{x}), (V^+\mathbb{P}_n(\mathbf{x}))^{\mathsf{T}}\right]^{\mathsf{T}},$$

which consists of a basis of $\Pi_n^d(V)$. Then, $P \in \Pi_n^d(V)$ can be written as $P = \mathbf{c}^\mathsf{T} Y$. Thus, the existence and the uniqueness of the interpolation polynomial are equivalent to that the linear system of equations $[Y(\mathbf{x}_1), \ldots, Y(\mathbf{x}_N)]\mathbf{c} = 0$ has only a trivial solution or that the coefficient matrix of the system is nonsingular. Assume otherwise and, without loss of generality, assume that

$$\operatorname{rank}[Y(\mathbf{x}_1), \ldots, Y(\mathbf{x}_N)] = m - 1, \qquad m \le N,$$

and the first $m-1$ columns of the matrix are linearly independent. It follows that there is $\mathbf{a} \in \mathbb{R}^m$ such that

$$\sum_{k=1}^{m} a_k Y(\mathbf{x}_k) = 0, \qquad \mathbf{a} = (a_1, \ldots, a_m) \neq 0. \tag{6.5}$$

In terms of the components of Y, the identity is equivalent to

$$\sum_{k=1}^{m} a_k \mathbb{P}_j(\mathbf{x}_k) = 0, \quad 0 \le j \le n-1, \quad \text{and} \quad \sum_{k=1}^{m} a_k V^+ \mathbb{P}_n(\mathbf{x}_k) = 0. \tag{6.6}$$

Since $U^\mathsf{T} \mathbb{P}_n(\mathbf{x}_k) = 0$ for $1 \le k \le N$ and, by (6.4), $\mathbb{P}_n = U^{+\mathsf{T}} U^\mathsf{T} \mathbb{P}_n + V V^+ \mathbb{P}_n$, the first identity also holds for $j = n$; that is, $\sum_{k=1}^m a_k \mathbb{P}_n(\mathbf{x}_k) = 0$. As a consequence of this observation and (6.6), it follows from the three-term relation (3.5) that

$$\sum_{k=1}^{m} a_k x_{k,i} \mathbb{P}_j(\mathbf{x}_k) = 0, \quad 0 \le j \le n-1, \quad 1 \le i \le d,$$

where we write $\mathbf{x}_k = (x_{k,1}, \ldots, x_{k,d})$. Moreover, since $\mathbb{Q}_{n+1}(\mathbf{x}_k) = 0$, $1 \le k \le N$, it follows from the three-term relation that

$$x_{k,i} \mathbb{P}_n(\mathbf{x}_k) = A_{n,i}[-\Gamma_1 \mathbb{P}_n(\mathbf{x}_k) - \Gamma_2 \mathbb{P}_{n-1}(\mathbf{x}_k)] + B_{n,i} \mathbb{P}_n(\mathbf{x}_k) + A_{n-1,i}^\mathsf{T} \mathbb{P}_{n-1}(\mathbf{x}_k),$$

and, as a consequence, we also obtain

$$\sum_{k=1}^{m} a_k x_{k,i} \mathbb{P}_n(\mathbf{x}_k) = \sum_{k=1}^{m} a_k \left[B_{n,i}^* \mathbb{P}_n(\mathbf{x}_k) + A_{n-1,i}^{*\mathsf{T}} \mathbb{P}_{n-1}(\mathbf{x}_k) \right] = 0.$$

Together, we have established that

$$\sum_{k=1}^{m} a_k x_{k,i} Y(\mathbf{x}_k) = 0, \qquad 1 \le i \le d. \tag{6.7}$$

The identity (6.6) for $j = 0$ gives $\sum_{k=1}^m a_j = 0$, which shows that there are at least two a_js that are nonzero. We may assume one of the two is a_1. Combining the identities (6.5) and (6.7), we obtain

$$\sum_{k=1}^{m-1} a_k (x_{m,i} - x_{k,i}) Y(\mathbf{x}_k) = x_{m,i} \sum_{k=1}^{m} a_k Y(\mathbf{x}_k) - \sum_{k=1}^{m} a_k x_{k,i} Y(\mathbf{x}_k) = 0.$$

Sine $Y(\mathbf{x}_1),\ldots,Y(\mathbf{x}_{m-1})$ are linearly independent, this shows that

$$a_k(x_{m,i}-x_{k,i})=0, \quad 1\le k\le m-1, \quad 1\le i\le d.$$

But $a_1\ne 0$, it follows that $x_{m,i}=x_{1,i}$, $1\le i\le d$, so that $\mathbf{x}_1=\mathbf{x}_m$, which contradicts the assumption that all nodes are distinct. This completes the proof. □

Theorem 6.1.2. *If the polynomials in* $\mathcal{Q}_n=\{U^{\mathsf{T}}\mathbb{P}_n,\mathbb{Q}_{n+1}\}$ *have* $\dim\Pi_{n-1}^d+\sigma_0$ *simple, real common zeros, then these zeros are the nodes of a cubature rule of degree* $2n-1$ *for the integral* $\mathcal{I}(f)$.

Proof The interpolation polynomial in $\Pi_n^d(V)$ can be written as

$$L_nf(\mathbf{x})=\sum_{k=1}^{N}f(\mathbf{x}_k)\ell_k(\mathbf{x}), \qquad \ell_k\in\Pi_n^d(V),$$

where ℓ_k are fundamental interpolation polynomials that satisfy $\ell_k(\mathbf{x}_j)=\delta_{k,j}$. Integrating L_n shows that the cubature rule

$$\mathcal{I}(f)=\sum_{k=1}^{N}\lambda_kf(\mathbf{x}_k)=:\mathsf{C}(f), \qquad \lambda_k=\mathcal{I}(\ell_k),$$

is exact for $f\in\Pi_n^d(V)$, by the uniqueness of the interpolation polynomial. Moreover, if $f\in\mathcal{Q}_n$, then $\mathcal{I}(f)=\mathsf{C}(f)$ since both sides are zero. Thus, the cubature rule is exact for $f\in\Pi_{n+1}^d$. Since $\mathcal{Q}_n$ contains $(2n-1)$-orthogonal polynomials, we can follow the proof of Theorem 4.1.1 to conclude that the cubature rule is of degree $2n-1$. □

The theorem shows that the main task in studying the cubature rule of degree $2n-1$ with $\dim\Pi_{n-1}^d+\sigma_0$ nodes is to study the zeros of $\mathcal{Q}_n$. What is conspicuously missing in the statement, and the proof, is the positivity of the cubature rule since there is no guarantee that the integration of $\lambda_k=\mathcal{I}(\ell_k)$ is nonnegative. In the next section, we give a set of sufficient conditions for $\mathcal{Q}_n$ to have N real and simple zeros, which is also necessary if the cubature rule based on the zeros of $\mathcal{Q}_n$ is positive.

We end this introductory section with a characterization of cubature rules that attain Möller's lower bound (5.3).

Corollary 6.1.3. *A cubature rule of degree* $2n-1$ *attains Möller's lower bound (5.3) if, and only if, its nodes are common zeros of* $r_n^d-\sigma_0$ *linearly independent orthogonal polynomials of degree n, where* $\sigma_0=\frac{1}{d}\operatorname{rank}\left[\Xi_{A_{n-1}}(A^{\mathsf{T}})_{n-1}\right]$.

Proof If a cubature rule attains the lower bound (5.3), then the inequality (5.5) in the proof of Theorem 5.2.1 becomes equality. In particular, it follows that $\operatorname{rank}U=r_n^d-\sigma_0$ with $\sigma_0=\frac{1}{d}\operatorname{rank}\left[\Xi_{A_{n-1}}(A^{\mathsf{T}})_{n-1}\right]$ and $\operatorname{span}\{x_iU^{\mathsf{T}}\mathbb{P}_n\}=$

r^d_{n+1}, so that the nodes of the cubature rules are zeros of $U^{\mathsf{T}}\mathbb{P}_n$ and, by the three-term relation, there is a $\mathbb{Q}_{n+1}$ that vanishes on the zeros of $U^{\mathsf{T}}\mathbb{P}_n$. In the other direction, the proof follows from applying Theorem 6.1.2. □

The cubature rules that attain Möller's lower bound are always positive, as we will see in the main theorem in Section 6.2.

6.2 Characterization of Positive Cubature Rules

We assume $\mathbb{P}_n$ consists of a basis of orthonormal polynomials relative to a square positive linear functional $\mathcal{I}$ or a nonnegative integral, normalized so that $\mathcal{I}1 = 1$ and, hence, $\mathbb{P}_0 = 1$. The main result of this section is the following characteristic theorem, given in terms of the coefficient matrices of the three-term relations, for cubature rules of degree $2n-1$ with $\dim \Pi^d_{n-1} + \sigma_0$ nodes that are common zeros of $\mathcal{Q}_n$ in (6.1).

Theorem 6.2.1. *The polynomials in $\mathcal{Q}_n = \langle U^{\mathsf{T}}\mathbb{P}_n, \mathbb{Q}_{n+1}\rangle$ have $\dim \Pi^d_{n-1} + \sigma_0$ real, simples zeros that are nodes of a positive cubature rule of degree $2n-1$ if, and only if, U and Γ_i are determined as follows: there is a nonnegative definite matrix $M : r^d_n \times r^d_n$ with $\operatorname{rank} M = \sigma_0$ such that*

(1) *the matrix U of $\mathcal{Q}_n$ satisfies $MU = 0$;*
(2) *the matrix Γ_1 of $\mathcal{Q}_n$ satisfies, for $1 \le i, j \le d$,*

$$A_{n-1,i}(M-I)A^{\mathsf{T}}_{n-1,j} = A_{n-1,j}(M-I)A^{\mathsf{T}}_{n-1,i}, \tag{6.8}$$

$$B^*_{n,i}M = MB^{*\mathsf{T}}_{n,i}, \tag{6.9}$$

$$MA^{\mathsf{T}}_{n-1,i}A_{n-1,j}M + B^*_{n,i}MB^{*\mathsf{T}}_{n,j} = MA^{\mathsf{T}}_{n-1,j}A_{n-1,i}M + B^*_{n,j}MB^{*\mathsf{T}}_{n,i}, \tag{6.10}$$

where $B^*_{n,i} := B_{n,i} - A_{n,i}\Gamma_1$;
(3) *the matrix Γ_2 of $\mathcal{Q}_n$ satisfies*

$$A_{n,i}\Gamma_2 = (I-M)A^{\mathsf{T}}_{n-1,i}, \qquad 1 \le i \le d. \tag{6.11}$$

The proof of this theorem is of interest in itself. In the sufficient direction, we need to show that if M and Γ_1 solve (6.8)–(6.10), then the set $\mathcal{Q}_n$ has $N = \dim \Pi^2_{n-1} + \sigma_0$ real and simple zeros. To show that they lead to a positive cubature rule of degree $2n-1$, we need a formula for the Lagrange interpolation polynomial, based on these zeros, so that the integral of the fundamental interpolation polynomial is positive, and this requires that the subspace $\Pi^d_n(V)$, halfway between Π^d_{n-1} and Π^d_n, possesses an analog of the Christoffel–Darboux formula.

The proof is long, and it will be divided into several subsections.

6.2.1 Common Zeros of $\mathcal{Q}_n$

We first study zeros of $\mathcal{Q}_n$ via the structural matrices of orthogonal polynomials. Let $\mathcal{Z}(\mathcal{Q}_n)$ be the set of real common zeros of $\mathcal{Q}_n$; that is,

$$\mathcal{Z}(\mathcal{Q}_n) = \left\{\mathbf{x} \in \mathbb{R}^d : U^{\mathsf{T}}\mathbb{P}_n(\mathbf{x}) = 0, \mathbb{Q}_{n+1}(\mathbf{x}) = 0\right\}.$$

We start with the necessary conditions for $\mathcal{Q}_n$ to have $\dim \Pi_{n-1}^d + \sigma_0$ real and simple common zeros. For $1 \le i \le d$, we define

$$A_{n-1,i}^* := A_{n-1,i} - \Gamma_2^{\mathsf{T}} A_{n,i}^{\mathsf{T}} \quad \text{and} \quad B_{n,i}^* := B_{n,i} - A_{n,i}\Gamma_1, \tag{6.12}$$

where $B_{n,i}^*$ is the same as stated in Theorem 6.2.1. Using these matrices, we can replace $\mathbb{P}_{n+1}$ in the three-term relation (3.5), with $m = n$, by $\mathbb{Q}_{n+1}$, so that the relation becomes

$$x_i\mathbb{P}_n = A_{n,i}\mathbb{Q}_{n+1} + B_{n,i}^*\mathbb{P}_n + A_{n-1,i}^{*\mathsf{T}}\mathbb{P}_{n-1}, \qquad 1 \le i \le d. \tag{6.13}$$

Proposition 6.2.2. *Let $1 \le \sigma_0 \le r_n^d$. Assume that $\mathcal{Q}_n$ in (6.1) has $N = \dim \Pi_{n-1}^d +$ σ_0 real and simple common zeros that are nodes of a cubature rule of degree $2n-1$. Then there is a matrix $M : r_n^d \times r_n^d$ such that (1)–(3) of Theorem 6.2.1 holds. Moreover, the matrix M is nonnegative definite if the cubature rule is positive.*

Proof Let the cubature rule of degree $2n-1$ be denoted $\mathsf{C}(f)$. Let

$$M = \mathsf{C}(\mathbb{P}_n\mathbb{P}_n^{\mathsf{T}}).$$

Since C is of degree $2n-1$, $\mathsf{C}(\mathbb{P}_n\mathbb{P}_{n-1}^{\mathsf{T}}) = \mathcal{I}(\mathbb{P}_n\mathbb{P}_{n-1}^{\mathsf{T}}) = 0$. Moreover, since $\mathbb{Q}_{n+1}$ vanishes on zeros of $\mathcal{Q}$, $\mathsf{C}(\mathbb{Q}_{n+1}g) = 0$ for all $g \in \Pi^d$. Therefore, using (6.13), we obtain $\mathsf{C}(x_i\mathbb{P}_n\mathbb{P}_{n-1}^{\mathsf{T}}) = A_{n-1,i}^{*\mathsf{T}}\mathcal{I}(\mathbb{P}_{n-1}\mathbb{P}_{n-1}^{\mathsf{T}}) = A_{n-1,i}^{*\mathsf{T}}$ and, furthermore, using the three-term relation on $x_i\mathbb{P}_{n-1}^{\mathsf{T}}$, we also have $\mathsf{C}(x_i\mathbb{P}_n\mathbb{P}_{n-1}^{\mathsf{T}}) = \mathsf{C}(\mathbb{P}_n\mathbb{P}_n^{\mathsf{T}})A_{n-1,i}^{\mathsf{T}} =$ $MA_{n-1,i}^{\mathsf{T}}$. Together, this gives $A_{n-1,i}^{*\mathsf{T}} = MA_{n-1,i}^{\mathsf{T}}$, which is (6.11). Similarly, using (6.13), we obtain $\mathsf{C}(x_i\mathbb{P}_n\mathbb{P}_n^{\mathsf{T}}) = B_{n,i}^*M$, which must be symmetric by its left-hand side, and it proves (6.9). Furthermore, using the three-term relation twice, we obtain

$$\mathsf{C}(x_ix_j\mathbb{P}_{n-1}\mathbb{P}_{n-1}^{\mathsf{T}}) = A_{n-1,i}MA_{n-1,j}^{\mathsf{T}} + B_{n-1,i}B_{n-1,j} + A_{n-2,i}^{\mathsf{T}}A_{n-2,j}^{\mathsf{T}},$$

and it holds when i and j are exchanged. The two ways of computing this quantity lead to an identity that yields (6.8) when combining with the third identity of (3.15). Finally, using $\mathsf{C}(x_j\mathbb{P}_n\mathbb{P}_n^{\mathsf{T}}) = MB_{n,j}^{*\mathsf{T}}$, we obtain

$$\mathsf{C}(x_j\mathbb{P}_{n+1}\mathbb{P}_n^{\mathsf{T}}) = -\Gamma_1\mathsf{C}(x_j\mathbb{P}_n\mathbb{P}_n^{\mathsf{T}}) - \Gamma_2\mathsf{C}(x_j\mathbb{P}_{n-1}\mathbb{P}_n^{\mathsf{T}}) = -\Gamma_1MB_{n,j}^{*\mathsf{T}} - \Gamma_2A_{n-1,j}M,$$

which can be used, together with (6.11), to deduce that

$$\begin{aligned}\mathsf{C}(x_ix_j\mathbb{P}_n\mathbb{P}_n^{\mathsf{T}}) &= A_{n,i}\mathsf{C}(x_j\mathbb{P}_{n+1}\mathbb{P}_n^{\mathsf{T}}) + B_{n,i}\mathsf{C}(x_j\mathbb{P}_n\mathbb{P}_n^{\mathsf{T}}) + A_{n-1,i}^{\mathsf{T}}\mathsf{C}(x_j\mathbb{P}_{n-1}\mathbb{P}_n^{\mathsf{T}})\\ &= A_{n,i}(-\Gamma_1 MB_{n,j}^{*\mathsf{T}} - \Gamma_2 A_{n-1,j}M) + B_{n,i}MB_{n,j}^{*\mathsf{T}} + A_{n-1,i}^{\mathsf{T}}A_{n-1,j}M\\ &= B_{n,i}^{*}MB_{n,j}^{*\mathsf{T}} - (A_{n-1,i}^{\mathsf{T}} - MA_{n-1,i}^{\mathsf{T}})A_{n-1,j}M + A_{n-1,i}^{\mathsf{T}}A_{n-1,j}M\\ &= B_{n,i}^{*}MB_{n,j}^{*\mathsf{T}} + MA_{n-1,i}^{\mathsf{T}}A_{n-1,j}M.\end{aligned}$$

The left-hand side shows that the identity holds if i and j exchange, from which (6.10) follows. This completes the proof. □

The main conditions in Theorem 6.2.1 are (6.8)–(6.10) since the condition (6.11) determines Γ_2 once M is known.

Proposition 6.2.3. *Equation* (6.11) *is equivalent to*

$$\Gamma_2 = \sum_{i=1}^{d} D_{n,i}^{\mathsf{T}}(I-M)A_{n-1,i}^{\mathsf{T}}, \tag{6.14}$$

where $D_{n,i}$ is the generalized inverse of $A_{n,i}$ defined in (3.12).

Proof Multiplying (6.11) by $D_{n,i}^{\mathsf{T}}$ and summing up, we obtain immediately (6.14). Assume now that Γ_2 is given by (6.14). Then, (6.8) is equivalent to $A_{n-1,i}Y_j = A_{n-1,j}Y_i$, where $Y_j = (I-M)A_{n-1,j}^{\mathsf{T}}$. By Lemma 6.2.4, this implies that $(Y_1^{\mathsf{T}},\dots,Y_d^{\mathsf{T}})^{\mathsf{T}}$ is in the null space of $I - A_nD_n^{\mathsf{T}}$, which leads to

$$(I-M)A_{n-1,i}^{\mathsf{T}} = A_{n,i}\sum_{j=1}^{d} D_{n,j}^{\mathsf{T}}(I-M)A_{n-1,j}^{\mathsf{T}} = A_{n,i}\Gamma_2,$$

and this is exactly (6.11). □

Lemma 6.2.4. *Let $Y = (Y_1^{\mathsf{T}},\dots,Y_d^{\mathsf{T}})^{\mathsf{T}}$ with $Y_i \in \mathbb{R}^{r_n^d}$. Then $(I - A_nD_n^{\mathsf{T}})Y = 0$ if and only if $A_{n-1,i}Y_j = A_{n-1,j}Y_i$, $1 \le i,j \le d$.*

Proof The system $A_{n-1,i}Y_j = A_{n-1,j}Y_i$, $1 \le i,j \le d$, is equivalent to $\Xi_{A_{n-1}}Y = 0$. By the first identity of (3.5), $\Xi_{A_{n-1}}A_n = 0$. Since $\operatorname{rank}\Xi_{A_{n-1}} = dr_n^d - r_{n+1}^d$ and $\operatorname{rank}A_n = r_{n+1}^d$ by (3.6), it follows that the columns of A_n consist of a basis of the null space of $\Xi_{A_{n-1}}$. Consequently, $Y = A_n\xi$ for some nonzero ξ. By (3.12), $(I - A_nD_n^{\mathsf{T}})A_n = 0$, so that $(I - A_nD_n^{\mathsf{T}})Y = 0$. The process is reversible if $\operatorname{rank}(I - A_nD_n^{\mathsf{T}}) = dr_n^d - r_{n+1}^d$. □

We now give a sufficient condition for $\mathcal{Q}_n$ to have $\dim\Pi_n^{d-1} + \sigma_0$ common zeros.

Theorem 6.2.5. *The polynomial ideal $\langle\mathcal{Q}_n\rangle = \langle U^{\mathsf{T}}\mathbb{P}_n, \mathbb{Q}_{n+1}\rangle$ has $\dim\Pi_{n-1}^d + \sigma_0$ real, simple zeros if there is a nonnegative definite matrix $M : r_n^d \times r_n^d$ with* rank$M = \sigma_0$, *such that* M *and* Γ_1 *satisfy the system of equations* (6.8)–(6.10), *and the matrix* $U : r_n^d \times (r_n^d - \sigma_0)$ *is determined by* $MU = 0$ *and* Γ_2 *is given by* (6.14).

Proof Assume (6.8)–(6.11). Since M has rank σ_0, the matrix U is determined by $MU = 0$ up to an invertible matrix. Furthermore, since M is nonnegative definite, there is a matrix V

$$M = VV^{\mathsf{T}}, \qquad V : r_n^d \times \sigma_0 \quad \text{and} \quad \operatorname{rank} V = \sigma_0. \tag{6.15}$$

Moreover, since $V^{\mathsf{T}}V$ is invertible, we see that $MU = 0$ if and only if $U^{\mathsf{T}}V = 0$.

We define matrices $S_{n,i}$ of size $N \times N$ by

$$S_{n,i} = \begin{bmatrix} B_{0,i} & A_{0,i} & & & \bigcirc \\ A_{0,i}^{\mathsf{T}} & B_{1,i} & A_{1,i} & & \\ & \ddots & \ddots & \ddots & \\ & & A_{n-2,i}^{\mathsf{T}} & B_{n-1,i} & A_{n-1,i}V \\ \bigcirc & & & V^{+}A_{n-1,i}^{*\mathsf{T}} & V^{+}B_{n,i}^{*}V \end{bmatrix}, \qquad 1 \le i \le d.$$

We first claim that the joint eigenvalues of $S_{n,1}, \ldots, S_{n,d}$ are common zeros of $\mathcal{Q}_n$. An element $\lambda = (\lambda_1, \ldots, \lambda_d) \in \mathbb{C}^d$ is a joint eigenvalue if there is a nonzero vector ξ such that

$$S_{n,i}\xi = \lambda_i \xi, \qquad 1 \le i \le d.$$

We write $\xi = \{\xi_0, \xi_1, \ldots, \xi_n\}$ as a column vector, so that $\xi_j \in \mathbb{R}^{r_j^d}$ for $1 \le j \le n-1$ and $\xi_n \in \mathbb{R}^{\sigma_0}$. Then $S_{n,i}\xi = \lambda_i\xi$ is equivalent to

$$\begin{aligned} B_{0,i}\xi_0 + A_{0,i}\xi_1 &= \lambda_i \xi_0 \\ A_{j-1}^{\mathsf{T}}\xi_{j-1} + B_{j,i}\xi_j + A_{j,i}\xi_{j+1} &= \lambda_i \xi_j, \quad 1 \le j \le n-2 \\ A_{n-2,i}^{\mathsf{T}}\xi_{n-2} + B_{n-1,i}\xi_{n-1} + A_{n-1,i}V\xi_n &= \lambda_i \xi_{n-1}, \\ V^{+}A_{n-1,i}^{*\mathsf{T}}\xi_{n-1} + V^{+}B_{n,i}^{*}\xi_n &= \lambda_i \xi_n. \end{aligned}$$

The first n equations are exactly the three-term relations satisfied by the orthogonal polynomials $\mathbb{P}_j(\lambda)$.

We claim that ξ_0 is nonzero. Indeed, if $\xi_0 = 0$, then we apply the generalized inverse $D_{j,i}$ of $A_{j,i}$ and (3.12) on the first n equations of $S_{n,i}\xi = \lambda_i\xi$, recursively from $j = 0$ to $j = n-1$, which shows that $\xi_1 = 0, \ldots, \xi_{n-1} = 0$ and $V\xi_n = 0$. Since $V\xi_n = 0$ implies $\xi_n = 0$ by $V^{+}V = I_{\sigma_0}$, we obtain $\xi = 0$, a contradiction. Since $\xi_0 \ne 0$, we can assume, without loss of generality, $\xi_0 = 1$. Thus, in particular, $\xi_0 = \mathbb{P}_0(\lambda)$. Let

$$\eta_j = \xi_j - \mathbb{P}_j(\lambda), \quad 1 \le j \le n-1, \quad \text{and} \quad \eta_n = V\xi_n - \mathbb{P}_n(\lambda).$$

Then $\eta_0,\ldots,\eta_n$ satisfies the first n equations of the three-term relations (3.5). Since $\eta_0 = 0$, following the same argument that shows $\xi_0 \neq 0$ given above, we obtain that $\eta_n = V\xi_n - \mathbb{P}_n(\lambda) = 0$ or $V\xi_n = \mathbb{P}_n(\lambda)$. By $U^{\mathsf{T}}V = 0$, we conclude that $U^{\mathsf{T}}\mathbb{P}_n(\lambda) = U^{\mathsf{T}}V\xi_n = 0$. Thus, λ is a zero of $U^{\mathsf{T}}\mathbb{P}_n$.

We remain to show that λ is a zero of $\mathbb{Q}_n$. Since $U^{\mathsf{T}}\mathbb{P}_n(\lambda) = 0$, by (6.4),

$$VV^{+}\mathbb{P}_n(\lambda) = U^{+\mathsf{T}}U^{\mathsf{T}}\mathbb{P}_n(\lambda) + VV^{+}\mathbb{P}_n(\lambda) = \mathbb{P}_n(\lambda).$$

Moreover, by $V^{+}V = I_{\sigma_0}$ and $V\xi_n = \mathbb{P}_n(\lambda)$, we also have $\xi_n = V^{+}\mathbb{P}_n(\lambda)$. Hence, multiplying the last of the n equations in $S_{n,i}\xi = \lambda_i\xi$ by V and using $\xi_n = V^{+}\mathbb{P}_n(\lambda)$, we obtain

$$VV^{+}A_{n-1}^{*\mathsf{T}}\mathbb{P}_{n-1}(\lambda) + VV^{+}B_{n,i}^{*}V\mathbb{P}_n(\lambda) = \lambda_i V\xi_n = \lambda_i VV^{+}\mathbb{P}_n(\lambda) = \lambda_i\mathbb{P}_n(\lambda).$$

Combining with the identity (6.13), we conclude that

$$A_{n,i}\mathbb{Q}_{n+1}(\lambda) = (VV^{+} - I_{\sigma_0})B_n^{*}VV^{+}\mathbb{P}_n(\lambda) + (VV^{+} - I_{\sigma_0})A_{n-1}^{*\mathsf{T}}\mathbb{P}_{n-1}(\lambda).$$

We claim that the right-hand side is zero. Indeed, since $M = VV^{\mathsf{T}}$ and $V^{+}V = I_{\sigma_0}$, Eq. (6.9) gives $B_{n,i}^{*}V = MB_{n,i}^{*\mathsf{T}}(V^{+})^{\mathsf{T}}$ and Eq. (6.11) is equivalent to $A_{n-1}^{*\mathsf{T}} = MA_{n-1,i}^{\mathsf{T}}$, so that the claim follows from

$$(VV^{+} - I_{\sigma_0})M = (VV^{+} - I_{\sigma_0})VV^{\mathsf{T}} = VV^{\mathsf{T}} - VV^{\mathsf{T}} = 0.$$

Thus, $A_{n,i}\mathbb{Q}_{n+1}(\lambda) = 0$ for $1 \le i \le n$. By (3.12), we conclude that $\mathbb{Q}_{n+1}(\lambda) = 0$. Hence, every joint eigenvalue of $S_{n,1},\ldots,S_{n,d}$ is a common zero of $\mathcal{Q}_n$.

Since $S_{n,i}$ is of size $N \times N$ with $N = \dim \Pi_{n-1}^d + \sigma_0$, this shows that $\mathcal{Q}_n$ has at most N zeros. We claim that $S_{n,i}$ is symmetric, which holds if

$$V^{+}A_{n-1}^{*\mathsf{T}} = V^{\mathsf{T}}A_{n-1}^{\mathsf{T}} \quad \text{and} \quad V^{+}B_{n,i}^{*}V = V^{\mathsf{T}}B_{n,i}^{*\mathsf{T}}(V^{+})^{\mathsf{T}}. \tag{6.16}$$

While the first identity follows from rewriting the assumed identity (6.11) as $A_{n-1,i}^{*\mathsf{T}} = VV^{\mathsf{T}}A_{n-1,i}^{\mathsf{T}}$ and then multiplying it from the left by V^{+}, the second identity follows from multiplying V^{+} from the left of (6.9) and using $V^{+}M = V^{+}VV^{\mathsf{T}} = V^{\mathsf{T}}$. Thus, the matrix $S_{n,i}$ is symmetric and, as a consequence, the joint eigenvalues of $S_{n,1},\ldots,S_{n,d}$ are real, which shows that the zeros of $\mathcal{Q}_n$ are real. We now show that $S_{n,1},\ldots,S_{n,i}$ consist of a commuting family of matrices. Indeed, using the relations (3.15) and (6.8), it is straightforward to see that $S_{n,i}S_{n,j} = S_{n,j}S_{n,i}$ if we can confirm

$$\begin{aligned} A_{n-1,i}(VV^{+}B_{n,j}^{*} - B_{n,j})V &= A_{n-1,i}(VV^{+}B_{n,j}^{*} - B_{n,j})V, \\ V^{\mathsf{T}}A_{n-1,i}^{\mathsf{T}}A_{n-1,j}V + V^{+}B_{n,i}^{*}VV^{+}B_{n,j}^{*}V &= V^{\mathsf{T}}A_{n-1,j}^{\mathsf{T}}A_{n-1,i}V + V^{+}B_{n,j}^{*}VV^{+}B_{n,i}^{*}V. \end{aligned}$$

Combining the identities (6.9) and (6.16), we obtain

$$B_{n,i}^{*}V = VV^{\mathsf{T}}B_{n,i}^{*\mathsf{T}}V^{+\mathsf{T}} = VV^{+}B_{n,i}^{*}V. \tag{6.17}$$

Using this identity and the definition of $A^*_{n-1,i}$ at (6.12), the first equation to be confirmed is equivalent to $A_{n-1,i}A_{n,j}\Gamma_1 V = A_{n-1,j}A_{n,i}\Gamma_1 V$, which follows immediately from the first identity of (3.15). By $V^+V = I_{\sigma_0}$, the validity of the second equation to be confirmed does not change if we multiply it from the left by V and from the right by V^T, and since

$$VV^+B^*_{n,i}VV^+B^*_{n,j}VV^\mathsf{T} = VV^+B^*_{n,i}V(VV^+B^*_{n,j}V)^\mathsf{T} = B^*_{n,i}VV^\mathsf{T}B^{*\mathsf{T}}_{n,j},$$

by (6.16) and (6.17), it follows that the validity is equivalent to (6.10) as $M = VV^\mathsf{T}$. Thus, the matrices $S_{n,1},\ldots,S_{n,d}$ commute, and as a consequence, they have N joint eigenvalues. This proves that $\mathbb{Q}_n$ has N real zeros. We remain to show that all zeros are simple, which will be given right after Theorem 6.2.6 in Subsection 6.2.2. □

6.2.2 An Analog of the Christoffel–Darboux Formula

Let M be the matrix in the statement of Theorem 6.2.5. Let V be the matrix $M = VV^\mathsf{T}$ with $\operatorname{rank} V = \sigma_0$ as in (6.15). We consider the subspace $\Pi^d_n(V)$ defined at (6.2), which has dimension $\dim \Pi^d_{n-1} + \sigma_0$.

The space $\Pi^d_n(V)$ has an orthogonal basis that consists of $\mathbb{P}_{n-1}$ and $V^+\mathbb{P}_n$. We define a kernel of the space by

$$\mathbf{K}^*_n(\mathbf{x},\mathbf{y}) = \mathbf{K}_{n-1}(\mathbf{x},\mathbf{y}) + \left[V^+\mathbb{P}_n(\mathbf{x})\right]^\mathsf{T} V^+\mathbb{P}_n(\mathbf{y}). \tag{6.18}$$

If $\mathsf{C}(f)$ is the positive cubature rule of degree $2n-1$, then the matrix M is defined in the proof of Proposition 6.2.2 by $M = \mathsf{C}(\mathbb{P}_n\mathbb{P}_n^\mathsf{T})$, which is nonnegative definite. Writing $M = VV^\mathsf{T}$ and $V^+V = I_{\sigma_0}$, we then have

$$I = V^+MV^{+\mathsf{T}} = V^+\mathsf{C}(\mathbb{P}_n\mathbb{P}_n^\mathsf{T})V^{+\mathsf{T}} = \mathsf{C}\big(V^+\mathbb{P}_n(V^+\mathbb{P}_n)^\mathsf{T}\big),$$

which shows that if we regard $\mathsf{C}(f)$ as a linear functional defined on Π^d_{2n-1}, then $V^+\mathbb{P}_n$ is orthonormal with respect to C, so is $\mathbb{P}_{n-1}$ by its orthogonality since the components of $\mathbb{P}_{n-1}\mathbb{P}_{n-1}^\mathsf{T}$ are of degree $2n-2$. This shows, in particular, that the kernel $\mathbf{K}^*_n(\cdot,\cdot)$ is the reproducing kernel of the space $\Pi^d_n(V)$ with respect to the linear functional C.

We show that this kernel satisfies an analog of the Christoffel–Darboux formula.

Theorem 6.2.6. *Assume* $M = VV^\mathsf{T}$ *with* $\operatorname{rank} V = \sigma_0$ *satisfies* (6.8)–(6.11). *Then, the kernel* K^*_n *satisfies*

$$\begin{aligned}(x_i - y_i)\mathbf{K}^*_n(\mathbf{x},\mathbf{y}) &= \mathbb{P}_n(\mathbf{x})^\mathsf{T}UU^+A^\mathsf{T}_{n-1,i}\mathbb{P}_{n-1}(\mathbf{y}) - \mathbb{P}_n(\mathbf{y})^\mathsf{T}UU^+A^\mathsf{T}_{n-1,i}\mathbb{P}_{n-1}(\mathbf{x})\\ &\quad + [V^+\mathbb{P}_n(\mathbf{y})]^\mathsf{T}V^+A_{n,i}\mathbb{Q}_{n+1}(\mathbf{x}) - [V^+\mathbb{P}_n(\mathbf{x})]V^+A_{n,i}\mathbb{Q}_{n+1}(\mathbf{y})\\ &\quad + [V^+\mathbb{P}_n(\mathbf{y})]^\mathsf{T}V^+B^*_{n,i}U^{+\mathsf{T}}U^\mathsf{T}\mathbb{P}_n(\mathbf{x}) - [V^+\mathbb{P}_n(\mathbf{x})]^\mathsf{T}V^+B^*_{n,i}U^{+\mathsf{T}}U^\mathsf{T}\mathbb{P}_n(\mathbf{y}).\end{aligned}$$

Furthermore, with $\partial_i = \frac{\partial}{\partial x_i}$ *and for* $1 \le i \le d$,

$$\begin{aligned}\mathbf{K}_n^*(\mathbf{x},\mathbf{x}) = &\left[\partial_i U^\mathsf{T}\mathbb{P}_n(\mathbf{x})\right]^\mathsf{T} U^+ A_{n-1,i}^\mathsf{T}\mathbb{P}_{n-1}(\mathbf{x}) - \left[U^\mathsf{T}\mathbb{P}_n(\mathbf{x})\right]^\mathsf{T} U^+ A_{n-1,i}^\mathsf{T}\partial_i\mathbb{P}_{n-1}(\mathbf{x})\\ &+ \mathbb{P}_n(\mathbf{x})^\mathsf{T} V^{+\mathsf{T}} V^+\left[A_{n,i}\partial_i\mathbb{Q}_{n+1}(\mathbf{x}) + B_{n,i}^* U^{+\mathsf{T}}\partial_i(U^\mathsf{T}\mathbb{P}_n(\mathbf{x}))\right]\\ &- \partial_i\mathbb{P}_n(\mathbf{x})^\mathsf{T} V^{+\mathsf{T}} V^+\left[A_{n,i}\mathbb{Q}_{n+1}(\mathbf{x}) + B_{n,i}^* U^{+\mathsf{T}} U^\mathsf{T}\mathbb{P}_n(\mathbf{x})\right].\end{aligned}$$

Proof Using $V^+A_{n-1,i}^{*\mathsf{T}} = V^\mathsf{T} A_{n-1,i}^\mathsf{T}$, which follows from (6.11), we deduce from (6.13) that

$$x_i V^+\mathbb{P}_n = V^+A_{n,i}\mathbb{Q}_{n+1} + V^+B_{n,i}^*\mathbb{P}_n + V^\mathsf{T}A_{n-1,i}\mathbb{P}_{n-1}.$$

Using this identity and (6.4), we obtain

$$\begin{aligned}&(x_i - y_i)[V^+\mathbb{P}_n(\mathbf{x})]^\mathsf{T} V^+\mathbb{P}_n(\mathbf{y})\\ &\quad= [V^+\mathbb{P}_n(\mathbf{y})]^\mathsf{T} V^+A_{n,i}\mathbb{Q}_{n+1}(\mathbf{x}) - [V^+\mathbb{P}_n(\mathbf{x})]V^+A_{n,i}\mathbb{Q}_{n+1}(\mathbf{y})\\ &\quad+ \left\{[V^+\mathbb{P}_n(\mathbf{y})]^\mathsf{T} V^+B_{n,i}^* U^{+\mathsf{T}}U^\mathsf{T}\mathbb{P}_n(\mathbf{x}) - [V^+\mathbb{P}_n(\mathbf{x})]^\mathsf{T} V^+B_{n,i}^* U^{+\mathsf{T}}U^\mathsf{T}\mathbb{P}_n(\mathbf{y})\right\}\\ &\quad+ \left[\mathbb{P}_n(\mathbf{y})^\mathsf{T} V^{+\mathsf{T}}V^\mathsf{T}A_{n-1,i}^\mathsf{T}\mathbb{P}_{n-1}(\mathbf{x}) - \mathbb{P}_n(\mathbf{x})^\mathsf{T} V^{+\mathsf{T}}V^\mathsf{T}A_{n-1,i}^\mathsf{T}\mathbb{P}_{n-1}(\mathbf{y})\right].\end{aligned}$$

The last term of the right-hand side can be written as, using $V^{+\mathsf{T}}V^\mathsf{T} = I - UU^+$ at (6.4) and the Christoffel–Darboux formula (3.9),

$$(y_i - x_i)\mathbf{K}_{n-1}(\mathbf{x},\mathbf{y}) - \left[\mathbb{P}_n(\mathbf{y})^\mathsf{T} UU^+A_{n-1,i}^\mathsf{T}\mathbb{P}_{n-1}(\mathbf{x}) - \mathbb{P}_n(\mathbf{x})^\mathsf{T} UU^+A_{n-1,i}^\mathsf{T}\mathbb{P}_{n-1}(\mathbf{y})\right],$$

from which the stated formula for $\mathbf{K}_n^*(\mathbf{x},\mathbf{y})$ follows readily. Dividing $x_i - y_i$ and writing the right-hand side formula for $\mathbf{K}_n^*(\mathbf{x},\mathbf{y})$ as appropriate differences, we can take limit $y_i \to x_i$ to derive the formula for $\mathbf{K}_n^*(\mathbf{x},\mathbf{x})$ as in (3.10). □

If $\mathbf{x}_k$ is a common zero of $\mathcal{Q}_n$, then $U^\mathsf{T}\mathbb{P}_n(\mathbf{x}_k) = 0$ and $\mathbb{Q}_{n+1}(\mathbf{x}_k) = 0$, so that the formula for $\mathbf{K}_n^*(\mathbf{x},\mathbf{x})$ with $\mathbf{x} = \mathbf{x}_k$ becomes

$$\begin{aligned}\mathbf{K}_n^*(\mathbf{x}_k,\mathbf{x}_k) = &\left[\partial_i U^\mathsf{T}\mathbb{P}_n(\mathbf{x}_k)\right]^\mathsf{T} U^+A_{n-1,i}^\mathsf{T}\mathbb{P}_{n-1}(\mathbf{x}_k)\\ &+ \mathbb{P}_n(\mathbf{x}_k)^\mathsf{T} V^{+\mathsf{T}}V^+\left[A_{n,i}\partial_i\mathbb{Q}_{n+1}(\mathbf{x}_k) + B_{n,i}^* U^{+\mathsf{T}}\partial_i U^\mathsf{T}\mathbb{P}_n(\mathbf{x}_k)\right].\end{aligned}$$

In particular, since $\mathbf{K}_n^*(\mathbf{x},\mathbf{x}) > 0$, it follows that at least one of $\partial_i U^\mathsf{T}\mathbb{P}_n(\mathbf{x}_k)$ and $\partial_i\mathbb{Q}_{n+1}(\mathbf{x}_k)$ is not zero, which shows that $\mathbf{x}_k$ must be a simple zero of $\mathcal{Q}_n$. This finishes the last piece for the proof of Theorem 6.2.5.

As one consequence of the Christoffel–Darboux formula, we obtain an expression for the fundamental interpolation polynomials in $\Pi_n^d(V)$ discussed in Proposition 6.1.1.

Theorem 6.2.7. *Let* $\mathcal{Z}(\mathcal{Q}_n) = \{\mathbf{x}_1,\dots,\mathbf{x}_N\}$, $N = \dim\Pi_{n-1}^d + \sigma_0$, *be the set of real and simple zeros of* $\mathcal{Q}_n$ *specified in Theorem 6.2.5. Then, the polynomial*

$$L_n^* f(\mathbf{x}) = \sum_{k=1}^N f(\mathbf{x}_k)\ell_k(\mathbf{x}) \quad \textit{with} \quad \ell_k(\mathbf{x}) = \frac{\mathbf{K}_n^*(\mathbf{x},\mathbf{x}_k)}{\mathbf{K}_n^*(\mathbf{x}_k,\mathbf{x}_k)} \tag{6.19}$$

is the unique Lagrange interpolation polynomial in $\Pi_n^d(V)$ *that satisfies*

$$L_n^* f(\mathbf{x}_k) - f(\mathbf{x}_k), \qquad k = 1,2,\ldots,N.$$

Proof From the Christoffel–Darboux formula for $\mathbf{K}_n^*(\mathbf{x},\mathbf{y})$ that we established earlier, it follows readily that $\mathbf{K}_n(\mathbf{x}_k,\mathbf{x}_j) = 0$ for $k \neq j$. Consequently, $\ell_k(\mathbf{x}_j) = \delta_{k,j}$, so that $L_n^* f(\mathbf{x}_k) = f(\mathbf{x}_k)$ for $1 \le k \le N$. The uniqueness is established in Proposition 6.1.1. □

6.2.3 Proof of Theorem 6.2.1

If a positive cubature rule $\mathsf{C}(f)$ of degree $2n-1$ exists, then its nodes are zeros of a $\mathcal{Q}_n$ by Proposition 5.1.5. The necessary conditions of (6.8)–(6.10) are satisfied with $M = \mathsf{C}(\mathbb{P}_n \mathbb{P}_n^{\mathsf{T}})$, as shown in Proposition 6.2.2. Since $\mathsf{C}(f)$ is positive, the matrix M is nonnegative positive definite. Moreover, the columns of U consist of a basis of the kernel space of M. Thus, M has rank σ_0.

Assume now that the nonnegative definite matrix M exists and satisfies, together with Γ_1 and Γ_2, (6.8)–(6.11). By Theorem 6.2.5, the ideal $\mathcal{Q}_n$ has $\dim \Pi_{n-1}^d + \sigma_0$ real simple zeros, and there are nodes of a unique interpolation polynomial in $\Pi_n^d(V)$ by Theorem 6.2.7. Integrating the polynomial given in (6.19) gives a cubature rule

$$\mathcal{I}(f) = \sum_{k=1}^{N} \lambda_k f(\mathbf{x}_k) =: \mathsf{C}(f), \qquad f \in \Pi_n^d(V),$$

by the uniqueness of the interpolation polynomial. This is the cubature rule of degree $2n-1$ established in the proof of Theorem 6.1.2. We remain to show $\lambda_k > 0$, so that $\mathsf{C}(f)$ is positive. Let ℓ_k be the fundamental interpolation polynomial defined in (6.19). Then, ℓ_k is of degree n and $\ell_k(\mathbf{x}_j) = \delta_{k,j}$. Applying the cubature rule on ℓ_j, we obtain

$$\lambda_k = \mathsf{C}(\ell_k) = \mathcal{I}(\ell_k) = \frac{1}{\mathbf{K}_n^*(\mathbf{x}_j,\mathbf{x}_j)} > 0, \tag{6.20}$$

where we have used $\mathcal{I}(\mathbf{K}_n^*(\cdot,\mathbf{x}_k)) = 1$, which follows from the orthogonality of $\mathbb{P}_j$. This completes the proof. □

6.3 Cubature Rules That Attain Möller's Lower Bound

The characterization in Theorem 6.2.1 is given in terms of the matrix M that is required to satisfy the system of equations (6.8)–(6.10). The existence of such a matrix, however, is not warranted since the system is nonlinear. Moreover, the requirement that M is nonnegative definite puts a further stringent restraint on its existence. We will give several positive examples in the next section. Here, we discuss several special cases that are of interest.

The number σ_0 in the characterization can be any nonnegative integer less than or equal to r_n^d. By (5.14), the cubature rules that attain Möller's lower bound (5.12) satisfy $\sigma_0 < r_n^d$, so that they are covered under Theorem 6.2.1. We show that the characterization can be substantially simplified in the centrally symmetric setting.

Let us first consider the case $\sigma_0 = 0$. In this case, $M = 0$ as $\operatorname{rank} M = 0$, so that the theorem degenerates to Theorem 4.1.5 for the Gauss cubature rule of degree $2n-1$, for which the nodes of the cubature rule are zeros of $\mathbb{P}_n$, and it follows from the three-term relation (3.5) and (3.12) that

$$\mathbb{Q}_{n+1}(\mathbf{x}) = \sum_{i=1}^{d} D_{n,i}^{\mathsf{T}} x_i \mathbb{P}_n(\mathbf{x}) = \mathbb{P}_{n+1}(\mathbf{x}) + \Gamma_1 \mathbb{P}_n(\mathbf{x}) + \Gamma_2 \mathbb{P}_n(\mathbf{x}),$$

vanishes on the zeros of $\mathbb{P}_n$ trivially. Thus, $\mathbb{Q}_{n+1}$ belongs to the ideal $\langle \mathbb{P}_n \rangle$.

More generally, we are interested in when $\mathbb{Q}_{n+1}$ belongs to the ideal generated by $U^{\mathsf{T}} \mathbb{P}_n$.

Proposition 6.3.1. *Let* $\mathrm{C}(f)$ *be a positive cubature rule of degree* $2n-1$ *and its nodes are zeros of* $\{U^{\mathsf{T}}\mathbb{P}_n, \mathbb{Q}_{n+1}\}$, *where* $U : r_n^d \times (r_n^d - \sigma_0)$ *and* $\operatorname{rank} U = r_n^d - \sigma_0$. *Define*

$$\sigma(n,d) := r_n^d - \frac{1}{d} r_{n+1}^d.$$

In particular, $\sigma(n,2) = \frac{n}{2}$. *Then the following holds:*

(i) *If* $\sigma_0 \le \sigma(n,d)$, *then* $\mathcal{Q}_n = \langle U^{\mathsf{T}}\mathbb{P}_n \rangle$;

(ii) *If* $\sigma_0 > \sigma(n,d)$, *then* $\mathcal{Q}_n$ *contains* $r_{n+1}^d - d(r_n^d - \sigma_0)$ *linearly independent polynomials that do not belong to* $\langle U^{\mathsf{T}}\mathbb{P}_n \rangle$.

Moreover, if (i) *holds and the integral is centrally symmetric, then* $\Gamma_1 = 0$ *in* $\mathbb{Q}_{n+1}$.

Proof Let $A_n : dr_n^d \times r_{n+1}^d$ be the matrix defined in (3.6). We define the matrix $H : d(r_n^d - \sigma_0) \times r_{n+1}^d$ by

$$H = \operatorname{diag}\{U^{\mathsf{T}}, \ldots, U^{\mathsf{T}}\} A_n = \begin{bmatrix} U^{\mathsf{T}} A_{n,1} \\ \vdots \\ U^{\mathsf{T}} A_{n,d} \end{bmatrix}, \quad \operatorname{diag}\{U^{\mathsf{T}}, \ldots, U^{\mathsf{T}}\} = \begin{bmatrix} U^{\mathsf{T}} & & \bigcirc \\ & \ddots & \\ \bigcirc & & U^{\mathsf{T}} \end{bmatrix},$$

so that $\operatorname{diag}\{U^{\mathsf{T}}, \ldots, U^{\mathsf{T}}\}$ is of size $d(r_n^d - s_0) \times dr_n^d$. We first consider the case

$\sigma_0 \le \sigma(n,d)$, which is equivalent to $d(r_n^d - \sigma_0) \ge r_{n+1}^d$, so that the matrix H has more rows than columns if it is not square. We determine the rank of H. Let $\xi \in \mathbb{R}^{d(r_n^d - \sigma_0)}$ be a vector in the null space of H^T. Then,

$$0 = H^\mathsf{T}\xi = A_n^\mathsf{T}\eta \quad \text{with} \quad \zeta = \operatorname{diag}\{U^\mathsf{T},\ldots,U^\mathsf{T}\}\xi.$$

Since A_n is of full rank, $\operatorname{rank} A_n = r_{n+1}^d$ by (3.6), this shows that ζ is an element of the null space of A_n^T and the latter has dimension $dr_n^d - r_{n+1}^d$. Let Z be the matrix of size $dr_n^d \times (dr_n^d - r_{n+1}^d)$, so that its columns consist of a basis for the null space of A_n^T. Then, the relation between η and ξ shows that there is a matrix Ξ such that

$$\operatorname{diag}\{U^\mathsf{T},\ldots,U^\mathsf{T}\}\Xi = Z, \qquad \Xi : d(r_n^d - \sigma_0) \times (dr_n^d - r_{n+1}^d),$$

and the columns of Ξ are elements of the null space of H^T. The above identity shows that $\operatorname{rank}\Xi \le dr_n^d - r_{n+1}^d$, by the size of Ξ, and

$$dr_n^d - r_{n+1}^d = \operatorname{rank} Z \ge \operatorname{rank}\operatorname{diag}\{U^\mathsf{T},\ldots,U^\mathsf{T}\} + \operatorname{rank}\Xi - d(r_n^d - \sigma_0) = \operatorname{rank}\Xi,$$

so that $\operatorname{rank}\Xi = dr_n^d - r_{n+1}^d$. Thus, the null space of H^T has the dimension $dr_n^d - r_{n+1}^d$. Consequently, we conclude that $\operatorname{rank} H = \operatorname{rank} H^\mathsf{T} = r_{n+1}^d$. In particular, H has a generalized inverse $H^+ = (H_1^+,\ldots,H_d^+)$, so that

$$I = H^+H = \sum_{i=1}^{d} H_i^+ H_i, \qquad H_i = U^\mathsf{T} A_{n,i}.$$

Thus, by the three-term relation (3.5), we obtain

$$\mathbb{P}_{n+1} = H^+H\mathbb{P}_{n+1} = \sum_{k=1}^{d} H_k^+ U^\mathsf{T} A_{n,i}\mathbb{P}_{n+1} = \sum_{i=1}^{d} H_i^+ U^\mathsf{T}(x_i\mathbb{P}_n - B_{n,i}\mathbb{P}_n - A_{n-1,i}\mathbb{P}_{n-1}),$$

which can be rewritten as

$$\sum_{i=1}^{d} x_i H_i^+ U^\mathsf{T}\mathbb{P}_n = \mathbb{P}_{n+1} + \Gamma_1\mathbb{P}_n + \Gamma_2\mathbb{P}_{n-1} = \mathbb{Q}_{n+1},$$

where the matrices Γ_1 and Γ_2 are given by

$$\Gamma_1 = \sum_{i=1}^{d} H_i^+ U^\mathsf{T} B_{n,i} \quad \text{and} \quad \Gamma_2 = \sum_{i=1}^{d} H_i^+ U^\mathsf{T} A_{n-1,i}^\mathsf{T}.$$

Thus, $\mathbb{Q}_{n+1}$ belongs to the ideal generated by $U^\mathsf{T}\mathbb{P}_n$. This completes the proof of (i). Moreover, if the integral is centrally symmetric, then $B_{n,i} = 0$, so that $\Gamma_1 = 0$ by its representation given above.

The case $\sigma_0 > \sigma(n,d)$ is easier since H has more columns than rows. From

$\operatorname{rank} A_{n,i} = r_n^d$ (3.6), it follows that $\operatorname{rank} U^\top A_{n,i} = r_n^d - \sigma_0$ and, consequently, $\operatorname{rank} H = d(r_n^d - \sigma_0)$. Thus, by the three-term relation (3.5), we obtain

$$\begin{bmatrix} x_1 U^\top \mathbb{P}_n \\ \vdots \\ x_d U^\top \mathbb{P}_n \end{bmatrix} = H\mathbb{P}_{n+1} + F_{n,1}\mathbb{P}_n + F_{n,2}\mathbb{P}_{n-1}, \; F_{n,1} = \begin{bmatrix} U^\top B_{n,1} \\ \vdots \\ U^\top B_{n,d} \end{bmatrix}, \; F_{n,2} = \begin{bmatrix} U^\top A_{n-1,1}^\top \\ \vdots \\ U^\top A_{n-1,d}^\top \end{bmatrix}.$$

Since H has a generalized inverse H^+ such that $HH^+ = I$, the right-hand side of the above identity can be written as

$$H\left(\mathbb{P}_{n+1} + H^+ F_{n,1}\mathbb{P}_n + H^+ F_{n,2}\mathbb{P}_{n-1}\right),$$

and it vanishes on all nodes of the cubature rule. Thus, it consists of the portion of $\mathbb{Q}_{n+1}$ that is generated by $U^\top \mathbb{P}_n$, and there are $d(r_n^d - \sigma_0)$ linearly independent of such polynomials. This completes the proof. □

The above proposition provides an upper bound $\sigma(n,d)$ for σ_0 that warrants $\mathbb{Q}_{n+1}$ belonging to the ideal generated by $U^\top \mathbb{P}_n$. The upper bound is satisfied for cubature rules attaining Möller's lower bound. If the integral for the cubature rule is centrally symmetric, then $\Gamma_1 = 0$, so that the characterization of positive cubature rules in Theorem 6.2.1 is substantially simplified. In particular, it yields the following characterization of cubature rules that attain Möller's lower bound.

Theorem 6.3.2. *For a centrally symmetric integral $\mathcal{I}$, a positive cubature rule of degree $2n-1$ attains Möller's lower bound* (5.11) *if, and only if, there is a nonnegative definite matrix $M : r_n^d \times r_n^d$ with $\operatorname{rank} M = \sigma_0$ and satisfies*

$$A_{n-1,i}(M-I)A_{n-1,j}^\top = A_{n-1,j}(M-I)A_{n-1,i}^\top, \qquad 1 \le i,j \le d, \tag{6.21}$$

and

$$M\left(A_{n-1,i}^\top A_{n-1,j} - A_{n-1,j}^\top A_{n-1,i}\right)M = 0, \qquad 1 \le i,j \le d. \tag{6.22}$$

The nodes of the cubature rule are zeros of $U^\top \mathbb{P}_n$, where U is a matrix of $r_n^d \times (r_n^d - \sigma_0)$ with full rank and satisfies $MU = 0$.

Proof For the lower bound $N_{n,d}$ in (5.12), $\sigma_0 = \dim \mathscr{P}_{n-1}^{d-1} - \delta_n$, and it is bounded by, using (5.10),

$$\sigma_0 \le \frac{1}{2}\sum_{\ell=1}^{d-1} \dim \Pi_{n-1}^\ell = \frac{1}{2}\sum_{\ell=1}^{d-1} \dim \binom{n+\ell-1}{\ell} = \frac{d}{n}\binom{n+d-1}{d} - 1,$$

which is less than $r_n^d - \frac{1}{d} r_{n+1}^d$ as can be seen by a simple calculation. This shows that σ_0 for Möller's lower bound satisfies $\sigma_0 \le \sigma(n,d)$ and, by Proposition

6.3.1, the cubature rules attaining the lower bound are generated by $U^T\mathbb{P}_n$. Furthermore, since $\mathcal{I}$ is centrally symmetric, $B_{n,i} = 0$ by Theorem 3.2.5. Hence, by the proof of Proposition 6.3.1, $\Gamma_1 = 0$ and, consequently, $B^*_{n,i} = 0$. Thus, Eq. (6.9) holds trivially and Eq. (6.10) simplifies to (6.22), whereas (6.10) is the same as (6.21). The stated result follows as a simplified version of Theorem 6.2.1. □

Remark 6.3.3. *The above proof shows that the upper bound $\sigma(n,d)$ is larger than the corresponding σ_0 in Möller's lower bound* (5.3). *In the case of $d = 2$, the upper and lower bounds agree, and $\sigma_0 = \lfloor \frac{n}{2} \rfloor$.*

Compared to the characterization of Theorem 6.2.1, the simplification of the above theorem comes from $\Gamma_1 = 0$, which holds for cubature rules attaining Möller's lower bound for centrally symmetric integrals and is deduced from the relation

$$x_i U^{\mathsf{T}}\mathbb{P}_n = U^{\mathsf{T}}A_{n,i}\mathbb{P}_{n+1} + U^{\mathsf{T}}A^{\mathsf{T}}_{n-1,i}\mathbb{P}_{n-1}, \quad 1 \le i \le d,$$

that follows from the three-term relation (3.5) for the centrally symmetric integrals. When σ_0 satisfies the upper bound $\sigma(n,d)$, the rank $U = r^d_n - \sigma_0$ is large enough to ensure that $\mathbb{Q}_n$ comes entirely from $x_i U^{\mathsf{T}}\mathbb{P}_n$. Our next proposition gives a lower bound of σ_0 under which Γ_1 cannot be zero.

Proposition 6.3.4. *Let $\mathcal{I}$ be a centrally symmetric integral. Assume the polynomial ideal $\mathcal{Q}_n = \langle U^{\mathsf{T}}\mathbb{P}_n, \mathbb{Q}_{n+1}\rangle$ has $\dim \Pi^d_n + \sigma_0$ zeros that are nodes of a positive cubature rule of degree $2n-1$. Define*

$$\sigma_*(n,d) := r^d_n - \frac{1}{\binom{d}{2}+1}\begin{cases} r^d_n & n \text{ odd} \\ r^d_n - 1 & n \text{ even}\end{cases}; \tag{6.23}$$

in particular, $\sigma_*(n,2) = \lfloor \frac{n}{2} \rfloor + 1$. If $\Gamma_1 = 0$ in $\mathbb{Q}_n = \mathbb{P}_{n+1} + \Gamma_1\mathbb{P}_n + \Gamma_2\mathbb{P}_{n-1}$, then $\sigma_0 \le \sigma_*(n,d)$. Equivalently, if $\sigma_0 > \sigma_*(n,d)$, then $\Gamma_1 \ne 0$.

Proof By the characterization in Theorem 6.2.5, if $\Gamma_1 = 0$, then the identities in (6.10) become

$$M\left(A^{\mathsf{T}}_{n-1,i}A_{n-1,j} - A^{\mathsf{T}}_{n-1,j}A_{n-1,i}\right)M = 0, \qquad 1 \le i < j \le d.$$

Written in terms of the matrix Ξ_C defined at (4.17), they are equivalent to

$$\operatorname{diag}\{M,\ldots,M\}\mathcal{A}_n M = 0, \qquad \mathcal{A}_n = \Xi_{A^{\mathsf{T}}_{n-1}} A_{n-1},$$

where $\operatorname{diag}\{M,\ldots,M\}$ is a square block diagonal matrix of size $\binom{d}{2}r^d_n$. This shows that M belongs to the null space of $\operatorname{diag}\{M,\ldots,M\}\mathcal{A}_n$. Hence,

$$\sigma_0 \le \dim \operatorname{null}(\operatorname{diag}\{M,\ldots,M\}\mathcal{A}_n) = r^d_n - \operatorname{rank}\left[\operatorname{diag}\{M,\ldots,M\}\mathcal{A}_n\right].$$

Since $\mathcal{A}_n = \Xi_{A_n}(A^\top)_n$ by the third identity of (3.15), it follows from (4.19)

$$\operatorname{rank}[\operatorname{diag}\{M,\dots,M\}\mathcal{A}_n] \ge \binom{d}{2}\sigma_0 + \operatorname{rank}\mathcal{A}_n - \binom{d}{2}r_n^d$$
$$= \binom{d}{2}(\sigma_0 - r_n^d) + \begin{cases} r_n^d & n \text{ odd} \\ r_n^d - 1 & n \text{ even} \end{cases}.$$

Putting together, the last two displayed inequality leads to an upper bound σ_0, as stated. □

Note that the proposition states that $\Gamma_1 = 0$ implies $\sigma \le \sigma_*(n,d)$, but not vice versa. The case $d = 2$ is particularly interesting because of the examples given later in this chapter. Compared to Proposition 6.3.1, we state the following corollary for future reference.

Corollary 6.3.5. *Let $d = 2$ and let $\mathcal{I}$ be a centrally symmetric integral. Assume that $\mathcal{Q}_n$ has $\dim\Pi_n^2 + \sigma_0$ zeros, and they are nodes of a positive cubature rule of degree $2n-1$. If $\Gamma_1 = 0$ in $\mathbb{Q}_n = \mathbb{P}_{n+1} + \Gamma_1\mathbb{P}_n + \Gamma_2\mathbb{P}_{n-1}$, then $\sigma_0 = \lfloor \frac{n}{2} \rfloor$ or $\sigma_0 = \lfloor \frac{n}{2} \rfloor + 1$.*

Remark 6.3.6. *The case $\sigma_0 = \lfloor \frac{n}{2} \rfloor$ is when Möller's lower bound is attained, for which $\mathcal{Q}_n = \langle U^\top \mathbb{P}_n \rangle$. In the case $\sigma_0 = \lfloor \frac{n}{2} \rfloor + 1$, $\mathbb{Q}_{n+1}$ contains exactly one polynomial not in the ideal $\langle U^\top \mathbb{P}_n \rangle$ when n is odd and exactly two polynomials not in the ideal $\langle U^\top \mathbb{P}_n \rangle$ when n is even. Since the statement in Proposition 6.3.4 is one direction, Γ_1 may or may not be zero if $\sigma_0 = \lfloor \frac{n}{2} \rfloor + 1$. However, if the system of equations* (6.21) *and* (6.22) *has an appropriate solution M, then the cubature rule of degree $2n-1$ exists by Theorem 6.2.1, which is a solution with $\Gamma_1 = 0$. Moreover, there is no need to calculate $\mathbb{Q}_{n+1}$, since it is determined by $\Gamma_1 = 0$ and Γ_2 given by* (6.14).

Remark 6.3.7. *In the characterization of Theorem 6.2.1, we allow $\sigma_0 = r_n^d$, for which $U = 0$ and $\mathcal{Q}_n = \langle \mathbb{Q}_{n+1} \rangle$. If $\mathcal{I}$ is centrally symmetric, Proposition 6.3.4 implies that $\Gamma_1 \ne 0$ in this case. It should be mentioned that if a cubature rule of degree $2n-1$ exists and its nodes are zeros of $\mathcal{Q}_n$ alone, then it has $\dim\Pi_{n-1}^d + r_n^d = \dim\Pi_n^d$ nodes. By Theorem 6.2.7, there is a unique polynomial of total degree n that interpolates on these nodes, which is of interest from the point of view of polynomial interpolation in high dimensions.*

For the existence of the cubature rule, we need to find orthogonal polynomials $U^\top \mathbb{P}_n$ that have N nodes. If n is odd, then all $\mathbb{P}_n$ vanishes on the origin by symmetry. Furthermore, in the case $d = 2$, two orthogonal polynomials that vanish on all nodes of the cubature rules are known.

Proposition 6.3.8. *Let $d = 2$ and let $\mathcal{I}$ be centrally symmetric. Let $\mathrm{C}_{2n-1}(f)$ be a cubature rule of degree $2n-1$ that attains Möller's lower bound (5.4). If n is odd and η is the unique solution, up to a constant multiple, of*

$$\left(A_{n,1}^{\mathsf{T}} A_{n,2} - A_{n,2}^{\mathsf{T}} A_{n,1}\right)\eta = 0, \qquad \eta \in \mathbb{R}^{n+1},$$

then the two polynomials $\eta^{\mathsf{T}} A_{n,1}^{\mathsf{T}} \mathbb{P}_n$ and $\eta^{\mathsf{T}} A_{n,2}^{\mathsf{T}} \mathbb{P}_n$ vanish on all nodes of C_{2n-1}.

Proof Since $\mathcal{I}$ is centrally symmetric, by the third identity of (3.15) with $B_{n,i} = 0$ and (4.20), we obtain for odd n

$$\operatorname{rank}\left(A_{n,1}^{\mathsf{T}} A_{n,2} - A_{n,2}^{\mathsf{T}} A_{n,1}\right) = \operatorname{rank}\left(A_{n+1,1} A_{n+1,2}^{\mathsf{T}} - A_{n+1,2} A_{n+1,1}^{\mathsf{T}}\right) = n+1.$$

Consequently, up to a constant multiple, the null space of $A_{n,1}^{\mathsf{T}} A_{n,2} - A_{n,2}^{\mathsf{T}} A_{n,1}$ is one dimensional. Let $\eta \neq 0$ be an element in the null space. Let $\zeta_1 = A_{n,2}\eta$ and $\zeta_2 = -A_{n,1}\eta$. Then, it follows readily that ζ_1 and ζ_2 satisfy (5.7). Since the cubature rule attains the lower bound (5.4), the proof of Theorem 5.2.1 shows that $\zeta_i = U\xi_i$ for some ξ; hence, $\zeta_i^{\mathsf{T}} \mathbb{P}_n$ vanishes on zeros of the cubature rules. □

This proposition shows that $A_{n,1}\eta$ and $A_{n,2}\eta$ are parts of U. In particular, M satisfies $MA_{n,i}\eta = 0$, an additional information that can help in determining the solvability of (6.21) and (6.21). The proposition does not hold for $d > 2$ since the above argument relies on the sharpness of the lower bound (5.3), which however is weaker than (5.11), hence not sharp, for $d \geq 3$.

The system of equations (6.8)–(6.10) is nonlinear; the first equation can be reformulated as a structure of the matrix M. Indeed, using (3.7), Eq. (6.8) is equivalent to

$$L_{n-1,i} G_n^{-1}(M-I) G_n^{-\mathsf{T}} L_{n-1,j} = L_{n-1,j} G_n^{-1}(M-I) G_n^{-\mathsf{T}} L_{n-1,i}, \qquad 1 \leq i,j \leq d,$$

where G_n is the leading coefficient matrix of $\mathbb{P}_n$, which holds if and only if, by Lemma 3.5.2, that the matrix $G_n^{-1}(M-I)G_n^{-\mathsf{T}}$ is a Hankel matrix; that is, M satisfies

$$M = I + G_n H_n G_n^{\mathsf{T}}, \qquad H_n = \left[h_{\alpha+\beta}\right]_{|\alpha|=n,|\beta|=n}. \tag{6.24}$$

In particular, for $d = 2$, $H_n = \left[h_{i+j}\right]_{i=0}^{n}{}_{j=0}^{n}$ that has $2n+1$ variables. The identity (6.24) provides an expression for M that solves (6.8). Thus, if $\Gamma_1 = 0$ and the integral is centrally symmetric, we only need to solve (6.10) with M given by (6.24). However, two challenging problems persist in finding a solution. The first one is evident: the system is nonlinear, and there are more equations than variables. The second one is hidden: We need the matrix M to be nonnegative definite and satisfy $\operatorname{rank} M = \sigma_0 \geq \frac{n}{2}$.

6.4 Minimal Cubature Rules on the Square

We consider the product Chebyshev weight functions on the square $[-1,1]^2$

$$W_0(\mathbf{x}) = w_0(x_1)w_0(x_2) \quad \text{and} \quad W_1(\mathbf{x}) = w_1(x_1)w_1(x_2),$$

of the first and second kind, respectively, where w_0 and w_1 are defined by

$$w_0(t) = \frac{1}{\pi\sqrt{1-t^2}} \quad \text{and} \quad w_1(t) = \frac{2}{\pi}\sqrt{1-t^2},$$

both normalized, so that their integrals over $[-1,1]$ are 1. The orthogonal polynomials in two variables are the product Chebyshev polynomials,

$$P_{k,n}(\mathbf{x}) = \hat{T}_{n-k}(x_1)\hat{T}_k(x_2) \quad \text{or} \quad P_{k,n}(\mathbf{x}) = U_{n-k}(x_1)U_k(x_2),$$

for $0 \le k \le n$, so that $\{P_{k,n} : 0 \le k \le n\}$ is an orthonormal basis for the space $\mathcal{V}_n^2(W_0)$ or $\mathcal{V}_n^2(W_1)$, respectively, where

$$T_n(t) = \cos n\theta \quad \text{and} \quad U_n(t) = \frac{\sin(n+1)\theta}{\sin\theta}, \qquad t = \cos\theta,$$

as usual. While U_n are orthonormal polynomials for w_1, the polynomials $\hat{T}_0(t) = 1$ and $\hat{T}_n(t) = \sqrt{2}T_n(t)$ for $n \ge 1$ are orthonormal for w_0. Both these families of polynomials satisfy the three-term relations

$$\begin{aligned} t\,p_0(t) &= \tfrac{1}{2}a\,p_1(t), \quad t\,p_1(t) = \tfrac{1}{2}p_2(t) + \tfrac{1}{2}a\,p_0(t), \\ t\,p_n(t) &= \tfrac{1}{2}p_{n+1}(t) + \tfrac{1}{2}p_{n-1}(t), \qquad n \ge 2, \end{aligned}$$

where $a = \sqrt{2}$ if $p_n = \hat{T}_n$ and $a = 1$ if $p_n = U_n$. The three-term relations (3.5) for the vectors $\mathbb{P}_n = (P_{0,n}, \ldots, P_{n,n})^{\mathsf{T}}$ are given in Subsection 3.4.1, in which the coefficient matrices are of the form

$$A_{n,1} = \frac{1}{2}\begin{bmatrix} 1 & & & \bigcirc & 0 \\ & \ddots & & & \vdots \\ & & 1 & & 0 \\ \bigcirc & & & a & 0 \end{bmatrix}, \quad A_{n,2} = \frac{1}{2}\begin{bmatrix} 0 & a & & & \bigcirc \\ 0 & & 1 & & \\ \vdots & & & \ddots & \\ 0 & \bigcirc & & & 1 \end{bmatrix}, \tag{6.25}$$

where $a = \sqrt{2}$ if $P_{k,n}(x) = \hat{T}_{n-k}(x_1)\hat{T}_k(x_2)$ and $a = 1$ if $P_{k,n}(x) = U_{n-k}(x_1)U_k(x_2)$.

6.4.1 Minimal Cubature Rule of Degree $2n-1$ for Even n

These cubature rules attain Möller's lower bound. To state our cubature rules, we define

$$t_{k,n} = \cos\frac{k\pi}{n}, \quad 0 \le k \le n, \quad \text{and} \quad u_{k,n} = \cos\frac{k\pi}{n+2}, \quad 0 \le k \le n+2. \tag{6.26}$$

Then, the zeros of $T_n(t)$ are $t_{2k-1,2n}$ and the zeros of $U_n(t)$ are $u_{2k,2n}$ for $1 \le k \le n$.

Theorem 6.4.1. *Let $n = 2\ell$ with $\ell \in \mathbb{N}$. Then, the minimal cubature rules of degree $2n-1$ exist for W_0 and W_1. More precisely, a minimal cubature rule of degree $2n-1$ for W_0 is given by*

$$\frac{1}{\pi^2}\int_{[-1,1]^2} f(\mathbf{x})W_0(\mathbf{x})d\mathbf{x} = \frac{1}{2\ell^2}\sum_{j=0}^{\ell}\lambda_j\sum_{k=1}^{\ell}\left[f(t_{2j,n},t_{2k-1,n}) + f(t_{2k-1,n},t_{2j,n})\right],$$

where $\lambda_j = 1$ if $1 \le j \le \ell-1$ and $\lambda_0 = \lambda_\ell = \frac{1}{2}$, and a minimal cubature rule of degree $2n-1$ for W_1 is given by

$$\frac{4}{\pi^2}\int_{[-1,1]^2} f(\mathbf{x})W_1(\mathbf{x})d\mathbf{x} = \frac{1}{2(\ell+1)^2}\sum_{j=1}^{\ell}\sum_{k=1}^{\ell+1}\left(\sin\tfrac{(2k-1)\pi}{n+2}\right)^2\left(\sin\tfrac{2j\pi}{n+2}\right)^2 \times\left[f(u_{2j,n},u_{2k-1,n}) + f(u_{2k-1,n},u_{2j,n})\right].$$

Both these cubature rules have $N_{\min} = \dim\Pi_{n-1}^2 + \lfloor\frac{n}{2}\rfloor = 2\ell(\ell+1)$ nodes.

Proof With $n = 2\ell$, we define the matrix of M_n of size $(2\ell+1)\times(2\ell+1)$ by

$$M_n = \begin{bmatrix} a^2 & & & & 0 & & & & -a^2 \\ & 1 & & & & & & -1 & \\ & & \ddots & & \vdots & & \iddots & & \\ & & & 1 & & -1 & & & \\ 0 & & \cdots & & 0 & & \cdots & & 0 \\ & & & -1 & & 1 & & & \\ & & \iddots & & \vdots & & \ddots & & \\ & -1 & & & & & & 1 & \\ -a^2 & & & & 0 & & & & a^2 \end{bmatrix}.$$

It is straightforward to verify M_n satisfies (6.21) and (6.22) for both Chebyshev weight functions of the first and second kind. Moreover, it is obvious that rank $M_n = \ell = \frac{n}{2}$. Furthermore, a quick verification shows that the matrix V defined by

$$V^{\mathsf{T}} = \begin{bmatrix} a & & & 0 & & & -a \\ & 1 & & 0 & & -1 & \\ & & \ddots & \vdots & \iddots & & \\ & & 1 & 0 & -1 & & \end{bmatrix}$$

satisfies $M_n = VV^{\mathsf{T}}$, which shows that M_n is nonnegative definite. Hence, by Theorem 6.3.2, there is a cubature rule of degree $2n-1 = 4\ell-1$ with $N = \dim\Pi_{2\ell-1}^2 + \ell$ nodes for both W_0 and W_1, and its nodes are the zeros of $U^{\mathsf{T}}\mathbb{P}_n$, where the matrix $U : (2\ell+1)\times(\ell+1)$ satisfies $MU = 0$ and rank$U = \ell+1$. It is easy to check that the matrix U defined by

$$U^{\mathsf{T}} = \begin{bmatrix} 1 & & & & 0 & & & & 1 \\ & 1 & & & 0 & & & 1 & \\ & & \cdots & & \vdots & & \iddots & & \\ & & & 1 & 0 & 1 & & & \\ & & & & \sqrt{2} & & & & \end{bmatrix}$$

is such a matrix. Writing explicitly, the polynomials in $U^{\mathsf{T}}\mathbb{P}_n$ are $\ell+1$ polynomials

$$P_{2\ell-k}(x_1)P_k(x_2) + P_k(x_1)P_{2\ell-k}(x_2), \quad 0 \le k \le \ell-1, \quad \text{and} \quad P_\ell(x_1)P_\ell(x_2),$$

where $P_k(t) = T_k(t)$ or $U_k(t)$. With $t_{k,n}$ and $u_{k,n}$ in (6.26), the zeros of $T_\ell(t)$ are $t_{2k-1,2\ell}$, $1 \le k \le \ell$, and the zeros of $U_\ell(t)$ are $u_{2k,2\ell}$, $1 \le k \le \ell$. The zeros of $U^{\mathsf{T}}\mathbb{P}_n$ consist of the set

$$X_{2\ell} = \{(z_{2j,2\ell}, z_{2k-1,2\ell}), (z_{2k-1,2\ell}, z_{2j,2\ell}) : 1 \le j \le \ell, 1 \le k \le \ell+1\},$$

where $z_{k,n} = t_{k,n}$ for the Chebyshev weight of the first kind and $z_{k,n} = u_{k,n}$ for the Chebyshev weight of the second kind. These are the nodes of the minimal cubature rules. By (6.20), the weight of the cubature rule for the note $\mathbf{z}_{k,j}$ is given by $[\mathbf{K}_n^*(\mathbf{z}_{k,j}, \mathbf{z}_{k,j})]^{-1}$, whereas its value is determined in the next proposition. $\square$

The nodes of the minimal cubature rule of degree $2n-1 = 39$ or $n = 20$ for the Chebyshev weight W_0 are depicted in Figure 6.1.

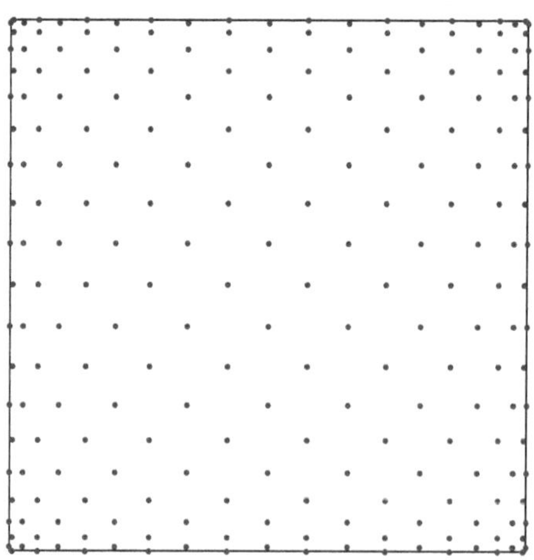

Figure 6.1 The 180-node minimal cubature rule of degree 39.

We determine the weights of the cubature rules and an explicit formula for the Lagrange interpolation (6.19). Let $z_{j,n} = t_{j,n}$ in the case of W_0 and $z_{j,n} = u_{j,n}$ in the case of W_1. Denote

$$\mathbf{z}_{2j,2k-1} = (z_{2j,n}, z_{2k-1,n}) \quad \text{and} \quad \mathbf{z}_{2k-1,2j} = (z_{2k-1,n}, z_{2j,n}),$$

so that $\mathbf{z}_{\mu,\nu}$ are nodes of the cubature rules when μ and ν have different parity. We will also use the notation $\mathbf{t}_{\mu,\nu}$ and $\mathbf{u}_{\mu,\nu}$ when needed.

Proposition 6.4.2. *Let $n = 2\ell$ and let μ and ν be positive integers with different parity. Then*

(i) *For the weight function W_0,*

$$\mathbf{K}_n^*(\mathbf{x}, \mathbf{t}_{\mu,\nu}) = \frac{1}{2}\left[\mathbf{K}_n(\mathbf{x}, \mathbf{t}_{\mu,\nu}) + \mathbf{K}_{n-1}(\mathbf{x}, \mathbf{t}_{\mu,\nu})\right] - \frac{1}{2}(-1)^{\mu}\left[T_n(x_1) + T_n(x_2)\right]$$

and, for $1 \le k \le \ell$ and $0 \le j \le \ell$,

$$\mathbf{K}_n^*(\mathbf{t}_{2j,2k-1}, \mathbf{t}_{2j,2k-1}) = \mathbf{K}_n^*(\mathbf{t}_{2k-1,2j}, \mathbf{t}_{2k-1,2j}) = 2\ell^2 \begin{cases} 1 & 1 \le j \le \ell - 1, \\ 2 & j = 0 \text{ or } j = \ell. \end{cases}$$

(ii) *For the weight function W_1,*

$$\mathbf{K}_n^*(\mathbf{x}, \mathbf{u}_{\mu,\nu}) = \frac{1}{2}\left[\mathbf{K}_n(\mathbf{x}, \mathbf{u}_{\mu,\nu}) + \mathbf{K}_{n-1}(\mathbf{x}, \mathbf{u}_{\mu,\nu})\right]$$

and, for $1 \le k \le \ell + 1$ and $1 \le j \le \ell$,

$$\mathbf{K}_n^*(\mathbf{u}_{2j,2k-1}, \mathbf{u}_{2j,2k-1}) = \mathbf{K}_n^*(\mathbf{u}_{2k-1,2j}, \mathbf{u}_{2k-1,2j}) = \frac{2(\ell+1)^2}{\left(\sin\frac{(2k-1)\pi}{n+2}\right)^2\left(\sin\frac{2j\pi}{n+2}\right)^2}.$$

Proof By the definition of $\mathbf{K}_n^*(\mathbf{x}, \mathbf{y})$, we need the generalized inverse $V^+ : (2\ell+1) \times \ell$ that satisfies $V^+V = I$, which is given by

$$V^+ = (V^\mathsf{T} V)^{-1} V^\mathsf{T} = \frac{1}{2}\operatorname{diag}\{a^{-2}, 1, \ldots, 1\} V^\mathsf{T},$$

where diag$\{\ldots\}$ denotes the diagonal matrix. Using their explicit formulas, we can verify that

$$(V^+)^\mathsf{T} V^+ + \frac{1}{4} U U^\mathsf{T} = \frac{1}{2} I_{2\ell+1} - \frac{1}{4}(1 - a^{-2}) \begin{bmatrix} 1 & & -1 \\ & \bigcirc & \\ -1 & & 1 \end{bmatrix}.$$

In particular, since $\mathbf{z}_{\mu,\nu}$ is a zero of $U^\mathsf{T}\mathbb{P}_n$, it follows that

$$\begin{aligned} &[V^+\mathbb{P}_n(\mathbf{x})]^\mathsf{T} V^+\mathbb{P}_n(\mathbf{z}_{\mu,\nu}) \\ &= \tfrac{1}{2}\mathbb{P}_n(\mathbf{x})^\mathsf{T}\mathbb{P}_n(\mathbf{z}_{\mu,\nu}) - \tfrac{1}{4}(1 - a^{-2})[P_n(x_1) - P_n(x_2)]\left[P_n(z_{\mu,n}) - P_n(z_{\nu,n})\right]. \end{aligned}$$

Since $\mathbb{P}_n^{\mathsf{T}}(\mathbf{x})\mathbb{P}_n^{\mathsf{T}}(\mathbf{y}) = \mathbf{K}_n(\mathbf{x},\mathbf{y}) - \mathbf{K}_{n-1}(\mathbf{x},\mathbf{y})$, it follows that, for $n = 2\ell$,

$$\mathbf{K}_n^*(\mathbf{x},\mathbf{z}_{\mu,\nu}) - \tfrac{1}{2}\left[\mathbf{K}_n(\mathbf{x},\mathbf{z}_{\mu,\nu}) + \mathbf{K}_{n-1}(\mathbf{x},\mathbf{z}_{\mu,\nu})\right] \\ -\tfrac{1}{4}(1-a^{-2})[P_n(x_1) - P_n(x_2)]\left[P_n(z_{\mu,n}) - P_n(z_{\nu,n})\right].$$

For the weight function W_0, the constant $1 - a^{-2} = \frac{1}{2}$ and $P_n(t) = \sqrt{2}T_n(t)$, so that $P_n(z_{\mu,n}) - P_n(z_{\nu,n}) = 2\sqrt{2}(-1)^{\mu}$, and for the weight W_1, $1 - a^{-2} = 0$. Thus, we have established the stated formulas for $\mathbf{K}_n^*(\mathbf{x},\mathbf{z}_{\mu,\nu})$.

To calculate the value of $\mathbf{K}_n^*(\mathbf{z}_{\mu,\nu},\mathbf{z}_{\mu,\nu})$, we need a closed-form formula for the quantity $\mathbf{K}_n(\mathbf{x},\mathbf{x})$. For the weight W_1, such a formula is stated in (4.16). For the weight W_0, we have the explicit formula

$$\mathbf{K}_{n-1}(\mathbf{x},\mathbf{x}) = \sum_{k=0}^{n-1}\sum_{j=0}^{k} \hat{T}_j^2(x_1)\hat{T}_{k-j}^2(x_2) = \frac{1}{4}L_n(\mathbf{x}) + \frac{1}{4}L_{n-1}(\mathbf{x}),$$

where $L_n(\mathbf{x}) = L_n(x_1,x_2)$ is given by

$$L_n(\mathbf{x}) = n^2 + U_{n-1}^2(x_1) + U_{n-1}^2(x_2) + \frac{T_n^2(x_1) - T_n^2(x_2)}{x_1^2 - x_2^2},$$

which can be verified by working with trigonometric sums or by using the closed-form formula of $\mathbf{K}_{n-1}(x,y)$ given in [168]. With $\mathbf{x} = \mathbf{y} = \mathbf{z}_{\mu,\nu}$, these formulas lead to closed-form formulas for $\mathbf{K}_n^*(\mathbf{z}_{\mu,\nu},\mathbf{z}_{\mu,\nu})$, from which its value can be calculated for each node. This completes the proof. □

As a consequence of the above proposition, we obtain a closed-form formula for the Lagrange interpolation polynomial (6.19) given in Theorem 6.2.7, which is the unique polynomial in the space $\Pi_n^d(V)$, defined in (6.2), that agrees with f on the nodes of the minimal cubature rule of degree $2n-1$.

6.4.2 Minimal Cubature Rule of Degree $2n-1$ for Odd n

The case $n = 2\ell - 1$ of the minimal cubature rule turns out to be more involved. We retain the notation $t_{k,n} = \cos\frac{k\pi}{n}$ and $u_{k,n} = \cos\frac{k\pi}{n+2}$ in (6.26).

Theorem 6.4.3. *Let $n = 2\ell - 1$ with $\ell \in \mathbb{N}$. Then, the minimal cubature rules of degree $2n-1$ exist for W_0 and W_1. More precisely,*

(i) *Let $h_{k,n} \in (-1,1)$, $1 \le k \le n$, be the zeros of $(n+1)U_n(t) - (n-1)U_{n-2}(t)$. Then, a minimal cubature rule of degree $2n-1$ for W_0 has the nodes*

$$X_{n,T} = \left\{(t_{2j,n},t_{2k,n}),(t_{n-2j,n},t_{n-2k,n}) : 0 \le j \ne k \le \ell - 1\right\} \cup \{(h_{k,n},h_{k,n}) : 1 \le k \le n\},$$

(ii) *Let* $j_{k,n} \in (-1,1)$, $1 \le k \le n$ *be zeros of the Gegenbauer polynomial* C_n^2. *Then, a minimal cubature rule of degree* $2n-1$ *for* W_1 *has the nodes*

$$X_{n,U} = \{(u_{2j,n}, u_{2k,n}), (u_{2j-1,n}, u_{2k-1,n}) : 1 \le j \ne k \le \ell\} \cup \{(j_{k,n}, j_{k,n}) : 1 \le k \le n\},$$

Both these cubature rules have $N_{\min} = \dim \Pi_{n-1}^2 + \lfloor \frac{n}{2} \rfloor = 2\ell^2 - 1$ *nodes.*

Proof With $n = 2\ell - 1$, we define the matrix of M_n of size $2\ell \times 2\ell$ by

$$M_n = D_a \left(\begin{bmatrix} -1 & \cdots & -1 \\ \vdots & \cdots & \vdots \\ -1 & \cdots & -1 \end{bmatrix} + \begin{bmatrix} \ell & & & & & \ell \\ & \ddots & & & \iddots & \\ & & \ell & \ell & & \\ & & \ell & \ell & & \\ & \iddots & & & \ddots & \\ \ell & & & & & \ell \end{bmatrix} \right) D_a,$$

where D_a is the diagonal matrix $D_a = \mathrm{diag}\{a, 1, \ldots, 1, a\}$. A careful computation verifies that M_n satisfies (6.21) and (6.22) for both Chebyshev weight functions of the first and second kind. Moreover, by symmetry, it is easy to see that the rank of M is equal to that of the sub-matrix of ℓ at the upper left corner. Moreover, the sum of each row of the sub-matrix is zero, from which it follows that the sub-matrix has rank $\ell - 1$. Consequently, $\mathrm{rank} M_n = \ell - 1 = \lfloor \frac{n}{2} \rfloor$. Furthermore, denote by Z the matrix

$$\begin{bmatrix} 0 & 1 & & & & -1 & -1 & & & & 1 & 0 \\ 0 & -1 & 2 & & & -1 & -1 & & & 2 & -1 & 0 \\ \vdots & \vdots & \ddots & \ddots & & \vdots & \vdots & & \iddots & \iddots & \vdots & \vdots \\ 0 & -1 & \cdots & -1 & \ell-2 & -1 & -1 & \ell-2 & -1 & \cdots & -1 & 0 \\ (\ell-1)a & -1 & \cdots & -1 & -1 & -1 & -1 & -1 & -1 & \cdots & -1 & (\ell-1)a \end{bmatrix}.$$

A careful computation shows that the matrix $V : 2\ell \times \ell - 1$ defined by

$$V^{\mathsf{T}} = Y_a Z, \qquad Y_a = \mathrm{diag}\left\{ \frac{1}{\sqrt{2}}, \frac{1}{\sqrt{2 \cdot 3}}, \ldots, \frac{1}{\sqrt{(\ell-1)\ell}} \right\}$$

satisfies $M = VV^{\mathsf{T}}$, so that M is nonnegative definite. Hence, by Theorem 6.3.2, there is a cubature rule of degree $2n - 1 = 4\ell - 3$ for both W_0 and W_1. Now, the matrix $U : 2\ell \times (\ell + 1)$ defined by

$$U^{\mathsf{T}} = \begin{bmatrix} a^{-1} & & & & & & -a^{-1} \\ & 1 & & & & -1 & \\ & & \ddots & & \iddots & & \\ & & & 1 \quad -1 & & & \\ a^{-1} & 1 & \cdots & 1 \quad 1 & \cdots & 1 & a^{-1} \end{bmatrix}$$

satisfies $M_n U = 0$ and $\operatorname{rank} U = \ell + 1$. Thus, the zeros of $U^{\mathsf{T}}\mathbb{P}_n$ are the nodes of the minimal cubature rules. Writing explicitly, these nodes are zeros of $\ell + 1$ polynomials

$$P_{n-k}(x_1)P_k(x_2) - P_k(x_1)P_{n-k}(x_2), \ 0 \le k \le \ell - 1,$$

where $P_k(t) = T_k(t)$ or $U_k(t)$, and

$$S_n(x_1, x_2) = P_n(x_1) + a^2 \sum_{k=1}^{n-1} P_{n-k}(x_1)P_k(x_2) + P_n(x_2).$$

For $P_k(t) = T_k(t)$ and $a = \sqrt{2}$, it is easy to verify that

$$S_n(x_1, x_2) = \frac{(1 - x_1^2)U_{n-1}(x_1) - (1 - x_2^2)U_{n-1}(x_2)}{x_2 - x_1}.$$

With $n = 2\ell - 1$ and $t_{k,n} = \cos\frac{k\pi}{n}$, $0 \le k \le n$, it follows readily that $S_n(t_{j,n}, t_{k,n}) = 0$ whenever $0 \le k \ne j \le n$. Furthermore, using symmetry, we can also verify that, for $0 \le k \le \ell - 1$, $T_{n-k}(x_1)T_k(x_2) - T_{n-k}(x_2)T_k(x_1)$ vanishes on $(t_{2j,n}, t_{2k,n})$ for $j \ne k$ and $0 \le j, k \le \ell - 1$, and on $(t_{2j-1,n}, t_{2k-1,n})$ for $j \ne k$ and $1 \le j, k \le \ell$. Moreover, using $(1 - x^2)U_{n-1}(x) = \frac{1}{2}(T_{n-1}(x) - T_{n+1}(x))$ and taking limit, we see that

$$S_n(x, x) = \frac{n+1}{2}U_n(x) - \frac{n-1}{2}U_{n-2}(x),$$

which is a quasi-orthogonal polynomial that has n real distinct zeros $h_{j,n}$, $1 \le j \le n$, in $(-1, 1)$. Furthermore, since $T_{n-k}(x_1)T_k(x_2) - T_{n-k}(x_2)T_k(x_1)$ is zero whenever $x_1 = x_2$, they vanish on $(h_{j,n}, h_{j,n})$ as well. This completes the proof that $X_{n,T}$ is the set of zeros of $U^{\mathsf{T}}\mathbb{P}_n$ for the weight W_0.

For $P_k(t) = U_k(t)$ and $a = 1$, the polynomial S_n satisfies the compact form

$$S_n(x_1, x_2) = \frac{U_{n+1}(x_1) - U_{n+1}(x_2)}{2(x_1 - x_2)}.$$

With $n = 2\ell - 1$ and $u_{k,n} = \cos\frac{k\pi}{n+2}$, $0 \le k \le n + 1$, it follows that $S_n(u_{j,n}, u_{k,n}) = 0$ when $j \ne k$. We can also verify that, for $0 \le k \le \ell - 1$, $U_{n-k}(x_1)U_k(x_2) - U_{n-k}(x_2)U_k(x_1)$ vanishes on $(u_{2j,n}, u_{2k,n})$ and $(u_{2j-1,n}, u_{2k-1,n})$ for $j \ne k$ and $1 \le j, k \le \ell$. Moreover, we also have $S_n(t, t) = C_n^2(t)$, the Gegenbauer polynomial with $\lambda = 2$, which has roots $j_{k,n}$, $1 \le j \le n$, in $(-1, 1)$. Hence, as in the case of $P_k = T_k$, $X_{n,U}$ is the set of zeros of $U^{\mathsf{T}}\mathbb{P}_n$ for the weight W_1. $\square$

The weights of the cubature rule are again given by $1/\mathbf{K}_n^*(\mathbf{z}_{j,k}, \mathbf{z}_{j,k})$. The formula of $\mathbf{K}_n^*$ depends on the matrix $V^+ : 2\ell \times \ell - 1$, and the latter is given by

$$(V^+)^{\mathsf{T}} = E_a Z, \qquad E_a = \frac{1}{4}\mathrm{dian}\left\{1, \frac{1}{\binom{3}{2}}, \ldots, \frac{1}{\binom{\ell-1}{2}}, \frac{2}{(\ell-1)(1+(\ell-1)a^2)}\right\}.$$

Hence, one can compute the weights of the cubature by (6.20). However, a closed-form formula for the kernel $\mathbf{K}_n^*$ is not easily available, so neither are the cubature weights.

The nodes of the minimal cubature rule of degree $2n-1=41$ or $n=21$ for the Chebyshev weight W_0 are depicted in Figure 6.2. It contains n nodes on the diagonal $x_1=x_2$.

Figure 6.2 The 241-node minimal cubature rule of degree 41.

6.4.3 Near-Minimal Cubature Rule of Degree $2n-1$ for Odd n

For even n, the minimal cubature rules of degree $2n-1$ in Subsection 6.4.1 are particularly appealing since all its nodes and weights are explicitly given. For odd n, in comparison, the minimal cubature rules in Subsection 6.4.2 are not nearly as explicit. In this subsection, we give another set of cubature rules of $2n-1$ for odd n that are comparable to the minimal cubature rules for even n but have $N_{n,2}+1$ nodes, one more than Möller's lower bound $N_{n,2}$.

This is the case discussed in Proposition 6.3.4 and in Remark 6.2.5, in which we have $\Gamma_1=0$, and there is one polynomial in $\mathbb{Q}_{n+1}$ that is not in the ideal $\langle U^{\mathsf{T}}\mathbb{P}_n\rangle$.

Theorem 6.4.4. *Let $n=2\ell-1$ with $\ell\in\mathbb{N}$. Then, a cubature rule of degree $2n-1$*

for W_0 *is given by*

$$\frac{1}{\pi^2}\int_{[-1,1]^2} f(\mathbf{x})W_0(\mathbf{x})\mathrm{d}\mathbf{x} = \frac{2}{n^2}\sum_{j=0}^{\ell-1}\sum_{k=0}^{\ell-1}\lambda_{j,k}\left[f(t_{2j,n},t_{2k,n}) + f(t_{n-2j,n},t_{n-2k,n})\right],$$

where $\lambda_j = 1$ *if* $1 \le j,k \le \ell$, $\lambda_{0,k} = \lambda_{j,0} = \frac{1}{2}$ *if* $1 \le j,k \le \ell$, *and* $\lambda_{0,0} = \frac{1}{4}$, *and a cubature rule of degree* $2n-1$ *for* W_1 *is given by*

$$\frac{4}{\pi^2}\int_{[-1,1]^2} f(\mathbf{x})W_1(\mathbf{x})\mathrm{d}\mathbf{x} = \frac{8}{(n+2)^2}\sum_{j=1}^{\ell}\sum_{k=1}^{\ell}\Bigg[\left(\sin\tfrac{2k\pi}{n+2}\right)^2\left(\sin\tfrac{2j\pi}{n+2}\right)^2 f(u_{2k,n},u_{2j,n})$$
$$+\left(\sin\tfrac{(2k-1)\pi}{n+2}\right)^2\left(\sin\tfrac{(2k-1)j\pi}{n+2}\right)^2 f(u_{2k-1,n},u_{2j-1,n})\Bigg].$$

Both these cubature rules have $N = \dim\Pi_{n-1}^2 + \lfloor\frac{n}{2}\rfloor + 1 = 2\ell^2$ *nodes.*

Proof With $n = 2\ell-1$, it is easy to verify that the matrix $M_n : 2\ell\times 2\ell$ defined by

$$M_n = \begin{bmatrix} a^2 & & & & & & & a^2 \\ & 1 & & & & & 1 & \\ & & \ddots & & & \iddots & & \\ & & & 1 & 1 & & & \\ & & & 1 & 1 & & & \\ & & \iddots & & & \ddots & & \\ & 1 & & & & & 1 & \\ a^2 & & & & & & & a^2 \end{bmatrix}$$

satisfies (6.21) and (6.22) for both Chebyshev weight functions of the first and second kind, which implies also that $\Gamma_1 = 0$. It is clear that $\operatorname{rank} M_n = \ell = \frac{n}{2} + 1$. Furthermore, a quick verification shows that the matrix $V : 2\ell\times\ell$ defined by

$$V^{\mathsf{T}} = \begin{bmatrix} a & & & & & & a \\ & 1 & & & & 1 & \\ & & \ddots & & \iddots & & \\ & & & 1 \quad 1 & & & \end{bmatrix}$$

satisfies $M = VV^{\mathsf{T}}$, which shows that M is nonnegative definite. Hence, by Theorem 6.3.2, there is a cubature rule of degree $2n-1 = 4\ell-1$ with $N = \dim\Pi_{2\ell-2}^2 + \ell + 1 = 2\ell^2$ nodes for both W_0 and W_1, and its nodes are the zeros of $U^{\mathsf{T}}\mathbb{P}_n$, where the matrix $U : 2\ell\times\ell$ satisfies $M_nU = 0$ and $\operatorname{rank} U = \ell+1$.

Moreover, we can choose the matrix U as

$$U^{\mathsf{T}} = \begin{bmatrix} 1 & & & & & & -1 \\ & 1 & & & & -1 & \\ & & \ddots & & \iddots & & \\ & & & 1 \quad -1 & & & \end{bmatrix}$$

for both W_0 and W_1. Writing explicitly, $U^{\mathsf{T}}\mathbb{P}_n$ consists of ℓ polynomials

$$P_{n-k}(x_1)P_k(x_2) - P_k(x_1)P_{n-k}k(x_2), \quad 0 \le k \le \ell - 1,$$

where $P_k(t) = T_k(t)$ or $U_k(t)$. With $t_{k,n}$ and $u_{k,n}$ in (6.26), the zeros of $T_\ell(t)$ are $t_{2k-1,2\ell}$, $1 \le k \le \ell$, and the zeros of $U_\ell(t)$ are $u_{2k,2\ell}$, $1 \le k \le \ell$. It is easy to verify that the zeros of $U^{\mathsf{T}}\mathbb{P}_n$ for the Chebyshev weight of the first kind consist of

$$X_{n,T} = \{(t_{2j,n}, t_{2k,n}), (t_{n-2j,n}, t_{n-2k,n}) : 0 \le j \le \ell - 1, 0 \le k \le \ell - 1\},$$

and the zeros of $U^{\mathsf{T}}\mathbb{P}_n$ for the Chebyshev weight of the first kind consist of

$$X_{n,U} = \{(u_{2j,n}, u_{2k,n}), (u_{2j-1,n}, u_{2k-1,n}) : 1 \le j \le \ell, 1 \le k \le \ell\}.$$

These are the nodes of the minimal cubature rules. The weights of the cubature rules can again be determined by (6.20) and a closed-form formula of $\mathbf{K}_n^*(\mathbf{z}_{k,j}, \mathbf{z}_{k,j})$, whereas the latter can be derived as in Proposition 6.4.2 as explained. □

The matrices M_n, U, and V resemble those for the minimal cubature rule in Subsection 6.4.2. Indeed, a further computation shows that

$$(V^+)^{\mathsf{T}} V^+ + \tfrac{1}{4} U U^{\mathsf{T}} = \frac{1}{2} I_{2\ell+1} + \frac{1}{4}(-1 + a^{-2}) \begin{bmatrix} 1 & & 1 \\ & \bigcirc & \\ 1 & & 1 \end{bmatrix},$$

which is the same as the one in the proof of Proposition 6.4.2. Hence, we can follow the same proof there to deducc a closed-form formula for $\mathbf{K}_n^*(x, \mathbf{z}_{\mu,\nu})$. We state the result only for the Chebyshev weight of the first kind. Let

$$\mathbf{t}_{2j,2k} = (t_{2j,n}, t_{2k,n}) \quad \text{and} \quad \mathbf{t}_{n-2j,n-2k} = (t_{n-2j,n}, t_{n-2k,n}),$$

so that $\mathbf{t}_{\mu,\nu}$ is a node of the cubature rule for W_0 in Theorem 6.4.4 if μ and ν have the same parity.

Proposition 6.4.5. *Let $n = 2\ell - 1$. Then,*

$$\mathbf{K}_n^*(\mathbf{x}, \mathbf{t}_{\mu,\nu}) = \tfrac{1}{2}\left[\mathbf{K}_n(\mathbf{x}, \mathbf{z}_{\mu,\nu}) + \mathbf{K}_{n-1}(\mathbf{x}, \mathbf{z}_{\mu,\nu})\right] - \tfrac{1}{2}(-1)^{\mu}\left[T_n(x_1) + T_n(x_2)\right]$$

and, for $0 \le k \le \ell - 1$ *and* $0 \le j \le n-1$,

$$\mathbf{K}_n^*\left(t_{2j,n}, t_{2k,n}\right) = \mathbf{K}_n^*\left(t_{n-2j,n}, t_{n-2k,n}\right) = \frac{n^2}{2}\begin{cases} 1 & 1 \le j,k \le \ell-1, \\ 2 & j = 0 \; or \; k = 0, \\ 4 & j = 0 \; or \; k = 0. \end{cases}$$

Comparing these near-minimal cubature rules with the minimal cubature rule of the same degree in Theorem 6.4.3, we see that their nodes differ only on the diagonal $\{(x,x) : -1 \le x \le 1\}$. The minimal cubature rules have n nodes in the interior of the diagonal, whereas the near-minimal rules have $n+1$, with two being the endpoints of the diagonal. The near-minimal cubature rules have the advantage that their weights are given explicitly and so are the interpolation polynomials based on the zeros of the cubature rules, as given by (6.19) and the above proposition.

6.5 Example: Padua Points on the Square

We consider the case $\sigma_0 = r_n^d$, for which $U = 0$ and the nodes of the cubature rule of degree $2n-1$ are the zeros of $\mathbb{Q}_{n+1} = \mathbb{P}_{n+1} + \Gamma_1 \mathbb{P}_n + \Gamma_2 \mathbb{P}_{n-2}$. By Proposition 6.3.4, $\Gamma_1 \ne 0$. In particular, the matrix $B_{n,i}^* = -A_{n,i}\Gamma_1$ in Theorem 6.2.5 is nonzero. In this case, the number of nodes is equal to $\dim \Pi_n^d$.

As we have observed in Remark 6.3.7, this case is of particular interest for polynomial interpolation since the unique Lagrange interpolation polynomial based on the nodes of such a cubature rule is in Π_n^d instead in $\Pi_n^d(V)$ as in the case of Subsections 6.4.1 and 6.4.3.

Theorem 6.5.1. *A cubature rule of degree* $2n-1$ *with* $N = \dim \Pi_n^d$ *nodes is*

$$\frac{1}{\pi^2}\int_{[-1,1]^2} f(\mathbf{x}) W_0(\mathbf{x}) d\mathbf{x} = \sum_{(\xi_j,\eta_k) \in \mathrm{Pad}_n} \lambda_{j,k} f(\xi_j, \eta_k),$$

where the set Pad_n *has cardinality* $\dim \Pi_n^d$ *and is given by, with* $t_{k,n} = \cos \frac{k\pi}{n}$,

$$\begin{aligned} \mathrm{Pad}_n = & \left\{ \left(t_{n-2j,n}, t_{n-2k,n+1}\right) : 0 \le j \le \lfloor \tfrac{n}{2} \rfloor, 0 \le k \le \rfloor \tfrac{n}{2} \rfloor \right\}, \\ & \bigcup \left\{ \left(t_{n-1-2j,n}, t_{n+1-2k,n+1}\right) : 0 \le j \le \lfloor \tfrac{n-1}{2} \rfloor, 0 \le k \le \rfloor \tfrac{n}{2} \rfloor + \delta_n \right\}, \end{aligned}$$

in which $\delta_n = 1$ *if* n *is odd and* $\delta_n = 0$ *if* n *is even, and* $\lambda_{j,k} = 1$ *if the nodes are in the interior of* $[-1,1]^2$, $\lambda_{j,k} = \frac{1}{2}$ *if the nodes are on the boundary but not at the corners of* $[-1,1]^2$ *and* $\lambda_{j,k} = \frac{1}{4}$ *if the nodes are on the corners of* $[-1,1]^2$.

Proof Let $M : (n+1) \times (n+1)$, $\Gamma_1 : (n+2) \times (n+1)$ and $\Gamma_2 : (n+2) \times n$ be defined by

$$M = \begin{bmatrix} 2 & & & \\ & 1 & & \\ & & \ddots & \\ & & & 1 \end{bmatrix}, \quad \Gamma_1 = \begin{bmatrix} 0 & \cdots & \cdots & 0 \\ & & & -\sqrt{2} \\ & & -1 & \\ & \iddots & & \\ -1 & & & \end{bmatrix}, \quad \Gamma_2 = \begin{bmatrix} -1 & & & \\ & 0 & & \\ & & \ddots & \\ & & & 0 \end{bmatrix}.$$

It is straightforward to verify that M and Γ_1 satisfy Eqs. (6.8)–(6.10). Furthermore, the matrix Γ_2 satisfies $(M-I)A_{n-1,i}^{\mathsf{T}} = -A_{n,i}\Gamma_2$ for $i = 1,2$, which implies that Γ_2 satisfies (6.14). Hence, by Theorem 6.2.5, there is a cubature rule of degree $2n-1$ with $\sigma_0 = \operatorname{rank} M = n+1$ or $\dim \Pi_{n-1}^d + \sigma_0 = \dim \Pi_n^d$ nodes, and the nodes are zeros of $\mathbb{Q}_{n+1}$. Writing explicitly in terms of the product Chebyshev polynomials, the nodes are zeros of

$$T_{n+1}(x_1) - T_{n-1}(x_1), \quad T_{n-k}(x_1)T_{k+1}(x_2) - T_k(x_1)T_{n-k}(x_2), \quad 0 \le k \le n.$$

The zeros of these polynomials can be determined from these polynomials, which consist of Pad_n. The cubature weights can be verified by working with (6.20), where a closed-form formula of K_n^* is given in Theorem 6.5.2. □

The number of Padua points Pad_n equals the dimension of Π_n^d. By Theorem 6.2.7, there is a unique polynomial $L_n^* f$ of degree n that interpolates f on the points in Pad_n.

Theorem 6.5.2. *The unique interpolation polynomial $L_n^* f$ of degree n that interpolates f on Pad_n is given by*

$$L_n^* f(\mathbf{x}) = \sum_{(u_j,v_k)\in \mathrm{Pad}_n} \lambda_{j,k} \left(\mathbf{K}_n(\mathbf{x},(u_j,v_k)) - \frac{1}{2} T_n(x_1) T_n(u_j) \right),$$

where $\lambda_{j,k}$ are the cubature weights specified in Theorem 6.5.1.

Proof Let $V = \operatorname{diag}\{\sqrt{2}, 1, \ldots, 1\}$. Then, $M = VV^{\mathsf{T}}$. Furthermore, $V^+ = V^{-1}$. Hence, it follows that

$$[V^+ \mathbb{P}_n(\mathbf{x})]^{\mathsf{T}} V^+ \mathbb{P}_n(\mathbf{y}) = \mathbb{P}_n(\mathbf{x})^{\mathsf{T}} \mathbb{P}_n(\mathbf{y}) - \frac{1}{2} T_n(x_1) T_n(y_1).$$

Consequently, we obtain

$$\mathbf{K}_n^*(\mathbf{x},\mathbf{y}) = \mathbf{K}_{n-1}(\mathbf{x},\mathbf{y}) + [V^+ \mathbb{P}_n(\mathbf{x})]^{\mathsf{T}} V^+ \mathbb{P}_n(\mathbf{y}) = \mathbf{K}_n(\mathbf{x},\mathbf{y}) - \frac{1}{2} T_n(x_1) T_n(y_1).$$

The formula of $L_n^* f$ is given in terms of $\mathbf{K}_n^*(\mathbf{x},\mathbf{y})$ by (6.19). □

An interesting and intriguing feature of the Padua points is their generating curve, given by $T_{n+1}(x_2) - T_n(x_1)$, which is one of the polynomials in $\mathbb{Q}_{n+1}$ and vanishes on all points in Pad_n. Since $T_n(t) = \cos(n \arccos t)$, this polynomial vanishes on the parametric curve

$$\gamma_n(t) = (\cos(n+1)t, \cos nt), \qquad 0 \le t \le \pi,$$

which is a special case of the Lissajous curves, a family of self-intersecting curves. It turns out that the Padua points Pad_n is precisely the set of the self-intersection points of this generating curve. In other words, the set consists of distinct points of $\gamma_n(\frac{j\pi}{n(n+1)})$ for $1 \le j \le n(n+1)$. Hence,

$$\mathrm{Pad}_n = \left\{ (t_{j,n}, t_{j,n+1}), \quad 1 \le n(n+1) \right\}.$$

The points in Pad_n and its generating curve γ_n are depicted in Figure 6.3. The

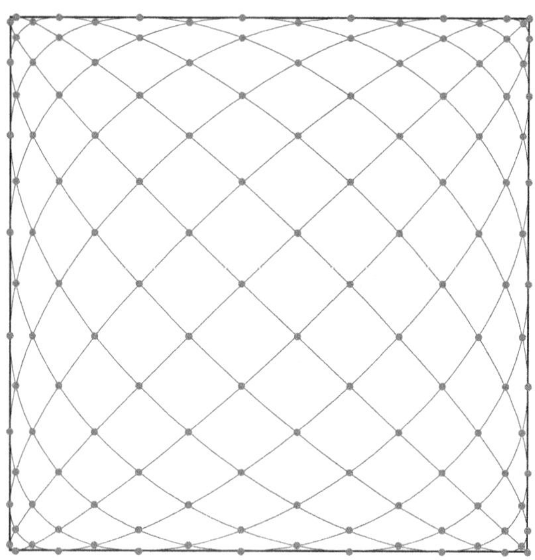

Figure 6.3 Padua nodes and the generating curve for $n = 15$.

Padua points are not unique. Changing the sign of x_1 and/or x_2 leads to three other sets of Padua points.

6.6 Chessboard Data

Given two sets of data $t_0 < t_1 < \cdots < t_n$ and $s_0 < s_1 < \cdots < s_m$, we consider the chessboard data defined by

$$\mathcal{N} = \{(t_i, s_j) : 0 \le i \le n, 0 \le j \le m, \, j - k \equiv \sigma \mod 2\},$$

where $\sigma = 0$ or 1. The points in the previous examples, including the Padua points, are examples of the chessboard data. We consider polynomial interpolations on such data.

We start with a proposition on the polynomial of one variable that is of interest in itself.

Proposition 6.6.1. *Given $t_0 < t_1 < \cdots < t_m$. There is a sequence of polynomials $p_0(t), \ldots, p_k(t)$ that satisfies*

$$p_{m-k}(t_i) = (-1)^i p_k(t_i), \qquad 1 \le i \le d, \quad 0 \le k \le m,$$

and there are sequences a_k and b_k, $0 \le k \le m$, such that

$$p_{k+1}(t) = (a_k t - b_k) p_k(t) - p_{k-1}(t), \quad 0 \le k \le m, \tag{6.27}$$

where $p_{-1}(t) = 0$ and $p_{m+1}(t) = (t - t_0) \ldots (t - t_m)$. Furthermore, the polynomials p_k are orthogonal with respect to the inner product

$$\langle f, g \rangle_m = \sum_{k=0}^{m} \lambda_i f(t_i) g(t_i), \quad \lambda_i = \frac{(-1)^{m-i}}{p'_{m+1}(t_i)} > 0.$$

Proof Let $L_m f(t)$ denote the unique Lagrange interpolation polynomial that satisfies $L_m f(t_i) = f(t_i)$, $1 \le i \le m$. Then,

$$L_m f(t) = \sum_{i=0}^{m} f(t_i) \ell_i(t), \qquad \ell_i(t) = \frac{p_{m+1}(t)}{(t - t_i) p'_{m+1}(t_i)},$$

and $L_m f = f$ if $f \in \Pi_m$. Let $f_0(t)$ be a function that satisfies $f_0(t_i) = (-1)^i$.

Let $p_{m+1}(t) = (t - t_0) \ldots (t - t_m)$ and $p_m = L_m f_0$. By the Newton form of the interpolation polynomials,

$$p_m(t) = f(t_0) + \sum_{k=1}^{m} \alpha_k \xi_k(t), \quad \xi_k(t) = (t - t_0) \ldots (t - t_{k-1}),$$

where $\alpha_k = f_0[t_0, t_1, \ldots, t_{k-1}]$ in terms of the kth divided difference of f_0. We show that there are polynomials $p_0, \ldots, p_{m-1}$ that satisfy (6.27) as well as

$$p_{m-k}(t) = p_k(t) p_m(t) + r_{k-1}(t) p_{m+1}(t), \qquad 1 \le k \le m - 1, \tag{6.28}$$

where r_k is some polynomial of degree $k-1$. To prove the claim, we assume

$$p_k(t) = \sum_{j=0}^{k} \alpha_{j,k}\xi_j(t), \quad 0 \le k \le m-1.$$

For $1 \le k \le m-1$, we consider the recursive relation

$$p_{m-k}(t) = \big(A_{m-k+1}(t-t_{m-k+1}) + C_{m-k+1}\big)p_{m-k+1}(t) - p_{m-k+2}(t), \tag{6.29}$$

which is equivalent to (6.27). We start with p_{m-1}. Let $A_m = \alpha_m^{-1}$ and $C_m = \alpha_{m-1}\alpha_m^{-2}$. Using $(t-t_k)\xi_k(t) = \xi_{k+1}(t)$, it is easy to verify that the right-hand side of (6.29) for $k=1$ is indeed a polynomial of degree $m-1$, which defines p_{m-1}. Moreover, define $p_1(t) = A_m(t-t_m)+C_m$. Then, (6.29) with $k=1$ shows that (6.28) holds for $k=1$ with $r_0(t)=1$. For $k \ge 2$, the same argument shows that there exist A_k and C_k that define p_k recursively in the downward order $k = m-2, m-3, \ldots$. Furthermore, assuming that (6.28) has been established for $k \le j-1$. Then, by (6.29),

$$\begin{aligned}
p_{m-j}(t) &= \big(A_{m-j+1}(t-t_{m-j+1}) + C_{m-j+1}\big)\big(p_{j-1}(t)p_m(t) - r_{j-1}(t)p_{m+1}(t)\big) \\
&\qquad - \big(p_{j-2}(t)p_m(t) + r_{j-2}(t)p_{m+1}(t)\big) \\
&= \big[\big(A_{j-1}(t-t_{m-j+1}) + C_{j-1}\big)p_{j-1}(t) - p_{j-2}(t)\big]p_m(t) + r_j(t)p_{m+1}(t),
\end{aligned}$$

where r_j is a polynomial of degree j, and it agrees with (6.28) if the expression in the square bracket in the right-hand side is $p_j(t)$, which is the case of $j = k+1$ of the relation

$$p_{k+1}(t) = \big(A_k(t-t_{m-k}) + C_k\big)p_k(t) - p_{k-1}(t), \tag{6.30}$$

where $A_j = A_{m-j}$ and $C_j = C_{m-j}$. Let $p_{-1}(t) = 0$ and $p_0(t) = 1$. We see that $p_1(t)$ agree with (6.30) for $k=0$. Thus, the polynomials $p_j(t)$ can also be defined by (6.30). In particular, the identity (6.28) holds, which implies, since $p_{m+1}(t_i) = 0$ for $0 \le i \le n$ and $p_m(t_i) = (-1)^i$, that $p_{m-k}(t_i) = (-1)^i p_k(t_i)$ for $0 \le i \le m$.

It remains to prove the orthogonality. It suffices to show, for each k with $1 \le k \le m$, that $\langle p_{n-k}, p_j\rangle_m = 0$ for $0 \le j < m-k$. By (6.28) and the formula of λ_i, we obtain

$$\begin{aligned}
\langle p_{m-k}, p_j\rangle_m &= (-1)^m \sum_{i=0}^{m} \frac{1}{p'_{m+1}(t_i)} p_k(t_i)p_j(t_i) \\
&= \frac{(-1)^m}{p_{m+1}(t)} \sum_{k=i}^{m} \ell_i(t)(t-t_i)p_k(t_i)p_j(t_i) \\
&= L_m Q(t) = Q(t) = 0,
\end{aligned}$$

where $Q(s) = (t-s)p_k(s)p_j(s)$ and $\deg Q = k+j+1 \le m$, and we have used

$L_m f = f$ whenever f is a polynomial of degree at most m. This completes the proof. □

Theorem 6.6.2. *Given $t_0 < t_1 < \cdots < t_n$ and $s_0 < s_1 < \cdots s_{n+1}$. Let*

$$\begin{aligned} \mathsf{V}_n = & \left\{ (t_{n-2j}, s_{n-2k}) : 0 \le j \le \lfloor \tfrac{n}{2} \rfloor, 0 \le k \le \rfloor \tfrac{n}{2} \rfloor \right\}, \\ & \bigcup \left\{ (t_{n-1-2j}, s_{n+1-2k}) : 0 \le j \le \lfloor \tfrac{n-1}{2} \rfloor, 0 \le k \le \rfloor \tfrac{n}{2} \rfloor + \delta_n \right\}, \end{aligned}$$

where $\delta_n = 1$ if n is odd and $\delta_n = 0$ if n is even. Then, $|\mathsf{V}_n| = \frac{1}{2}(n+1)(n+2)$, and there is a unique polynomial of degree n that interpolates on points in V_n.

Proof We use the polynomials defined in Proposition 6.6.1. Let $p_0, \ldots, p_n$ be polynomials that satisfy $p_{n-k}(t_i) = (-1)^i p_k(t_i)$, $0 \le t_i \le n$, and let $q_0, \ldots, q_{n+1}$ be polynomials that satisfy $q_{n+1-k}(s_j) = (-1)^j q_k(s_j)$, $0 \le j \le n+1$. It is easy to see that polynomials of degree $n+1$ defined by

$$Q_{k,n}(t,s) = p_{n-k}(t) q_{k+1}(s) - p_k(t) q_{n-k}(s), \qquad 0 \le k \le n, \tag{6.31}$$

vanishes on V_n. Together with $p_{n+1}(t) = (t-t_0)\ldots(t-t_n)$, this leads to $n+2$ independent polynomials of degree $n+1$ that vanish on V_n. By Proposition 2.2.4, V_n is not on an algebraic surface of degree n. Hence, by Theorem 2.2.3, there is a unique interpolation polynomial that interpolates on V_n. □

The interpolation polynomial can be given constructively using the orthogonality of the polynomials p_j and q_k. Let $h_{j,1} = \langle p_j, p_j \rangle_n$ and $h_{k,2} = \langle q_k, q_k \rangle_{n+1}$ denote the norm of p_j and q_k with respect to the inner products defined on $\{t_0, \ldots, t_n\}$ and $\{s_0, \ldots, s_{n+1}\}$, respectively. Then the polynomials

$$\hat{p}_k(t) = \frac{p_k(t)}{\sqrt{h_{k,1}}}, \quad 0 \le k \le n, \quad \text{and} \quad \hat{q}_k(t) = \frac{q_k(t)}{\sqrt{h_{k,2}}}, \quad 0 \le k \le n+1,$$

are orthonormal polynomials. We now define the inner product for polynomials of two variables,

$$\langle f, g \rangle = \sum_{i=0}^{n} \sum_{j=0}^{n+1} \lambda_i \mu_j f(t_i, s_j), \qquad f, g \in \Pi_n \times \Pi_{n+1},$$

where λ_i and μ_j are positive numbers defined by

$$\lambda_i = \frac{(-1)^{n-i}}{p'_{n+1}(t_i)} \qquad \text{and} \qquad \mu_j = \frac{(-1)^{n+1-j}}{q'_{n+2}(s_j)}$$

with $p_{n+1}(t) = (t-t_0)\cdots(t-t_n)$ and $q_{n+2}(s) = (s-s_0)\cdots(s-s_{n+1})$. Furthermore, if $\mathbf{z} = (t_i, s_j)$, then we denote $\Lambda_{\mathbf{z}} = \lambda_i \mu_j$. The polynomials

$$P_{j,k}(\mathbf{t}) = \hat{p}_j(t_1) \hat{q}_k(t_2), \qquad 0 \le j \le n, \quad 0 \le k \le n+1,$$

are orthonormal polynomials of two variables with respect to the inner product. Furthermore, for $\mathbf{t} = (t_1, t_2)$ and $\mathrm{b} = (s_1, s_2)$, the kernel

$$\mathbf{K}_n(\mathbf{t}, \mathbf{s}) = \sum_{j=0}^{n} \sum_{k=0}^{j} P_{j,k}(\mathbf{t}) P_{j,k}(\mathbf{s})$$

is the reproducing kernel of Π_n^2.

Proposition 6.6.3. *The unique interpolation polynomial $L_n f$ in Theorem 6.6.2 is given by*

$$L_n f(\mathbf{t}) = \sum_{\mathbf{z} \in \mathsf{V}_n} f(\mathbf{z}) \ell_{\mathbf{z}}(\mathbf{t}), \qquad \ell_{\mathbf{z}}(\mathbf{t}) = \frac{\mathsf{K}_n(\mathbf{t}, \mathbf{z})}{\mathsf{K}_n(\mathbf{z}, \mathbf{z})}.$$

Proof The polynomials $\hat{p}_j$ and $\hat{q}_k$ satisfy the three-term relations by orthogonality. Furthermore, let us define $\hat{p}_{n+1}(t) = p_{n+1}(t)$; then

$$\begin{aligned} t_1 \hat{p}_k(t_1) &= a_{k,1} \hat{p}_{k+1}(t_1) + b_{k,1} \hat{p}_k(t_1) + a_{k-1,1} \hat{p}_{k-1}(t_1), \quad 0 \le k \le n, \\ t_2 \hat{q}_k(t_2) &= a_{k,2} \hat{q}_{k+1}(t_2) + b_{k,2} \hat{q}_k(t_2) + a_{k-1,2} \hat{q}_{k-1}(t_2), \quad 0 \le k \le n, \end{aligned}$$

where $a_{k,i}$ and $b_{k,i}$ can be derived from orthonormality, which gives, for example,

$$a_{k,1} = \langle t_1 p_k, p_{k+1} \rangle_n \quad \text{and} \quad a_{k,2} = \langle t_2 q_k, q_{k+1} \rangle_{n+1},$$

except $a_{n,1}$, or they can be derived by a renormalization of (6.27). The latter shows, in particular, that $a_{n,1} = 1/(A_n \sqrt{h_{n,1}}) = f_0[t_0, t_1, \ldots, t_n] / \sqrt{h_n}$ by the proof of Proposition 6.6.1. The polynomials $P_{j,k}$ of two variables satisfy the three-term relations (3.5). Let $\mathbb{P}_k$ denote the set, or the column vector, of polynomials of degree n,

$$\mathbb{P}_k = \left(P_{k,0}, P_{k-1,1}, \ldots, P_{1,k-1}, P_{0,k}\right), \qquad 0 \le k \le n+1,$$

where we also define $P_{n+1,0}(t, s) = p_{n+1}(t)$. Then they satisfy the three-term relations given by

$$t_i \mathbb{P}_k(\mathbf{t}) = A_{k,i} \mathbb{P}_{k+1}(\mathbf{t}) + B_{k,i} \mathbb{P}_k(\mathbf{t}) + A_{k-1,i}^{\mathsf{T}} \mathbb{P}_{k-1}(\mathbf{t}), \quad i = 1, 2,$$

where the coefficient matrices of $A_{k,i}$ and $B_{k,i}$ are of the form given in Subsection 3.4.1 since $P_{j,k}$ are product orthogonal polynomials, and these relations hold for $0 \le k \le n$. In particular, the matrices $A_{n,1}$ and $A_{n,2}$ are given by

$$A_{n,1} = \begin{bmatrix} a_{n,1} & & \bigcirc & 0 \\ & \ddots & & \vdots \\ \bigcirc & & a_{0,1} & 0 \end{bmatrix} \quad \text{and} \quad A_{n,2} = \begin{bmatrix} 0 & a_{0,2} & & \bigcirc \\ \vdots & & \ddots & \\ 0 & \bigcirc & & a_{n,2} \end{bmatrix}.$$

Moreover, a renormalization of (6.31) shows that the polynomials

$$Q_{k+1,n+1}(\mathbf{t}) = P_{n-k,k+1}(\mathbf{t}) - \gamma_{n-k,k+1}P_{k,n}(\mathbf{t}), \quad \gamma_{k,n+1} = \frac{\sqrt{h_{k,1}h_{n-k,2}}}{\sqrt{h_{n-k,1}h_{k+1,2}}},$$

vanish on V_n for $0 \le k \le n$. In particular, define $Q_{0,n+1} = P_{n+1,0}$ and let

$$\mathbb{Q}_{n+1} = \mathbb{P}_{n+1} + \Gamma\mathbb{P}_n \quad \text{with} \quad \Gamma = \begin{bmatrix} 0 & \cdots & \cdots & 0 \\ \bigcirc & & & \gamma_{n,1} \\ & & \gamma_{n-1,2} & \\ & \iddots & & \\ \gamma_{0,n+1} & & & \bigcirc \end{bmatrix};$$

then $\mathbb{Q}_{n+1}$ vanish on V_n. The kernel $\mathbf{K}_n(\mathbf{t},\mathbf{s})$ satisfies the Christoffel–Darboux formula (3.9), which shows, using $\mathbb{Q}_{n+1}$, that

$$\begin{aligned}(t_i - s_i)\mathbf{K}_n(\mathbf{t},\mathbf{s}) &= [\mathbb{Q}_{n+1}(\mathbf{t}) + \Gamma\mathbb{P}_n(\mathbf{t})]^{\mathsf{T}} A_{n,i}^{\mathsf{T}}\mathbb{P}_n(\mathbf{s}) \\ &\quad - \mathbb{P}_n(\mathbf{s})^{\mathsf{T}} A_{n,i}[\mathbb{Q}_{n+1}(\mathbf{t}) + \Gamma\mathbb{P}_n(\mathbf{t})] \\ &= \mathbb{P}_n(\mathbf{s})^{\mathsf{T}} A_{n,i}\mathbb{Q}_{n+1}(\mathbf{t}) - \mathbb{P}_n(\mathbf{t})^{\mathsf{T}} A_{n,i}\mathbb{Q}_{n+1}(\mathbf{s}),\end{aligned}$$

since $\mathbb{P}_n(\mathbf{t})^{\mathsf{T}}\Gamma^{\mathsf{T}}A_{n,i}^{\mathsf{T}}\mathbb{P}_n(\mathbf{s})$, as a scaler, is equal to its transpose. If both $\mathbf{z}$ and $\mathbf{y}$ are elements of V_n and $\mathbf{z} \ne \mathbf{y}$, then $z_1 - y_1$ and $z_2 - y_2$ cannot both be 0. Hence, the displayed formula shows that $\mathbf{K}_n(\mathbf{z},\mathbf{y}) = 0$ since $\mathbb{Q}_{n+1}$ vanishes on V_n. Furthermore, $\mathbf{K}_n(\mathbf{z},\mathbf{z}) > 0$. Hence, $\ell_{\mathbf{z}}$ satisfies $\ell_{\mathbf{z}}(\mathbf{y}) = \delta_{\mathbf{z},\mathbf{y}}$ if $\mathbf{z},\mathbf{y} \in \mathsf{V}_n$. By the uniqueness of the interpolation polynomial, the proof is complete. □

The result is of interest for polynomial interpolation in two variables. The construction via interpolation also provides a cubature rule of degree $2n-1$ for a square positive linear functional $\mathcal{I}$, defined on Π_{2n-1}^2, that satisfies $\mathcal{I}(fg) = \langle f,g\rangle$ for $f,g \in \Pi_n \times \Pi_{n+1}$. Indeed, since the polynomials $P_{k,j}$ are orthogonal polynomials with respect to $\mathcal{I}$, by Theorem 9.2.2, V_n is the set of nodes of a cubature rule of degree $2n-1$ for $\mathcal{I}$. Moreover, integrating the Lagrange interpolation polynomial shows that the weights of the cubature rule are $1/\mathbf{K}_n(z,z)$, $z \in \mathsf{V}$, so that the cubature rule is positive.

However, from the point of view of numerical integration, such cubature rules are of interest only if the linear functional $\mathcal{I}$ has an integral expression with respect to a simple weight function. Due to its point distribution, the check-board data is most interesting for the rectangular domain, conformal to $[-1,1]^2$. For the Padua points, the linear functional $\mathcal{I}$ is given by the integral associated with the product Chebyshev weight W_0. Apart from the weight functions $(1-t^2)^{\pm\frac{1}{2}}(1-s^2)^{\pm\frac{1}{2}}$ on $[-1,1]^2$, however, we do not know any other candidate for the integral expression of $\mathcal{I}$.

We can also work with other check-board data sets. For example, given $t_0 < t_1 < \cdots < t_n$ and $s_0 < s_1 < \cdots < s_n$ for an even positive integer $n = 2\ell$, let

$$\mathsf{V}_n = \{(t_{2j}, s_{2k-1}), (t_{2k-1}, s_{2j}) : 0 \le j \le \ell, \quad 1 \le k \le \ell\}.$$

Then, $|\mathsf{V}_n| = 2\ell(\ell+1) = \dim \Pi_{n-1}^2 + \lfloor \frac{n}{2} \rfloor$. One example of such a data set is the nodes of a minimal cubature rule in Subsection 6.4.1. Let $p_j(t)$ and $q_k(s)$ be orthogonal polynomials in Proposition 6.6.1 with respect to $\{t_i : 0 \le i \le n\}$ and $\{s_j : 0 \le j \le n\}$, respectively. Then, it is easy to verify that the polynomials

$$P_k(t,s) = p_{2\ell-k}(t) q_k(s) - p_k(t) q_{2\ell-k}(s), \quad 0 \le k \le \ell - 1,$$

vanishes on V_n, and so is the polynomial

$$P_\ell(t,s) = (t - t_1)(t - t_3) \dots (t - t_{2\ell-1})(s - s_1)(s - s_3) \dots (s - s_{2\ell-1}).$$

Together, these polynomials consist of $\ell + 1$ polynomials of degree n. There is a unique polynomial $L_n f$ in the space $\Pi_n^2 \setminus \{P_0, \dots, P_\ell\}$ that interpolates on the points in V_n. This follows from Theorem 9.1.5 since V is the variety of the polynomial ideal $\mathcal{I} = \langle P_0, \dots, P_\ell \rangle$ and $\operatorname{codim} \mathcal{I} = \dim \Pi_{n-1}^2 + \ell = |\mathsf{V}_n|$.

6.7 Notes and Further Results

The proof of Möller's lower bound yields a characterization when the bound is sharp [107], which was stated in terms of the coefficients of recursive relations and the polynomial ideal. The characterization in Theorem 6.2.1 appeared in [167], with an amendment of the proof in [170], from which the characterization of cubature rules attaining Möller's lower bound in Theorem 6.3.2, follows as a special case. The analog of the Christoffel–Darboux formula is a useful byproduct of the proof, which gives a simple proof for the positivity of the cubature rule and an explicit construction of the Lagrange interpolation polynomial on the nodes of the cubature rule.

Although the characterization is satisfactory, it is applicable only if the number of nodes of the minimal cubature rules is less than $\dim \Pi_n^d$. This, however, is far from guaranteed. Indeed, according to linear programming lower bounds in Section 5.4, it is not applicable for a subset of classical weight functions on the unit ball and the simplex when n is large. Furthermore, it is likely to be less applicable when the dimension increases.

A nontrivial example of a cubature rule of degree 9 with 17 nodes is given in [107]. The first infinite family of minimal cubature rules attaining Möller's lower bound appeared in [110], where the cubature rule of degree $2n - 1$ for even n in Theorem 6.4.1 is determined explicitly for W_0. The minimal cubature rules for W_0 and W_1 are determined in [31, 167], and the near-minimal cubature rules in Subsection 6.4.3 are in [167]. The explicit construction of minimal rules for even n and near-minimal rules for odd n are also established later in [14] and [95] via different approaches. The minimal cubature rules for Chebyshev weights on the square were also studied in [139], where the matrix equations are derived from the monomial orthogonal basis rather than the orthonormal basis.

The polynomial interpolation based on the nodes of the minimal cubature rules of degree $2n-1$ when n is even and near-minimal cubature rule when n is odd was studied in [169], making use of the explicit construction of the Lagrange interpolation polynomial on the nodes of the cubature rule, where the weighted L^p convergence was established. The Lebesgue constant of the interpolation process was shown to be in the order of $(\log n)^2$ in [16], which is the optimal order for minimal projection [37, 149]. Subsequently, the Padua points were recognized in [22] and established in [18], via the generating curve, and in [19], via the common zeros of $(2n-1)$-orthogonal polynomials. The Lebesgue constant of the interpolation on the Padua points also has the optimal order $(\log n)^2$.

The Padua nodes are the intersection points of a Lissajous curve. There are generalizations of Lissajous curves in higher dimensions. The polynomial interpolations on such points are studied in [17] for $d = 3$ and in [40] for higher dimensions. However, they do not lead to cubature rules of fewer nodes for polynomials of total degree. The interpolation on the chessboard data is studied in [59, 60] and [23].

7
Further Minimal Cubature Rules

In this chapter, we give minimal cubature rules that attain Möller's lower bound for several families of weight functions, which arise from the curved domain that has appeared in Subsection 3.4.5 by a transformation. The weight functions are on two domains: One is the square $[-1,1]^2$, and the other is unbounded in the complement of the square. In the square, there are two families of fairly general weight functions: One contains the product Chebyshev weight of the first kind and the other contains the product Chebyshev weight of the second kind. The weight functions and their associated orthogonal polynomials are discussed in Section 7.1. Following the characterization of cubature rules of degree $2n-1$, we determine minimal ones for n even in Section 7.2 and near-minimal ones for n odd, for which the number of nodes is one more than the lower bound, in Section 7.3. The minimal cubature rules for n odd are also determined in Section 7.3 but only in the setting of the Jacobi weight. In Section 7.4, we extend the results to two more classes of functions by replacing the product Chebyshev weight of the first and the second kind with mixed type. Finally, minimal cubature rules of even degree are constructed explicitly for a family of weight functions composed with the Chebyshev polynomials in Section 7.5.

7.1 Weight Function and Orthogonal Structure

We start with the definition of the weight functions and the domain on which they are defined. For a weight function w defined on the real line, we denote by $s(w)$ the support set of w.

Definition 7.1.1. Let w be a weight function defined on $s(w)=[-1,1]$ or on $s(w)=[1,+\infty)$. Let $\gamma \geq -\frac{1}{2}$.

(i) If $s(w) = [-1,1]$, then we define W_γ on $s(w)^2 = [-1,1]^2$ by

$$W_\gamma(\mathbf{x}) = w(\cos(\theta_1-\theta_2))w(\cos(\theta_1+\theta_2))\left(1-x_1^2\right)^\gamma\left(1-x_2^2\right)^\gamma\left|x_1^2-x_2^2\right|,$$

where $\mathbf{x} = (x_1,x_2) = (\cos\theta_1,\cos\theta_2)$ with $0 \le \theta_1,\theta_2 \le \pi$.

(ii) If $s(w) = [1,+\infty)$, then we define W_γ on $s(w)^2 = [1,+\infty)^2$ by

$$W_\gamma(\mathbf{x}) = w(\cosh(t_1-t_2))w(\cosh(t_1+t_2))\left(x_1^2-1\right)^\gamma\left(x_2^2-1\right)^\gamma\left|x_1^2-x_2^2\right|,$$

where $\mathbf{x} = (x_1,x_2) = (\cosh t_1,\cosh t_2)$ with $t_1,t_2 \ge 0$.

These weight functions are closely related to the weight functions $\mathcal{W}_\gamma$ defined on the curved domain Ω in (3.40) that is the image of the triangular domain $\triangle = \{\mathbf{x} : x_1 < x_2,\ x \in s(w)^2\}$ under the change of variables (3.39), which we restate for easy reference

$$u_1 = x_1 + x_2 \quad \text{and} \quad u_2 = x_1 x_2, \tag{7.1}$$

Under this mapping, the weight function $\mathcal{W}_\gamma$ is defined by

$$\mathcal{W}_\gamma(\mathbf{u}) = w(x_1)w(x_2)\left(u_1^2-4u_2\right)^\gamma, \quad \mathbf{u} \in \Omega. \tag{7.2}$$

As it is shown in Theorem 4.2.2, the weight functions $\mathcal{W}_{\pm\frac{1}{2}}$ admit the Gauss cubature rules of degree $2n-1$ on the domain Ω. If w is supported on $[-1,1]$, the set Ω is given by

$$\Omega = \left\{\mathbf{u} : |u_1| \le 1+u_2 \le 2,\ 4u_2 \le u_1^2\right\}, \qquad s(w) = [-1,1],$$

which is depicted in Figure 3.1 of Subsection 3.4.5, whereas if w is supported on $[1,\infty)$, then

$$\Omega = \left\{\mathbf{u} : 2 \le u_1 \le 1+u_2,\ 4u_2 \le u_1^2\right\}, \qquad s(w) = [1,\infty).$$

Lemma 7.1.2. *If w is supported on $[-1,1]$ or $[1,+\infty]$, then*

$$W_\gamma(\mathbf{x}) := 4^\gamma \mathcal{W}_\gamma\left(2x_1x_2, x_1^2+x_2^2-1\right)\left|x_1^2-x_2^2\right|, \quad (x_1,x_2) \in s(w)^2, \tag{7.3}$$

so that W_γ is centrally symmetric. Moreover, for integrable function f on Ω,

$$\int_\Omega f(\mathbf{u})\mathcal{W}_\gamma(\mathbf{u})d\mathbf{u} = 4\int_{s(w)^2} f\left(2x_1x_2, x_1^2+x_2^2-1\right)W_\gamma(\mathbf{x})d\mathbf{x}. \tag{7.4}$$

Proof For $(x_1,x_2) \in [-1,1]^2$, we write $x_1 = \cos\theta_1$ and $x_2 = \cos\theta_2$, so that

$$\begin{aligned} 2x_1x_2 &= \cos(\theta_1-\theta_2) + \cos(\theta_1+\theta_2), \\ x_1^2+x_2^2-1 &= \cos(\theta_1-\theta_2)\cos(\theta_1+\theta_2). \end{aligned} \tag{7.5}$$

Likewise, for $(x_1,x_2) \in [1,+\infty)^2$, we write $x_1 = \cosh t_1$ and $x_2 = \cosh t_2$, then

$$\begin{aligned} 2x_1x_2 &= \cosh(t_1 - t_2) + \cosh(t_1 + t_2), \\ x_1^2 + x_2^2 - 1 &= \cosh(t_1 - t_2)\cosh(t_1 + t_2). \end{aligned} \tag{7.6}$$

These identities imply immediately the relation (7.3) and $(2x_1x_2, x_1^2 + x_2^2 - 1) \in \Omega$ for $(x_1,x_2) \in s(w)^2$. Moreover, the mapping $(x_1,x_2) \mapsto (u_1,u_2)$ defined by $u_1 = 2x_1x_2$ and $u_2 = x_1^2 + x_2^2 - 1$ is a bijection between Ω and the domain

$$\left\{(x_1,x_2) \in s(w)^2 : -x_2 < x_1 < x_2\right\},$$

which is a quarter of $s(w)^2$. Moreover, the change of variable has the Jacobian $du = 4|x_1^2 - x_2^2|dx$. Hence, writing the integral in the right-hand side of (7.4) as four times the integral over the quarter defined earlier and then changing variables $\mathbf{x} \mapsto \mathbf{u}$, we arrive at the integral in the left-hand side. The proof is complete. □

Example 7.1.3. If w is the Jacobi weight function $w = w_{\alpha,\beta}$ in (1.11), then, using the identity $(1 \pm \cos(\theta_1 - \theta_2))(1 \pm \cos(\theta_1 + \theta_2)) = (x_1 \pm x_2)^2$, the weight function W_γ becomes $W_{\alpha,\beta,\gamma}$ defined by

$$W_{\alpha,\beta,\gamma}(\mathbf{x}) = |x_1 - x_2|^{2\alpha+1}|x_1 + x_2|^{2\beta+1}\left(1 - x_1^2\right)^\gamma\left(1 - x_2^2\right)^\gamma$$

for $\mathbf{x} \in [-1,1]^2$, where $\alpha, \beta > -1$ and $\gamma \ge -\frac{1}{2}$. Correspondingly, the weight $\mathcal{W}_\gamma$ becomes the weight function $\mathcal{W}_{\alpha,\beta,\gamma}$ in Example 3.4.12.

Example 7.1.4. If w is the generalized Gegenbauer weight function $w = \omega_{\lambda,\mu}$ in (1.18) then, using $\cos(\theta_1 - \theta_2)\cos(\theta_1 + \theta_2) = x_1^2 - x_2^2 - 1$, the weight function W_γ becomes

$$W_{\lambda,\mu,\gamma}(\mathbf{x}) = \left|1 - x_1^2 - x_2^2\right|^{2\mu}\left|x_1^2 - x_2^2\right|^{2\lambda}\left(1 - x_1^2\right)^\gamma\left(1 - x_2^2\right)^\gamma$$

for $\mathbf{x} \in [-1,1]^2$, where $\lambda, \mu > -\frac{1}{2}$ and $\gamma \ge -\frac{1}{2}$.

Example 7.1.5. If w is the Laguerre weight function $w(t) = (t-1)^\alpha e^{-t+1}$ on $[1,+\infty)$, then the weight function W_γ becomes

$$W_{\alpha,\gamma}(\mathbf{x}) = |x_1 - x_2|^{2\alpha}\left(x_1^2 - 1\right)^\gamma\left(x_2^2 - 1\right)^\gamma\left|x_1^2 - x_2^2\right|e^{-2x_1x_2-2}$$

for $\mathbf{x} \in [1,+\infty)^2$, where $\alpha > -1$ and $\gamma \ge -\frac{1}{2}$.

Remark 7.1.6. *As a consequence of* (7.3)*, the weight function W_γ is symmetric under $x \to -x$, so that it can be extended to $(-\infty,-1]^2$ by $W_\gamma(-x_1,-x_2) = W_\gamma(x_1,x_2)$. The extended W_γ defined on $(-\infty,-1] \cup [1,+\infty) = -s(w)^2 \cup s(w)^2$ is centrally symmetric.*

Using the integral relation (7.4), the orthogonal polynomials associated with W_γ on $s(w)^2$ can be deduced in terms of those associated with $\mathcal{W}_\gamma$ and three other related weight functions on the domain Ω. While the weight function $\mathcal{W}_\gamma$ in (7.2) is defined via w, the other weight functions are defined via $(1-x^2)w(x)$, $(1-x)w(x)$, and $(1+x)w(x)$, respectively. More precisely, they are defined by

$$\begin{aligned} \mathcal{W}_\gamma^{(1,1)}(\mathbf{u}) &:= (1-u_1+u_2)(1+u_1+u_2)\mathcal{W}_\gamma(\mathbf{u}), \\ \mathcal{W}_\gamma^{(1,0)}(\mathbf{u}) &:= (1-u_1+u_2)\mathcal{W}_\gamma(\mathbf{u}), \\ \mathcal{W}_\gamma^{(0,1)}(\mathbf{u}) &:= (1+u_1+u_2)\mathcal{W}_\gamma(\mathbf{u}), \end{aligned} \tag{7.7}$$

for $\mathbf{u} \in \Omega$ since $1-u_1+u_2 = (1-x_1)(1-x_2)$ and $1+u_1+u_2 = (1+x_1)(1+x_2)$ under (7.1). Let $\{P_{k,n}^{(\gamma)} : 0 \le k \le n\}$ be an orthogonal basis for $\mathcal{V}_n(\mathcal{W}_\gamma)$ under the inner product

$$\langle f, g\rangle_{\mathcal{W}_\gamma} = \int_\Omega f(\mathbf{u})g(\mathbf{u})\mathcal{W}_\gamma(\mathbf{u})d\mathbf{u}.$$

We further denote by $P_{k,n}^{(\gamma),1,1}$, $P_{k,n}^{(\gamma),1,0}$, $P_{k,n}^{(\gamma),0,1}$ the orthogonal polynomials of degree n with respect to $\langle f,g\rangle_{\mathcal{W}}$ for $\mathcal{W} = \mathcal{W}_\gamma^{(1,1)}, \mathcal{W}_\gamma^{(1,0)}, \mathcal{W}_\gamma^{(0,1)}$, respectively.

Proposition 7.1.7. *An orthogonal basis of $\mathcal{V}_{2n}(W_\gamma)$ is given by*

$$\left\{{}_1Q_{k,2n}^{(\gamma)} : 0 \le k \le n\right\} \bigcup \left\{{}_2Q_{k,2n}^{(\gamma)} : 0 \le k \le n-1\right\},$$

where

$$\begin{aligned} {}_1Q_{k,2n}^{(\gamma)}(\mathbf{x}) &:= P_{k,n}^{(\gamma)}(2x_1x_2, x_1^2+x_2^2-1), && 0 \le k \le n, \\ {}_2Q_{k,2n}^{(\gamma)}(\mathbf{x}) &:= (x_1^2-x_2^2)P_{k,n-1}^{(\gamma),1,1}(2x_1x_2, x_1^2+x_2^2-1), && 0 \le k \le n-1; \end{aligned} \tag{7.8}$$

and an orthogonal basis of $\mathcal{V}_{2n+1}(W_\gamma)$ is given by

$$\left\{{}_1Q_{k,2n+1}^{(\gamma)} : 0 \le k \le n\right\} \bigcup \left\{{}_2Q_{k,2n+1}^{(\gamma)} : 0 \le k \le n\right\},$$

where

$$\begin{aligned} {}_1Q_{k,2n+1}^{(\gamma)}(\mathbf{x}) &:= (x_1+x_2)P_{k,n}^{(\gamma),0,1}(2x_1x_2, x_1^2+x_2^2-1), && 0 \le k \le n, \\ {}_2Q_{k,2n+1}^{(\gamma)}(\mathbf{x}) &:= (x_1-x_2)P_{k,n}^{(\gamma),1,0}(2x_1x_2, x_1^2+x_2^2-1), && 0 \le k \le n. \end{aligned} \tag{7.9}$$

Proof Let $\langle \cdot,\cdot\rangle_W$ denote the inner product associated with W on $s(w)^2$. It is easy to see that the polynomials $Q_{k,n}^{(\gamma)}$ are of degree n. Hence, it suffices to show that they are orthogonal with respect to $\langle \cdot,\cdot\rangle_W$. By Lemma 7.1.2, for $0 \le j \le m$ and $0 \le k \le n$,

$$\left\langle {}_1Q_{k,2n}^{(\gamma)}, {}_1Q_{j,2m}^{(\gamma)} \right\rangle_{W_\gamma} = \left\langle P_{k,n}^{(\gamma)}, P_{j,m}^{(\gamma)} \right\rangle_{\mathcal{W}_\gamma} = \delta_{k,j}\delta_{n,m}.$$

Moreover, under the mapping (7.1), $(x_1^2-x_2^2)^2 W_\gamma(\mathbf{x})d\mathbf{x}$ is equal to a constant multiple of $\mathcal{W}_\gamma^{(1,1)}(\mathbf{u})d\mathbf{u}$, it follows that

$$\left\langle {}_2Q_{k,2n}^{(\gamma)}, {}_2Q_{j,2m}^{(\gamma)} \right\rangle_{W_\gamma} = c\left\langle P_{k,n-1}^{(\gamma),1,1}, P_{j,m-1}^{(\gamma),1,1} \right\rangle_{\mathcal{W}_\gamma^{(1,1)}} = c\delta_{k,j}\delta_{n,m}.$$

Furthermore, setting $F = P_{k,n-1}^{(\gamma),1,1} P_{k,n}^{(\gamma)}$, we obtain

$$\left\langle {}_1Q_{k,2n}^{(\gamma)}, {}_2Q_{j,2m}^{(\gamma)} \right\rangle_{W_\gamma} = \int_{s(w)^2} (x_1^2-x_2^2)F(2x_1x_2, x_1^2+x_2^2-1)W_\gamma(\mathbf{x})d\mathbf{x} = 0,$$

since the integrant changes sign under $(x_1,x_2) \mapsto (x_2,x_1)$, while the domain remains unchanged. The same argument also shows $\langle {}_1Q_{k,2n}^{(\gamma)}, {}_1Q_{j,2m+1}^{(\gamma)} \rangle_{W_\gamma} = 0$ and $\langle {}_1Q_{k,2n}^{(\gamma)}, {}_2Q_{j,2m+1}^{(\gamma)} \rangle_{W_\gamma} = 0$. Together, we have proved the orthogonality of ${}_1Q_{k,2n}^{(\gamma)}$ and ${}_2Q_{k,2n}^{(\gamma)}$ of even degree. The proof for polynomials of odd degree follows similarly. □

For $\gamma = \pm\frac{1}{2}$, an explicit basis of $\mathcal{V}_n(\mathcal{W}_{\pm\frac{1}{2}})$ can be given in terms of orthogonal polynomials $p_n(w)$ of one variable, as shown in Proposition 3.4.11. Thus, by the relation in Proposition 7.1.7, we can state a basis for $W_{\pm\frac{1}{2}}$ in terms of orthogonal polynomials of one variable. Let

$$w^{i,j}(t) = (1-t)^i(1+t)^j w(t), \qquad (i,j) = (1,1),(1,0),(0,1).$$

Proposition 7.1.8. *Let w be a weight function with support $\sigma(w) = [-1,1]$. Let $\mathbf{x} = (\cos\theta_1, \cos\theta_2)$. Then, an orthogonal basis of $\mathcal{V}_{2n}(W_{-\frac{1}{2}})$ is given by, for $0 \le k \le n$ and $0 \le k \le n-1$, respectively,*

$$\begin{aligned} {}_1Q_{k,2n}^{(-\frac{1}{2})}(\mathbf{x}) &= p_n(w;\cos(\theta_1-\theta_2))p_k(w;\cos(\theta_1+\theta_2)) \\ &\quad + p_k(w;\cos(\theta_1-\theta_2))p_n(w;\cos(\theta_1+\theta_2)), \\ {}_2Q_{k,2n}^{(-\frac{1}{2})}(\mathbf{x}) &= (x_1^2-x_2^2)\left[p_{n-1}\left(w^{1,1};\cos(\theta_1-\theta_2)\right)p_k\left(w^{1,1};\cos(\theta_1+\theta_2)\right)\right. \\ &\quad \left. + p_k\left(w^{1,1};\cos(\theta_1-\theta_2)\right)p_{n-1}\left(w^{1,1};\cos(\theta_1+\theta_2)\right)\right], \end{aligned}$$

and an orthonormal basis of $\mathcal{V}_{2n+1}(W_{-\frac{1}{2}})$ is given by, for $0 \le k \le n$,

$$\begin{aligned} {}_1Q_{k,2n+1}^{(-\frac{1}{2})}(\mathbf{x}) &= (x_1+x_2)\left[p_n\left(w^{0,1};\cos(\theta_1-\theta_2)\right)p_k\left(w^{0,1};\cos(\theta_1+\theta_2)\right)\right. \\ &\quad \left. + p_k\left(w^{0,1};\cos(\theta_1-\theta_2)\right)p_n\left(w^{0,1};\cos(\theta_1+\theta_2)\right)\right], \\ {}_2Q_{k,2n+1}^{(-\frac{1}{2})}(\mathbf{x}) &= (x_1-x_2)\left[p_n\left(w^{1,0};\cos(\theta_1-\theta_2)\right)p_k\left(w^{1,0};\cos(\theta_1+\theta_2)\right)\right. \\ &\quad \left. + p_k\left(w^{1,0};\cos(\theta_1-\theta_2)\right)p_n\left(w^{1,0};\cos(\theta_1+\theta_2)\right)\right]. \end{aligned}$$

Moreover, if w is supported on $s(w) = [-1,1]^c$, then the abovementioned relations hold if we replace cos *by* cosh *and setting $(x_1,x_2) = (\cosh t_1, \cosh t_2)$.*

Using the relations (7.5) or (7.6), this proposition follows from the explicit basis (3.44) for $\mathcal{W}_{-\frac{1}{2}}$ and the relations in Proposition 3.4.11. Similarly, an explicit basis for $W_{\frac{1}{2}}$ can be deduced from the basis (3.44) for $\mathcal{W}_{\frac{1}{2}}$.

7.2 Minimal Cubature Rules of Degree $2n-1$: n Even

We show that the minimal cubature rules of degree $2n-1$ for even n exist for $W_{\pm\frac{1}{2}}$ defined in Definition 7.1.1, which is extended to the centrally symmetric setting as described in Remark 7.1.6 when $s(w)=[1,\infty)$, in Subsection 7.2.1. Examples are given in Subsection 7.2.2 using the Jacobi weight.

7.2.1 Minimal Cubature Rules

We consider cubature rules for the centrally symmetric domain denoted by

$$\mathfrak{S}(w) = -s(w)^2 \cup s(w)^2 = \begin{cases} [-1,1]^2, & s(w)=[-1,1], \\ (-\infty,-1]^2 \cup [1,\infty)^2, & s(w)=[1,\infty). \end{cases}$$

With $n=2m$, the number of nodes of the minimal cubature rule of degree $2n-1$ is $2m(m+1)$, which attains Möller's lower bound (5.9). The nodes are the common zeros of polynomials ${}_1Q_{k,2m}^{(\pm\frac{1}{2})}$, $0\le k\le m$, in $\mathcal{V}_n(W_{\pm\frac{1}{2}})$. To describe these zeros, let us write the zeros $t_{k,n}$ of the orthogonal polynomial $p_n(w)$ as

$$t_{k,m} = \begin{cases} \cos\theta_{k,m}, & \text{if } s(w)=[-1,1], \\ \cosh\theta_{k,m}, & \text{if } s(w)=[1,+\infty), \end{cases} \qquad 1\le k\le m, \tag{7.10}$$

where $\theta_{k,m}\in(0,\pi)$ if $s(w)=[-1,1]$ and $\theta_{k,m}>1$ if $s(w)=[1,+\infty)$. Moreover, for $1\le j,k\le m$, we denote

$$\begin{aligned} s_{j,k}^{(m)} &= \cos\frac{\theta_{j,m}-\theta_{k,m}}{2} \quad\text{and}\quad t_{j,k}^{(m)} = \cos\frac{\theta_{j,m}+\theta_{k,m}}{2}, \quad \text{if } s(w)=[-1,1], \\ s_{j,k}^{(m)} &= \cosh\frac{\theta_{j,m}-\theta_{k,m}}{2} \quad\text{and}\quad t_{j,k}^{(m)} = \cosh\frac{\theta_{j,m}+\theta_{k,m}}{2}, \quad \text{if } s(w)=[1,+\infty). \end{aligned} \tag{7.11}$$

Furthermore, we denote by $\lambda_{k,m}(w)$ the weights in the Gauss quadrature rule of degree $2m-1$ for the weight w,

$$\int_{s(w)} f(t)w(t)\mathrm{d}t = \sum_{k=1}^{m} \lambda_{k,m}(w) f(t_{k,m}), \qquad \deg f \le 2m-1.$$

Theorem 7.2.1. *Let $n=2m$ be an even positive integer. Then the weight function $W_{\pm\frac{1}{2}}$, defined in Definition 7.1.1, admit the minimal cubature rule of degree $2n-1=4m-1$ with $2m(m+1)$ nodes. More precisely,*

(i) *for* $W_{-\frac{1}{2}}$ *and* $f \in \Pi^2_{2n-1}$,

$$\int_{\mathbf{S}(w)} f(\mathbf{x}) W_{-\frac{1}{2}}(\mathbf{x}) d\mathbf{x} = \frac{1}{2} \sum_{k=1}^{m} {\sum_{j=1}^{k}}' \lambda_{j,k}^{(m)} \Big[f\left(s_{j,k}^{(m)}, t_{j,k}^{(m)}\right) + f\left(t_{j,k}^{(m)}, s_{j,k}^{(m)}\right) + f\left(-s_{j,k}^{(m)}, -t_{j,k}^{(m)}\right) + f\left(-t_{j,k}^{(m)}, -s_{j,k}^{(m)}\right) \Big], \tag{7.12}$$

where $\lambda_{j,k}^{(m)} = \lambda_{j,m}(w)\lambda_{k,m}(w)$ *and* $\sum'$ *means that the terms for* $j = k$ *are divided by 2.*

(ii) *For* $W_{\frac{1}{2}}$ *and* $f \in \Pi^2_{2n-1}$,

$$\int_{\mathbf{S}(w)} f(\mathbf{x}) W_{\frac{1}{2}}(\mathbf{x}) d\mathbf{x} = \frac{1}{2} \sum_{k=2}^{m+1} \sum_{j=1}^{k-1} \mu_{j,k}^{(m+1)} \Big[f\left(s_{j,k}^{(m+1)}, t_{j,k}^{(m+1)}\right) + f\left(t_{j,k}^{(m+1)}, s_{j,k}^{(m+1)}\right) + f\left(-s_{j,k}^{(m+1)}, -t_{j,k}^{(m+1)}\right) + f\left(-t_{j,k}^{(m+1)}, -s_{j,k}^{(m+1)}\right) \Big], \tag{7.13}$$

where $\mu_{j,k}^{(m+1)} = \lambda_{j,m}(w)\lambda_{k,m}(w)(t_{j,m+1} - t_{k,m+1})^2$.

Proof By (4.6), the common zeros of orthogonal polynomials $P_{k,m}^{(-\frac{1}{2})}$, $0 \le k \le m$, are $\{(t_{k,m} + t_{j,m}, t_{k,m} t_{j,m}) : 1 \le j \le k \le m\}$. Hence, by Proposition 7.1.8 with $\gamma = -\frac{1}{2}$ and (7.5), it follows readily that $\{{}_1Q_{k,2m}^{(-\frac{1}{2})} : 0 \le k \le m\}$ has common zeros

$$\left(s_{j,k}^{(m)}, t_{j,k}^{(m)}\right), \left(-s_{j,k}^{(m)}, -t_{j,k}^{(m)}\right), \left(t_{j,k}^{(m)}, s_{j,k}^{(m)}\right), \left(-t_{j,k}^{(m)}, -s_{j,k}^{(m)}\right),$$

for $1 \le j \le k \le m$, and the cardinality of the set is $4\frac{m(m+1)}{2} = \frac{1}{2}n(n+2)$. These are the nodes of the cubature (7.12) with some $\lambda_{j,k}^{(m)}$, and it is of degree $2n-1$ by Theorem 6.1.2. Moreover, it is easy to verify that

$$2 s_{j,k}^{(m)} t_{j,k}^{(m)} = t_{j,m} + t_{k,m} \quad \text{and} \quad \left(s_{j,k}^{(m)}\right)^2 + \left(t_{j,k}^{(m)}\right)^2 - 1 = t_{j,m} t_{k,m}. \tag{7.14}$$

Hence, for $f(x_1,x_2) = g(2x_1x_2, x_1^2 + x_2^2 - 1)$ for $g \in \Pi_{2m-1}$, the cubature rule (7.12) is equivalent to, by (7.4), the Gauss cubature rule of degree $2m-1$ for the curved domain Ω in (4.8). Consequently, $\lambda_{j,k}^{(m)} = \lambda_{k,n}(w)\lambda_{j,n}(w)$, and we have proved (7.12).

Similarly, by (4.7), the common zeros of orthogonal polynomials $P_{k,m}^{(\frac{1}{2})}$, $0 \le k \le m$, are $\{(t_{k,m+1} + t_{j,m+1}, t_{k,m+1} t_{j,m+1}) : 1 \le j \le k-1, 2 \le k \le m+1\}$. Hence, by (7.8) with $\gamma = \frac{1}{2}$ and (7.5), we see that $\{{}_1Q_{k,n}^{(\frac{1}{2})} : 0 \le k \le m\}$ has common zeros

$$\left(s_{j,k}^{(m+1)}, t_{j,k}^{(m+1)}\right), \left(-s_{j,k}^{(m+1)}, -t_{j,k}^{(m+1)}\right), \left(t_{j,k}^{(m+1)}, s_{j,k}^{(m+1)}\right), \left(-t_{j,k}^{(m+1)}, -s_{j,k}^{(m+1)}\right),$$

for $1 \le j \le k-1 \le m$. The set of these zeros again has the cardinality $\frac{1}{2}n(n+2)$, and it consists of the nodes of the cubature (7.13). The rest of the proof follows as in the case of $\gamma = -\frac{1}{2}$ but making use of the Gauss cubature rule (4.9). □

7.2.2 Example with *w* Being the Jacobi Weight

To illustrate the general result, we consider the example on the domain $[-1,1]^2$ that arises from the Jacobi weight.

Let w be the Jacobi weight function $w_{\alpha,\beta}(t) = (1-t)^\alpha(1+t)^\beta$, $\alpha,\beta > -1$. Then, the minimal cubature rules of the degree $2n-1 = 4m-1$ are for the weight functions

$$W_{\alpha,\beta,-\frac{1}{2}}(\mathbf{x}) = \frac{|x_1-x_2|^{2\alpha+1}|x_1+x_2|^{2\beta+1}}{\sqrt{1-x_1^2}\sqrt{1-x_2^2}}, \quad \alpha,\beta > -1, \tag{7.15}$$

and

$$W_{\alpha,\beta,\frac{1}{2}}(\mathbf{x}) = |x_1-x_2|^{2\alpha+1}|x_1+x_2|^{2\beta+1}\sqrt{1-x_1^2}\sqrt{1-x_2^2}, \quad \alpha,\beta > -1. \tag{7.16}$$

Their $2m(m+2)$ nodes are given in (7.12) and (7.13) with $t_{j,n}(w) = t_{j,n}^{(\alpha,\beta)}$ being the nodes of the Jacobi polynomial $P_n^{(\alpha,\beta)}$.

When $\alpha = \beta = -\frac{1}{2}$, these are minimal cubature rules for the Chebyshev weight functions discussed in Subsection 6.4.1. For $\alpha > -\frac{1}{2}$ and/or $\beta > -\frac{1}{2}$, the nodes are propelled away from the diagonal(s) of the square $[-1,1]^2$, as shown in Figure 7.1 for the weight functions

$$W_{\frac{1}{2},\frac{1}{2},-\frac{1}{2}}(\mathbf{x}) = \frac{|x_1^2-x_2^2|}{\sqrt{1-x_1^2}\sqrt{1-x_2^2}} \quad \text{and} \quad W_{\frac{1}{2},-\frac{1}{2},-\frac{1}{2}}(\mathbf{x}) = \frac{|x_1-x_2|}{\sqrt{1-x_1^2}\sqrt{1-x_2^2}}.$$

The diagonals are where the singularity appears in $|x-y|^{2\alpha+1}|x+y|^{2\beta+1}$ of the weight function $\mathcal{W}_{\alpha,\beta,-\frac{1}{2}}$. The nodes in Figure 7.1 display a region around the diagonal(s) that contains no nodes. The region becomes larger so that the nodes will be further away from the diagonal(s) if the values of α and β increase. We can precisely determine the curves that bound the region that contains no nodes. Let $\cos\theta_{k,m}^{(\alpha,\beta)}$ be the zeros of the Jacobi polynomial $P_m^{(\alpha,\beta)}$ as before with the order $0 < \theta_{1,m}^{(\alpha,\beta)} < \theta_{2,m}^{(\alpha,\beta)} < \cdots < \theta_{m,m}^{(\alpha,\beta)} < \pi$.

Proposition 7.2.2. *Let $\alpha,\beta > -1$. For $0 < \theta_{1,m} < \theta_{2,m} < \ldots < \theta_{m,m} < \pi$, define parametric curves*

$$S_1^{\pm} := \{\pm(\cos\theta, \cos(\theta-\theta_{1,m})) : \theta_{1,m} < \theta < \pi\},$$
$$S_2^{\pm} := \{\pm(\cos\theta, \cos(\theta-\theta_{m,m})) : 0 < \theta < \theta_{m,m}\}.$$

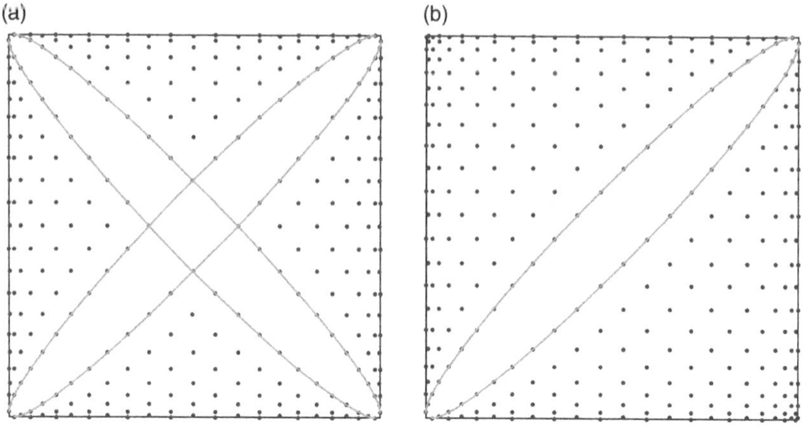

Figure 7.1 The 312-node minimal cubature rule of degree 47 for $W_{\frac{1}{2},\frac{1}{2},-\frac{1}{2}}$ (a) and $W_{\frac{1}{2},-\frac{1}{2},-\frac{1}{2}}$ (b).

Let $\theta_{j,m} = \theta_{j,m}^{(\alpha,\beta)}$. Then, all nodes of the minimal cubature rule of degree $4m-1$ for $W_{\alpha,\beta,-\frac{1}{2}}$ are in the region bounded by $S_1^{\pm}$ and $S_2^{\pm}$ and the boundary of $[-1,1]^2$.

Proof For $\theta_{k,m} = \theta_{k,m}^{(\alpha,\beta)}$, define $\xi_{j,k} = (\theta_{j,m} + \theta_{k,m})/2$, $\eta_{j,k} = (\theta_{j,m} - \theta_{k,m})/2$. Then, $t_{j,k} = \cos\xi_{j,k}$ and $s_{j,k} = \cos\eta_{j,k}$. Hence, the arguments of the nodes in $X_m^{(1)} := \{(t_{j,k}, s_{j,k}), 1 \le j \le k \le m\}$ consist of the set $\{(\xi_{j,k}, \eta_{j,k}) : 1 \le j \le k \le m\}$, which is roughly a triangle with $(\xi_{m,m}, \eta_{m,m}), \ldots, (\xi_{2,2}, \eta_{2,2}), (\xi_{1,1}, \eta_{1,1})$ as one side and $(\xi_{1,m}, \eta_{1,m})$ as the vertex opposite to that side. The two other sides of the triangle are $(\xi_{j,m}, \eta_{j,m})$, $1 \le j \le m$, and $(\xi_{1,k}, \eta_{1,k})$, $1 \le k \le m$, respectively. Since

$$\xi_{j,m} = (\theta_{j,m} + \theta_{m,m})/2 = (\theta_{j,m} - \theta_{m,m})/2 + \theta_{m,m} = \eta_{j,m} + \theta_{m,m},$$
$$\xi_{1,k} = (\theta_{1,m} + \theta_{k,m})/2 = -(\theta_{1,m} - \theta_{k,m})/2 + \theta_{1,m} = -\eta_{1,k} + \theta_{1,m},$$

we see that $X_m^{(1)}$ is bounded by the parametric curve $(\cos\theta, \cos(\theta - \theta_{m,m}))$ and $(\cos\theta, \cos(\theta - \theta_{1,m}))$, or the curves S_1^{+} and S_2^{+}. The other three subsets of the nodes can be considered similarly. □

The oval curves in Figure 7.1 are depicted with the help of the formulas of $S_1^{\pm}$ and $S_2^{\pm}$.

7.3 Minimal and Near-Minimal Rules of Degree $2n-1$: n Odd

In this section, we consider the cubature rules of degree $2n-1$ when n is odd. We choose $n=2m-1$ so that $2n-1=4m-3$. In Subsection 7.3.1, we construct explicitly near-minimal cubature rules with $2m^2$ nodes, just one more than the minimal cubature rule that attains Möller's lower bound (5.9), for the weight function $W_{\pm\frac{1}{2}}$ with a generic weight function w. In Subsection 7.3.2, we establish minimal cubature rules for $W_{\pm\frac{1}{2}}$ but only for w being a Jacobi weight since the argument requires further information pertinent to orthogonal polynomials.

7.3.1 Near-Minimal Cubature Rules for $W_{\pm\frac{1}{2}}$

In contrast to the case $n=2m$ that uses the Gauss quadrature rules, these cubature rules are derived from the Gauss–Radau quadrature rules (1.24) or (1.25).

Let w be a weight function defined on either $[-1,1]$ or $[1,+\infty)$. We define

$$w_1(t)=\begin{cases}(1-t)w(t), & s(w)=[-1,1],\\ (t-1)w(t), & s(w)=[1,+\infty).\end{cases}$$

Let $t_{k,m}$, $1\le k\le m$ be the zeros of $p_m(w_1,t)$, and we adopt the notation (7.10) and (7.11). Furthermore, we denote by $\mu_{k,m}(w)$ the weights in the Gauss–Radau quadrature rule of degree $2m-2$ for the weight w,

$$\int_{s(w)} f(t)w(t)\mathrm{d}t=\mu_{0,m}f(1)+\sum_{k=1}^{m}\mu_{k,m}(w)f(t_{k,m}),\qquad \deg f\le 2m-2.$$

To unify the notation, we define $t_{0,m}=1$.

Theorem 7.3.1. *Let $n=2m+1$ be an odd integer. Then, the weight function $W_{\pm\frac{1}{2}}$ of Definition 7.1.1 has near-minimal cubature rules of degree $4m+1$ that has $2(m+1)^2$ nodes, one more than that of the minimal cubature rule. More precisely,*

(i) *For $W_{-\frac{1}{2}}$ and $f\in\Pi^2_{2n-1}$,*

$$\int_{\mathbf{S}(w)} f(\mathbf{x})W_{-\frac{1}{2}}(\mathbf{x})\mathrm{d}\mathbf{x}=\frac{1}{2}\sum_{k=0}^{m}{\sum_{j=0}^{k}}'\mu_{k,m}\mu_{j,m}\tag{7.17}$$
$$\times\Big[f\big(s_{j,k}^{(m)},t_{j,k}^{(m)}\big)+f\big(t_{j,k}^{(m)},s_{j,k}^{(m)}\big)+f\big(-s_{j,k}^{(m)},-t_{j,k}^{(m)}\big)+f\big(-t_{j,k}^{(m)},-s_{j,k}^{(m)}\big)\Big],$$

where $\sum'$ means that the terms for $j=k$ are divided by 2.

(ii) *For* $W_{\frac{1}{2}}$ *and* $f \in \Pi^2_{2n-1}$,

$$\int_{\mathbf{S}(w)} f(\mathbf{x})W_{\frac{1}{2}}(\mathbf{x})d\mathbf{x} = \frac{1}{2}\sum_{k=1}^{m+1}\sum_{j=0}^{k-1}\mu_{k,m+1}\mu_{j,m+1}(\cos\theta_{j,n} - \cos\theta_{k,n})^2 \times\Big[f\Big(s^{(m+1)}_{j,k}, t^{(m+1)}_{j,k}\Big) + f\Big(t^{(m+1)}_{j,k}, s^{(m+1)}_{j,k}\Big) + f\Big(-s^{(m+1)}_{j,k}, -t^{(m+1)}_{j,k}\Big) + f\Big(-t^{(m+1)}_{j,k}, -s^{(m+1)}_{j,k}\Big)\Big]. \tag{7.18}$$

Proof We consider $W_{-\frac{1}{2}}$ first. We start with the product Gauss–Radau cubature rule

$$\int_{s(w)^2} f(\mathbf{x})w(x_1)w(x_2)d\mathbf{x} = \sum_{j=0}^{m}\sum_{k=0}^{m}\mu_{j,m}\mu_{k,m}f(t_{j,m}, t_{k,m}),$$

which holds for $f \in \Pi^2_{2m} \times \Pi^2_{2m}$ and apply it on the functions $f(x_1 + x_2, x_1x_2)$, which implies, by symmetry and changing variables $u_1 = x_1 + x_2$ and $u_2 = x_1x_2$ on $\{x \in s(w)^2 : x_1 < x_2\}$, that

$$\int_{\Omega} f(\mathbf{u})\mathcal{W}_{-\frac{1}{2}}(\mathbf{u})d\mathbf{u} = 2\sum_{k=0}^{m}\sum_{j=0}^{k}{}' \mu_{j,m}\mu_{k,m}f(u_{j,k}, v_{j,k}), \tag{7.19}$$

where $u_{j,k} = t_{j,m} + t_{k,m}$ and $v_{j,k} = t_{j,m}t_{k,m}$, for $\mathcal{W}_{-\frac{1}{2}}$ defined as in (7.2) and this cubature rule is of degree $2m - 2$. Consequently, by (7.4) and (7.14), we obtain

$$\int_{\mathbf{S}(w)} f\big(2x_1x_2, x_1^2 + x_2^2 - 1\big)W_{-\frac{1}{2}}(\mathbf{x})d\mathbf{x} = 2\sum_{k=0}^{m}\sum_{j=0}^{k}{}' \mu_{j,n}\mu_{k,n}f(s_{j,k}, t_{j,k}), \tag{7.20}$$

for all $f \in \Pi^2_{2m-1}$. By the Sobolev theorem on invariant cubature rules, and (7.5) or (7.6), it follows that the abovementioned cubature rule implies (7.17) for $f \in \Pi^2_{4m}$. Moreover, the cubature rule also holds for $f \in \Pi^2_{4m+1}$ since it is invariant under $\mathbf{x} \mapsto -\mathbf{x}$ and so is the weight function $W_{-\frac{1}{2}}$ since it is centrally symmetric. When $j = 0$, $s_{0,k} = t_{0,k}$, so that $(t_{0,k}, s_{0,k}) = (s_{0,k}, t_{0,k})$ for $0 \le k \le m$. Hence, the number of nodes of the cubature rule is

$$2m(m+1) + 2(m+1) = 2(m+1)^2 = N_{\min}(2m+1) + 1.$$

This completes the proof in the case of $W_{-\frac{1}{2}}$.

The proof for (7.18) can be carried out similarly. We start by applying the cubature rule (7.19) of degree $2n + 1$ on polynomials $f(\mathbf{u})(u_1^2 - 4u_2)$, where $f \in \Pi^2_{n-1}$, and using $\mathcal{W}_{\frac{1}{2}}(\mathbf{u}) = \mathcal{W}_{-\frac{1}{2}}(\mathbf{u})(u_1^2 - 4u_2)$ and $u_{j,k}^2 - 4v_{j,k} = (\cos\theta^{(m+1)}_{j,n} - \cos\theta^{(m+1)}_{k,n})^2$ to obtain an analog of (7.19) for $\mathcal{W}_{\frac{1}{2}}$ that has the summation index

$j \neq k$, which leads to, by (7.4) and (7.14) again, an analog of (7.20) for $W_{\frac{1}{2}}$. The rest of the proof follows exactly as in the case of $W_{-\frac{1}{2}}$. □

By Theorem 6.2.1, the nodes of the near-minimal cubature rules are zeros of $U^{\mathsf{T}}\mathbb{P}_n$ and $\mathbb{Q}_{n+1}$, and the latter contains exactly one polynomial not in $\langle U^{\mathsf{T}}\rangle$ as stated in Remark 6.3.6. The polynomials in $U^{\mathsf{T}}\mathbb{P}_n$ are determined in the next proposition.

Proposition 7.3.2. *For $n = 2m+1$, the nodes of the cubature rules of degree $2n-1 = 4m-1$ in (7.17) are zeros of the polynomial ideal generated by ${}_2Q_{k,2m+1}^{(-\frac{1}{2})}$, $0 \le k \le m$, defined in (7.9).*

The proof follows immediately from the formula of ${}_2Q_{k,2m+1}^{(-\frac{1}{2})}$ given in Proposition 7.1.8, where we need the factor $x_1 - x_2$ in the polynomials when $j = 0$ since $s_{0,k} = t_{0,k}$.

Example 7.3.3. We consider the case when w is the Jacobi weight $w_{\alpha,\beta}(t) = (1-t)^\alpha(1+t)^\beta$, $\alpha,\beta > -1$, so that $W_{-\frac{1}{2}}$ is for the weight functions $W_{\alpha,\beta,-\frac{1}{2}}$ defined in (7.15).

When $\alpha = \beta = -\frac{1}{2}$, these are near-minimal cubature rules for the Chebyshev weight functions discussed in Subsection 6.4.3. For $\alpha > -\frac{1}{2}$ and/or $\beta > -\frac{1}{2}$, the nodes are propelled away from the diagonal(s) of the square $[-1,1]^2$, as shown

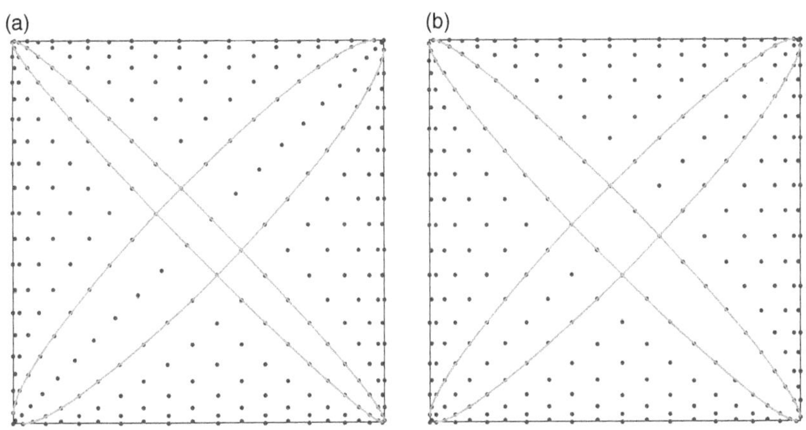

Figure 7.2 The 242-node near-minimal cubature rule of degree 41 for $W_{\frac{1}{2},-\frac{1}{2},-\frac{1}{2}}$ (a) and $W_{\frac{1}{2},0,-\frac{1}{2}}$ (b).

in Figure 7.2 for the weight functions

$$W_{\frac{1}{2},-\frac{1}{2},-\frac{1}{2}}(\mathbf{x}) = \frac{|x_1-x_2|^2}{\sqrt{1-x_1^2}\sqrt{1-x_2^2}} \quad \text{and} \quad W_{\frac{1}{2},0,-\frac{1}{2}}(\mathbf{x}) = \frac{|x_1-x_2|^2|x_1+x_2|}{\sqrt{1-x_1^2}\sqrt{1-x_2^2}},$$

where the curves are defined as in Proposition 7.2.2. Compared with the case of even n, the distributions of the nodes for these cubature rules are not mirrored if the values of α and β are switched. The nodes will be further away from the diagonals if the values of α and β increase, but there will always be n nodes on the diagonal $\{(x,x) : -1 \le x \le 1\}$.

7.3.2 Minimal Cubature Rules of Degree $2n-1$: n Odd

For the Chebyshev weight, minimal cubature rules of degree $2n-1$ for n odd are discussed in Subsection 6.4.2. Taking a cue from the structure there, one expects that the minimal cubature rule for $W_{\pm\frac{1}{2}}$ should have the same nodes as those of near-minimal cubature rules except on the diagonal $\{(x,x) : -1 \le x \le 1\}$. To identify the nodes on the diagonal, however, we need further information on orthogonal polynomials $p_n(w^{0,1})$ and $p_n(w^{(1,0)})$ in order to work with the basis in Proposition 7.1.8, which however may not hold for a generic w. Consequently, we will restrict to $w = w_{\alpha,\beta}$ being the Jacobi weight with $\alpha, \beta > -1$.

For $w = w_{\alpha,\beta}$, the weight function is $W_{\alpha,\beta,\pm\frac{1}{2}}$ on $[-1,1]^2$ defined in (7.15), as the example in the previous subsection. Its associate orthogonal polynomials are given in Proposition 7.1.8, with $p_n = P_n^{(\alpha,\beta)}$ and $p_n(w^{i,j}) = P_n^{(\alpha+i,\beta+j)}$. We denote the orthogonal polynomials on the cube by ${}_iQ_{k,n}^{\alpha,\beta,\pm\frac{1}{2}}$. By Theorem 5.2.3, the nodes of the minimal cubature rule of degree $2n-1$ with $n = 2m+1$ are zeros of $m+2$ orthogonal polynomials of degree n. We again expect that the off-diagonal nodes of the minimal cubature rule agree with those of the near-cubature rule and, as a consequence, they are common zeros of $\{{}_2Q_{k,2m+1}^{\alpha,\beta,\pm\frac{1}{2}} : 0 \le k \le m\}$ by Proposition 7.3.2. This collection of polynomials consists of $m+1$ polynomials, and its zeros are infinite since all vanish on the $x_2 = x_1$ diagonal. Thus, we need one more orthogonal polynomial of degree $n = 2m+1$ that vanishes on the off-diagonal nodes and also on $2m+1$ nodes on the diagonal.

Let $t_{k,m} = t_{k,m}^{\alpha+1,\beta}$ be the zeros of $P_m^{(\alpha+1,\beta)}$. We adopt the notation in (7.10) and (7.11). Furthermore, we define

$$S_m^{(\pm\frac{1}{2})}(t) = P_m^{(\alpha,\beta+1)}(1) P_m^{(\alpha+1,\beta)}(2t^2-1) \mp P_m^{(\alpha,\beta+1)}(2t^2-1) P_m^{(\alpha+1,\beta)}(1).$$

This polynomial has $2m$ distinct zeros in $(-1,1)$, as shown in Lemma 7.3.5.

Theorem 7.3.4. *Let $n = 2m+1$. Then, the minimal cubature rules of degree $2n-1$ exist for $W_{\alpha,\beta,\pm\frac{1}{2}}$. More precisely,*

(i) *Let* $\xi_{k,2m+1} \in (-1,1)$, $1 \le k \le 2m+1$, *be the zeros of* $tS_m^{(-\frac{1}{2})}(2t^2-1)$. *Then, a minimal cubature rule of degree* $2n-1$ *for* $W_{\alpha,\beta,-\frac{1}{2}}$ *has the nodes*

$$X_n^{(-\frac{1}{2})} = \Big\{\Big(s_{j,k}^{(m)}, t_{j,k}^{(m)}\Big), \Big(t_{j,k}^{(m)}, s_{j,k}^{(m)}\Big), \Big(-s_{j,k}^{(m)}, -t_{j,k}^{(m)}\Big), \Big(-t_{j,k}^{(m)}, -s_{j,k}^{(m)}\Big) : \\ 1 \le j \le k \le m\Big\} \bigcup \big\{(\xi_{k,2m+1}, \xi_{k,2m+1}) : 1 \le k \le 2m+1\big\}.$$

(ii) *Let* $\eta_{k,2m+1} \in (-1,1)$, $1 \le k \le 2m+1$, *be the zeros of* $S_{m+1}^{(\frac{1}{2})}(2t^2-1)$ *inside* $(-1,1)$. *A minimal cubature rule of degree* $2n-1$ *for* $W_{\alpha,\beta,\frac{1}{2}}$ *has the nodes*

$$X_n^{(\frac{1}{2})} = \Big\{\Big(s_{j,k}^{(m+1)}, t_{j,k}^{(m+1)}\Big), \Big(t_{j,k}^{(m+1)}, s_{j,k}^{(m+1)}\Big), \Big(-s_{j,k}^{(m+1)}, -t_{j,k}^{(m+1)}\Big), \\ \Big(-t_{j,k}^{(m+1)}, -s_{j,k}^{(m+1)}\Big) : \quad 0 \le j < k \le m+1\Big\} \\ \bigcup \big\{(\eta_{k,2m+1}, -\eta_{k,2m+1}) : 1 \le k \le 2m+1\big\}.$$

Both these cubature rules have $\dim \Pi_{n-1}^2 + \lfloor \frac{n}{2} \rfloor = 2(m+1)^2 - 1$ *nodes.*

Proof We first consider the case $\mathcal{W}_{\alpha,\beta,-\frac{1}{2}}$. For $t_{k,m} = t_{k,m}^{\alpha+1,\beta} = \cos\theta_{k,m}$, the orthogonal polynomials ${}_2Q_{k,2m+1}^{\alpha,\beta,-\frac{1}{2}}$, $0 \le k \le m$, of degree n vanish on $X_n^{(-\frac{1}{2})}$ and we need one more, which has to involve ${}_1Q_{k,2m+1}^{\alpha,\beta,-\frac{1}{2}}$. We can expand the Jacobi polynomial $P_n^{(\alpha,+1,\beta)}$ in terms of $P_k^{(\alpha,\beta+1)}$,

$$P_n^{(\alpha+1,\beta)}(t) = \sum_{k=0}^{n} d_k P_k^{(\alpha,\beta+1)}(t).$$

By Proposition 7.1.8 with $p_n(w^{1,0}) = P_n^{(\alpha+1,\beta)}$ and $p_n(w^{0,1}) = P_n^{(\alpha,\beta+1)}$, we see that

$$\begin{aligned} q_m(\mathbf{x}) &= \sum_{k=0}^{n} d_{k\,1} Q_{k,2m+1}^{\alpha,\beta,-\frac{1}{2}}(\mathbf{x}) \\ &= (x_1 + x_2)\Big[P_m^{(\alpha,\beta+1)}(\cos(\theta-\phi))P_m^{(\alpha+1,\beta)}(\cos(\theta+\phi)) \\ &\quad + P_m^{(\alpha,\beta+1)}(\cos(\theta+\phi))P_m^{(\alpha+1,\beta)}(\cos(\theta-\phi))\Big], \end{aligned}$$

so that q_m is an element of $\mathcal{V}_{2m+1}(W_{\alpha,\beta,-\frac{1}{2}})$. Since $P_n^{(\alpha+1,\beta)}(\cos(\theta \pm \phi))$ vanishes if $\theta = \pm\frac{\theta_{j,m}-\theta_{k,m}}{2}$ and $\phi = \pm\frac{\theta_{j,m}+\theta_{k,m}}{2}$, q_m vanishes on non-diagonal points of $X_m^{(-\frac{1}{2})}$. Moreover, as we shall see in Lemma 7.3.5, the polynomial

$$\begin{aligned} q_m(x,x) &= 2x\Big[P_m^{(\alpha,\beta+1)}(1)P_m^{(\alpha+1,\beta)}(\cos 2\theta) + P_m^{(\alpha,\beta+1)}(\cos 2\theta)P_m^{(\alpha+1,\beta)}(1)\Big] \\ &= 2x\Big[P_m^{(\alpha,\beta+1)}(1)P_m^{(\alpha+1,\beta)}(2x^2-1) + P_m^{(\alpha,\beta+1)}(2x^2-1)P_m^{(\alpha+1,\beta)}(1)\Big] \end{aligned}$$

has $2m+1$ distinct zeros, so that $q_m(x,y)$ has $2m+1$ zeros of the form $(\xi_{j,m},\xi_{j,m})$, of which one is $(0,0)$. Since ${}_2Q_{k,2m+1}^{\alpha,\beta,-\frac{1}{2}}$ contains a factor $x_1 - x_2$, we see it also has $\{(\xi_{j,m},\xi_{j,m}) : 0 \le j \le m\}$ as zeros. Consequently, there are $2m+2$ orthogonal polynomials of degree $n = 2m+1$ that vanish on $X_n^{(-\frac{1}{2})}$, and the cubature rules have $\left|X^{(-\frac{1}{2})}\right| = 2(m+1)^2 - 1$ elements. This completes the proof for $W_{\alpha,\beta,-\frac{1}{2}}$.

The proof for the weight function $W_{\alpha,\beta,\frac{1}{2}}$ is similar. For $s_{j,k}^{(m+1)}$ and $t_{j,k}^{(m+1)}$ defined in terms of $t_{k,m} = t_{k,m+1}^{(\alpha+1,\beta)}$ for $1 \le k \le m+1$, the orthogonal polynomials ${}_2Q_{k,2m+1}^{\alpha,\beta,\frac{1}{2}}$, $0 \le k \le m$, vanish on $X_n^{(\frac{1}{2})}$. The additional orthogonal polynomial that vanishes on the non-diagonal elements of $X_n^{(\frac{1}{2})}$ is

$$q_m(\mathbf{x}) = \frac{(x_1 - x_2)}{\sin\theta \sin\phi}\Big[P_{m+1}^{(\alpha,\beta+1)}(\cos(\theta-\phi))P_{m+1}^{(\alpha+1,\beta)}(\cos(\theta+\phi)) \\ -P_{m+1}^{(\alpha,\beta+1)}(\cos(\theta+\phi))P_{m+1}^{(\alpha+1,\beta)}(\cos(\theta-\phi))\Big],$$

which is a linear combination of ${}_1Q_{k,2m+1}^{\alpha,\beta,\frac{1}{2}}$ as in the case of $\gamma = -\frac{1}{2}$ and

$$q_m(t,-t) = \frac{2t}{(\sin\theta)^2} \\ \times\Big[P_{m+1}^{(\beta+1,\alpha)}(2t^2-1)P_{m+1}^{(\beta,\alpha+1)}(1) - P_{m+1}^{(\beta+1,\alpha)}(1)P_{m+1}^{(\beta,\alpha+1)}(2t^2-1)\Big],$$

where we have used $P_m^{(a,b)}(-t) = (-1)^m P_m^{(b,a)}(t)$. The nominator of $q_m(t,-t)$ has $2m+2$ zeros, but two are ± 1, which are not zeros of $q_m(t,-t)$ because of $\sin^2\theta = 1-x^2$ in the denominator. By Lemma 7.3.5, $q_m(t,-t)$ has $2m+1$ zeros, which is the diagonal part of $X_n^{(\frac{1}{2})}$. □

Lemma 7.3.5. *For $\alpha,\beta > -1$ and $m = 1,2,\ldots$,*

(1) The Jacobi polynomials $P_m^{(\alpha,\beta+1)}$ and $P_m^{(\alpha+1,\beta)}$ have no common zeros.
(2) The polynomial

$$P_m^{(\alpha,\beta+1)}(1)P_m^{(\alpha+1,\beta)}(2t^2-1) + P_m^{(\alpha+1,\beta)}(1)P_m^{(\alpha,\beta+1)}(2t^2-1)$$

has $2m$ simple zeros inside $(-1,1)$.

Proof We need two identities of the Jacobi polynomials [1, (22.7.18) and (22.7.19)]

$$\begin{aligned} P_m^{(\alpha+1,\beta+1)}(t) + \frac{m+\alpha+1}{m+\gamma}P_{m-1}^{(\alpha+1,\beta+1)}(t) &= \frac{2m+\gamma}{m+\gamma}P_m^{(\alpha+1,\beta)}(t), \\ P_m^{(\alpha+1,\beta+1)}(t) - \frac{m+\beta+1}{m+\gamma}P_{m-1}^{(\alpha+1,\beta+1)}(t) &= \frac{2m+\gamma}{m+\gamma}P_m^{(\alpha,\beta+1)}(t), \end{aligned} \tag{7.21}$$

where $\gamma = \alpha+\beta+2$. If x_0 is a common zero of $P_m^{(\alpha+1,\beta)}$ and $P_m^{(\alpha,\beta+1)}$, then the right-hand sides of these two identities are zero, which implies immediately, by linear combination, that x_0 is a common zero of $P_m^{(\alpha+1,\beta+1)}$ and $P_{m-1}^{(\alpha+1,\beta+1)}$, but consecutive Jacobi polynomials do not have common zeros.

For (2), let $q_m(t) := P_m^{(\alpha,\beta+1)}(1)P_m^{(\alpha+1,\beta)}(t) + P_m^{(\alpha+1,\beta)}(1)P_m^{(\alpha,\beta+1)}(t)$. Together with $P_m^{(\alpha+1,\beta)}(1) = \binom{m+\alpha+1}{m} = \frac{m+\alpha+1}{\alpha+1}P_m^{(\alpha,\beta+1)}(1)$, we can then rewrite $q_m(t)$ as

$$q_m(t) = \frac{(m+\gamma)(m+2\alpha+2)}{(2m+\gamma)\alpha+1}\binom{m+\alpha}{m}$$
$$\times\left[P_m^{(\alpha+1,\beta+1)}(t) - \frac{(m+\alpha+1)(m+\beta-\alpha)}{(m+\gamma)(m+2\alpha+2)}P_{m-1}^{(\alpha+1,\beta+1)}(t)\right].$$

By its definition, $q_m(1) > 0$ and $(-1)^m q_m(-1) > 0$, which are sufficient to imply that the quasi-orthogonal polynomial in the bracket has m simple zeros in $(-1,1)$ (see, e.g. [162, Theorem 5.3]). □

7.4 Cubature Rules for Other Chebyshev Weight Function

In this section, we consider cubature rules for the weight functions $W_{\pm\frac{1}{2},\pm\frac{1}{2}}$ that are defined by, similarly as in Definition 7.1.1,

$$W_{\pm\frac{1}{2},\pm\frac{1}{2}}(\mathbf{x}) = w(\cos(\theta_1-\theta_2))w(\cos(\theta_1+\theta_2))(1-x_1)^{\pm\frac{1}{2}}(1-x_2)^{\pm\frac{1}{2}}|x_1^2-x_2^2|$$

for $\mathbf{x} \in [-1,1]^2$, where w is supported on $s(w) = [-1,1]$. This consists of four classes of weight functions. While $W_{-\frac{1}{2},-\frac{1}{2}} = W_{-\frac{1}{2}}$ and $W_{\frac{1}{2},\frac{1}{2}} = W_{\frac{1}{2}}$ are considered in Section 7.3, the other two families, $W_{\pm\frac{1}{2},\mp\frac{1}{2}}$, are new. In particular, the two new weight functions are not centrally symmetric, so these are not of the form $W_{\pm\frac{1}{2}}$ in Definition 7.1.1.

Example 7.4.1. If $w = w_{\alpha,\beta}$ is the Jacobi weight, then, for $\alpha,\beta > -1$,

$$W_{-\frac{1}{2},\frac{1}{2}}(\mathbf{x}), = \frac{|x_1-x_2|^{2\alpha+1}|x_1+x_2|^{2\beta+1}\sqrt{1+x_1}\sqrt{1+x_2}}{\sqrt{1-x_1}\sqrt{1-x_2}},$$
$$W_{\frac{1}{2},-\frac{1}{2}}(\mathbf{x}), = \frac{|x_1-x_2|^{2\alpha+1}|x_1+x_2|^{2\beta+1}\sqrt{1-x_1}\sqrt{1-x_2}}{\sqrt{1+x_1}\sqrt{1+x_2}}.$$

Because the weights are not centrally symmetric, the number of nodes of a cubature rule of degree $2n-1$ for these weights may not obey Möller's lower bound if w is not a constant. If $w(t) = 1$, then Möller's lower bound is satisfied by Corollary 5.2.4, which suggests that the lower bound may as well hold for $W_{\pm\frac{1}{2},\mp\frac{1}{2}}$.

We first use the minimal cubature rules in Theorem 7.2.1 for $W_{-\frac{1}{2}}$ to construct cubature rules of degree $2n-1$ for $W_{\pm\frac{1}{2},\mp\frac{1}{2}}$ that have one node more than the lower bound. To simplify the statement, we state the result only for $W_{-\frac{1}{2},\frac{1}{2}}$. The notations $s_{j,k}^{(m)}$ and $t_{j,k}^{(m)}$ are the same as in (i) of Theorem 7.2.1.

Theorem 7.4.2. *Let $n = 2m-1$. Then, the weight function $W_{-\frac{1}{2},\frac{1}{2}}$ admits a cubature rule of degree $2n-1 = 4m-3$ that has $2m^2$ nodes, one more than the bound (5.9). More precisely, for $f \in \Pi_{2n-1}^2$,*

$$\int_{[1,1]^2} f(\mathbf{x}) W_{-\frac{1}{2},\frac{1}{2}}(\mathbf{x}) d\mathbf{x} = \frac{1}{2} \sum_{k=1}^{m} \sum_{j=1}^{k}{}' \lambda_{j,k}^{(m)} \left[f\left(s_{j,k}^{(m)}, t_{j,k}^{(m)}\right) + f\left(t_{j,k}^{(m)}, s_{j,k}^{(m)}\right) \right]$$

$$+ \frac{1}{2} \sum_{k=2}^{m} \sum_{j=1}^{k-1} \lambda_{j,k}^{(m)} \left[f\left(-s_{j,k}^{(m)}, -t_{j,k}^{(m)}\right) + f\left(-t_{j,k}^{(m)}, -s_{j,k}^{(m)}\right) \right],$$

where $\lambda_{j,k}^{(m)} = \lambda_{j,m}(w)\lambda_{k,m}(w)(1+s_{j,k}^{(m)})(1+t_{j,k}^{(m)})$.

Proof We start with the observation $W_{-\frac{1}{2},\frac{1}{2}}(\mathbf{x}) = (1+x_1)(1+x_2)W_{-\frac{1}{2}}(\mathbf{x})$. For $f \in \Pi_{4m-3}^d$, we apply the minimal cubature rule (7.12) on the polynomial $(1+x_1)(1+x_2)f(\mathbf{x})$ of degree $4m-1$ to obtain the stated cubature rule of degree $4m-3$ for $W_{-\frac{1}{2},\frac{1}{2}}$. Since $s_{k,k}^{(m)} = 1$, the cubature rules (7.12) has m nodes on each of the four boundary of $[-1,1]^2$ but none on the corner. Consequently, the polynomial $(1+x_1)(1+x_2)f(\mathbf{x})$ vanish on $(-1, -t_{k.k}^{(m)})$ and $(-t_{k.k}^{(m)}, -1)$ for $1 \le k \le m$, a total of $2m$ nodes, so that the cubature rule for $W_{-\frac{1}{2},\frac{1}{2}}$ has the nodes $2m(m+1) - 2m = 2m^2$ nodes. □

Next, we do the same but use the near-minimal cubature rules in Theorem 7.3.1 for $W_{-\frac{1}{2}}$ instead. Again, we state the result only for $W_{-\frac{1}{2},\frac{1}{2}}$. The notations $s_{j,k}^{(m)}$, $t_{j,k}^{(m)}$, and $\mu_{k,m}$ are the same as in (i) of Theorem 7.3.1.

Theorem 7.4.3. *Let $n = 2m$. Then, the weight function $W_{-\frac{1}{2},\frac{1}{2}}$ admits a cubature rule of degree $2n-1 = 4m-1$ that has $2m(m+1)$ nodes, exactly Möller's lower bound (5.9). More precisely, for $f \in \Pi_{2n-1}^2$,*

$$\int_{[1,1]^2} f(\mathbf{x}) W_{-\frac{1}{2},\frac{1}{2}}(\mathbf{x}) d\mathbf{x} = \frac{1}{2} \sum_{k=0}^{m} \sum_{j=0}^{k}{}' \mu_{j,k}^{(m)} \left[f\left(s_{j,k}^{(m)}, t_{j,k}^{(m)}\right) + f\left(t_{j,k}^{(m)}, s_{j,k}^{(m)}\right) \right]$$

$$+ \frac{1}{2} \sum_{k=2}^{m} \sum_{j=1}^{k-1} \mu_{j,k}^{(m)} \left[f\left(-s_{j,k}^{(m)}, -t_{j,k}^{(m)}\right) + f\left(-t_{j,k}^{(m)}, -s_{j,k}^{(m)}\right) \right],$$

where $\mu_{j,k}^{(m)} = \mu_{j,m}\mu_{k,m}(1+s_{j,k}^{(m)})(1+t_{j,k}^{(m)})$.

Proof The proof follows that of Theorem 7.4.2. For $f \in \Pi^d_{4m-1}$, we apply the near-minimal cubature rule (7.17) on the polynomial $(1+x_1)(1+x_2)f(\mathbf{x})$ of degree $4m+1$ to obtain the stated cubature rule of degree $4m-1$ for $W_{-\frac{1}{2},\frac{1}{2}}$. The cubature rule (7.17) has $m+1$ nodes in each side of the square $[-1,1]^2$, with two of them being the corner points, since $t_{0,m} = 1 = \cos 0$, it follows that $s^{(m)}_{0,0} = t^{(m)}_{0,0} = 1$. Thus, the number of the nodes becomes $2(m+1)^2 - 2(m+1) = 2m(m+1)$. □

7.5 Further Minimal Cubature Rules of $2n-1$ with n Even

Let w be a weight function defined on $[-1,1]$. We consider minimal cubature rules for a more general class of weight functions defined by, for $\ell = 0,1,2,\ldots$, and $x_1 = \cos\theta_1$ and $x_2 = \cos\theta_2$,

$$W^{(\ell)}_{-\frac{1}{2}}(\mathbf{x}) = w(\cos\ell(\theta_1-\theta_2))\,w(\cos\ell(\theta_1+\theta_2))\,\frac{\left|T_\ell(x_1)^2 - T_\ell(x_2)^2\right|}{\sqrt{1-x_1^2}\sqrt{1-x_2^2}},$$

where T_ℓ is the Chebyshev polynomial, which agrees with $W_{-\frac{1}{2}}$ when $\ell = 0$. For $\ell \ge 1$, this weight function agrees with $W_{-\frac{1}{2}}$ in (i) of Definition 7.1.1 by replacing w by $w^{(\ell)}$ defined by

$$w^{(\ell)}(t) = w(T_\ell(t))\,\frac{\sqrt{1-T^2_\ell(t)}}{\sqrt{1-t^2}}, \qquad -1 \le t \le 1. \tag{7.22}$$

Consequently, by Theorem 7.2.1, the minimal cubature rules of degree $2n-1$ with n even exist, and the nodes of the minimal cubature rule for $W^{(\ell)}_{-\frac{1}{2}}$ are based on zeros of $p_{n/2}(w^{(\ell)})$. For $n = 2\ell m$, we provide an explicit constructive later so that the nodes are given in terms of the zeros $t_{k,m}$ of $p_n(w)$ and the weights $\lambda_{k,m}(w)$ are those of the Gauss quadrature associated with w.

Let G_ℓ be the group defined by rotations of multiples of $2\pi/\ell$ about the origin. Denote the elements of G_ℓ by the angle of rotations,

$$G_\ell = \left\{\tau_{k,\ell} = \tfrac{2\pi k}{\ell} : 0 \le k \le \ell-1\right\}$$

and define the action of G_ℓ on $f : [-1,1] \mapsto \mathbb{R}$ by $\tau_{k,\ell} f(t) = f(\cos(\theta + \tau_{k,\ell}))$.

Lemma 7.5.1. *Let $\ell = 1,2,\ldots$. Then, the orthogonal polynomial of degree $n = \ell m$ associated with $w^{(\ell)}$ is*

$$p_{\ell m}\left(w^{(\ell)};t\right) = p_m(w;T_\ell(t)), \qquad m = 0,1,2,\ldots.$$

Proof We start with an elementary integral identity

$$\int_{-1}^{1} f(T_\ell(t)) \frac{\mathrm{d}t}{\sqrt{1-t^2}}\mathrm{d}t = \int_{-1}^{1} f(t)\frac{\mathrm{d}t}{\sqrt{1-t^2}}\mathrm{d}t. \tag{7.23}$$

This follows from a changing variable $t \mapsto \cos\theta$ so that

$$\begin{aligned}\int_{-1}^{1} f(T_\ell(t)) \frac{\mathrm{d}t}{\sqrt{1-t^2}}\mathrm{d}t &= \int_0^\pi f(\cos\ell\theta)\mathrm{d}\theta = \frac{1}{\ell}\int_0^{\ell\pi} f(\cos\theta)\mathrm{d}\theta \\ &= \frac{1}{\ell}\sum_{k=0}^{\ell-1}\int_{k\pi}^{(k+1)\pi} f(\cos\theta)\mathrm{d}\theta = \int_0^\pi f(\cos\theta)\mathrm{d}\theta,\end{aligned}$$

where the last step follows a change of variable $\theta \mapsto k\pi + \phi$ if k is even and $\theta \mapsto (k+1)\pi - \phi$ if k is odd. For $k = 1,2,\ldots$, we claim that

$$\int_{-1}^{1} f(T_\ell(t))\,T_k(t)\frac{\mathrm{d}t}{\sqrt{1-t^2}} = 0, \qquad \text{if } k \not\equiv 0 \mod \ell.$$

Indeed, changing variable $t \mapsto \cos\theta$, the integral in the left-hand side becomes

$$\begin{aligned}\int_0^\pi f(\cos\ell\theta)\cos k\theta\mathrm{d}\theta &= \frac{1}{2}\int_{-\pi}^{\pi} f(\cos\ell\theta)\cos k\theta\mathrm{d}\theta \\ &= \frac{1}{2\ell}\sum_{j=0}^{\ell-1}\int_{-\pi}^{\pi} f(\cos\theta)\cos k\left(\theta + \tfrac{2\pi j}{\ell}\right)\mathrm{d}\theta,\end{aligned}$$

by the 2π-periodicity. Writing $\cos k(\theta+\tau_j) = \cos k\theta\cos k\tau_j - \sin k\theta\sin k\tau_j$, the last integral is zero because

$$\sum_{j=0}^{\ell-1}\left(\cos k\tfrac{2\pi j}{\ell} + \mathrm{i}\sin\tfrac{2\pi j}{\ell}\right) = \sum_{j=0}^{\ell-1} \mathrm{e}^{\mathrm{i}\frac{2\pi k j}{\ell}} = 0 \quad \text{if } k \not\equiv 0 \mod \ell.$$

Let $\varpi(t) = w(t)\sqrt{1-t^2}$ so that $w^{(\ell)}(t) = \varpi(T_\ell(t))/\sqrt{1-t^2}$. Then, by (7.23) and $T_{\ell j}(t) = T_j(T_\ell(t))$,

$$\int_{-1}^{1} p_m(w;T_\ell(t))\,T_k(t)w^{(\ell)}(t)\mathrm{d}t = \int_{-1}^{1} p_m(w;t)T_j(t)w(t)\mathrm{d}t = 0,$$

if $k = \ell j$ and $0 \le j \le m-1$, whereas the integral on the left-hand side is zero if $k \not\equiv 0 \mod \ell$. This shows that $p_m(w;T_\ell(t))$ is an orthogonal polynomial of degree ℓm associated with $w^{(\ell)}$. □

Lemma 7.5.2. *Let $\ell = 1,2,\ldots$. Then, the integral*

$$\mathcal{I}^{(\ell)} f := \int_{[-1,1]^2} f(\mathbf{x})W_{-\frac{1}{2}}^{(\ell)}(\mathbf{x})\mathrm{d}\mathbf{x},$$

is invariant under the action of the group $G_\ell \times G_\ell$. Moreover,

$$\int_{[-1,1]^2} f\big(T_\ell(x_1), T_\ell(x_2)\big)W_{-\frac{1}{2}}^{(\ell)}(\mathbf{x})\mathrm{d}\mathbf{x} = \int_{[-1,1]^2} f(\mathbf{x})W_{-\frac{1}{2}}(\mathbf{x})\mathrm{d}\mathbf{x}. \tag{7.24}$$

Proof Changing variables $(x_1,x_2) \mapsto (\cos\theta_1, \cos\theta_2)$, we obtain

$$\mathcal{I}^{(\ell)} f = \int_0^\pi \int_0^\pi f(\cos\theta_1, \cos\theta_2) \varpi(\ell\theta_1, \ell\theta_2)\, \mathrm{d}\theta_1 \mathrm{d}\theta_2,$$

where the weight function ϖ is defined by

$$\varpi(\theta_1,\theta_2) = w(\cos(\theta_1 - \theta_2)) w((\cos(\theta_1 + \theta_2)) \left|\cos^2\theta_1 - \cos^2\theta_2)\right|$$

form, which we conclude that $\mathcal{I}^{(\ell)}$ is invariant, as $\varpi(\ell\theta_1, \ell\theta_2)$ is under $G_\ell \times G_\ell$. The identity (7.24) follows from applying (7.23) to both variables. □

Let $\ell = 1,2,3,\ldots$. For $s = \cos\theta$ and $t = \cos\phi$ with $0 \le s,t < \pi$, we define sets of points that will be used to define the nodes of the cubature rules. We need to consider two cases, depending on the parity of ℓ.

Case 1. ℓ even. Assume $\ell = 2,4,6,\ldots$, we define

$$X_\ell^-(s,t) = \bigcup \left\{ \left(\cos\tfrac{(2\nu_1+1)\pi\pm\theta}{\ell}, \cos\tfrac{(2\nu_2+1)\pi\pm\phi}{\ell} \right), 0 \le \nu_1,\nu_2 \le \frac{\ell}{2} - 1 \right\},$$

where the union is over all four sign changes and define

$$X_\ell^+(s,t) = \bigcup \left\{ \left(\cos\tfrac{2\nu_1\pi\pm\theta}{\ell}, \cos\tfrac{2\nu_2\pi\pm\phi}{\ell} \right), \delta_\pm \le \nu_1,\nu_2 \le \frac{\ell}{2} - 1 + \delta_\pm \right\},$$

where $\delta_+ = 0$ and $\delta_- = 1$.

Case 2. ℓ odd. Assume $\ell = 1,3,5,\ldots$, we define

$$X_\ell^-(s,t) = \bigcup \left\{ \left(\cos\tfrac{(2\nu_1+1)\pi\pm\theta}{\ell}, \cos\tfrac{(2\nu_2+1)\pi\pm\phi}{\ell} \right), 0 \le \nu_1,\nu_2 \le \frac{\ell-1}{2} - 1 + \delta_\pm \right\},$$

where $\delta_+ = 0$ and $\delta_- = 1$ and define, with the same $\delta_\pm$,

$$X_\ell^+(s,t) = \bigcup \left\{ \left(\cos\tfrac{2\nu_1\pi\pm\theta}{\ell}, \cos\tfrac{2\nu_2\pi\pm\phi}{\ell} \right), \delta_\pm \le \nu_1,\nu_2 \le \frac{\ell-1}{2} \right\}.$$

The choices of $\delta_\pm$ ensure that the angular argument of the points are all in $[0,\pi)$ so that all points in $X_\ell^\pm(s,t)$ are in $(-1,1]^2$. These sets are selected to have the property (7.25).

Lemma 7.5.3. *For $\ell = 2,3,\ldots$,*

$$\begin{aligned} \left(T_\ell(u), T_\ell(v)\right) &= (s,t), &&\text{if } (u,v) \in X_\ell^+(s,t), \\ \left(T_\ell(u), T_\ell(v)\right) &= (-s,-t), &&\text{if } (u,v) \in X_\ell^-(s,t). \end{aligned} \tag{7.25}$$

Moreover, the cardinality of $X_\ell^\pm(s,t)$ is given by

$$\left|X_\ell^\pm(s,t)\right| = \begin{cases} \ell^2, & s \ne 1, t \ne 1, \\ \frac{\ell(\ell+1)}{2}, & s = 1, ,t \ne 1 \, or \, s \ne 1, t = 1, \text{ if } \ell \text{ is odd}, \end{cases}$$

whereas if $s = 1$, $t \neq 1$ *or* $s \neq 1$, $t = 1$,

$$\left|X_\ell^-(s,t)\right| = \frac{\ell^2}{2} \quad \textit{and} \quad \left|X_\ell^+(s,t)\right| = \frac{\ell^2}{2} + \ell \quad \textit{if } \ell \textit{ is even}$$

$$\left|X_\ell^\pm(s,t)\right| = \frac{\ell(\ell+1)}{2}, \qquad \textit{if } \ell \textit{ is even.}$$

Proof The identity (7.25) is immediate and the reason that $X_\ell^\pm$ is defined. Since $s = 1$ or $t = 1$ is the same as $\theta = 0$ or $\phi = 0$, the cardinality of $X_\ell^\pm(s,t)$ is straightforwardly $4 \times (\ell/2)^2$ when $\theta \neq 0$ and $\phi \neq 0$. If $s = 0$ and $t \neq 0$, or $s \neq 0$ and $t = 0$, then the index for $\theta = 0$, say, is between 0 and $(\ell-1)/2$, so that the cardinality of $X_\ell^\pm(1,t)$ is $\ell \times (\ell+1)/2$, whereas if ℓ is even, then $X_\ell^-(s,t)$ is $2 \times (\ell/2)^2 = \ell^2/2$ and $X_\ell^+(1,t)$, say, contains $\ell/2+1$ distinct v_1, so that $|X_\ell^+(1,t)| = 2 \times (\ell/2+1)\ell/2 = \ell^2/2 + \ell$. □

Let w be a weight function on $[-1,1]$ and let $t_{k,m} = \cos\theta_{k,m}$ be the zeros of the polynomials $p_n(w)$. We retain the notation (7.11) but rewrite it further as

$$s_{j,k}^{(m)} = \cos\theta_{j,k}^{(m)} \quad \text{with} \quad \theta_{j,k}^{(m)} = \frac{\theta_{j,m} - \theta_{k,m}}{2},$$

$$t_{j,k}^{(m)} = \cos\phi_{j,k}^{(m)} \quad \text{with} \quad \phi_{j,k}^{(m)} = \frac{\theta_{j,m} + \theta_{k,m}}{2}.$$

We further define the set

$$X_\ell(j,k) := X_\ell^-\left(-s_{j,k}^{(m)}, -t_{j,k}^{(m)}\right) \bigcup X_\ell^-\left(-t_{j,k}^{(m)}, -s_{j,k}^{(m)}\right), \\ \bigcup X_\ell^+\left(s_{j,k}^{(m)}, t_{j,k}^{(m)}\right) \bigcup X_\ell^+\left(t_{j,k}^{(m)}, s_{j,k}^{(m)}\right).$$

As in Theorem 7.2.1, let $\lambda_{k,m}(w)$ denote the quadrature weights of the Gauss quadrature associated with w on $[-1,1]$.

Theorem 7.5.4. *Let* $n = 2\ell m$ *for* $\ell = 1,2,\ldots$. *Then, the weight function* $W_{-\frac{1}{2}}^{(\ell)}$ *admits the minimal cubature rule of degree* $2n-1 = 4\ell m - 1$ *with* $2\ell^2 m^2 + 2\ell m$ *nodes. More precisely,*

$$\int_{[-1,1]^2} f(\mathbf{x}) W_{-\frac{1}{2}}^{(\ell)}(\mathbf{x}) d\mathbf{x} = \frac{1}{2} \sum_{k=1}^{m} {\sum_{j=1}^{k}}' \mu_{k,m}\mu_{j,m} \frac{1}{|X_\ell(j,k)|} \sum_{(s,t)\in X_\ell(j,k)} f(s,t), \tag{7.26}$$

where $|X_\ell(j,k)| = 4\ell^2$ *if* $j \neq k$ *and* $|X_\ell(k,k)| = 2\ell^2 + 2\ell$.

Proof By the periodicity of the cosine function, the right-hand side of (7.26) is invariant under $G_\ell \times G_\ell$. Using the identity in (7.24) and applying the minimal cubature rule in Theorem 7.2.1, we obtain

$$\int_{[-1,1]^2} f(T_{\ell m}(x_1), T_{\ell m}(x_2))\, W^{(\ell)}_{-\frac12}(\mathbf{x})d\mathbf{x} = \int_{[-1,1]^2} f(x_1), x_2) W_{-\frac12}(\mathbf{x})d\mathbf{x}$$

$$= \frac12 \sum_{k=1}^{m} \sum_{j=1}^{k}{}' \lambda_{j,k} \Big[f\Big(s^{(m)}_{j,k}, t^{(m)}_{j,k}\Big) + f\Big(t^{(m)}_{j,k}, s^{(m)}_{j,k}\Big)$$

$$+ f\Big(-s^{(m)}_{j,k}, -t^{(m)}_{j,k}\Big) + f\Big(-t^{(m)}_{j,k}, -s^{(m)}_{j,k}\Big)\Big]$$

$$= \frac12 \sum_{k=1}^{m} \sum_{j=1}^{k}{}' \lambda_{j,m}(w)\lambda_{k,m}(w) \frac{1}{|X_\ell(j,k)|} \sum_{(u,v)\in X_\ell(j,k)} f(T_\ell(u), T_\ell(v)),$$

where $\lambda_{j,k} = \lambda_{j,m}(w)\lambda_{k,m}(w)$ and the last equal sign follows from (7.25), which holds for all polynomials $f \in \Pi^d_{4m-1}$. This establishes the cubature rule (7.26) for all polynomials of the form $f(T_\ell(x_1), T_\ell(x_2))$ of degree $4\ell m - 1$, which consist of all invariant polynomials under $G_\ell \times G_\ell$ of the degree less than $4\ell m$. By Sobolev's Theorem 2.5.2, this proves (7.26) for all polynomials of degree $4\ell m - 1$. Finally, by Lemma 7.5.3, we obtain $|X_\ell(j,k)| = 4\ell^2$ if $j \ne k$ and $|X_\ell(k,k)| = 2\ell(\ell+1)$. Since $s_{k,k} = \cos 0 = 1$, the number of nodes of the cubature rule is equal to

$$4\ell^2 \left(\frac{m(m+1)}{2} - m \right) + 2\ell(\ell+1)m = 2\ell^2 m^2 + 2\ell m,$$

which agrees with Möller's lower bound when $n = 2\ell m$. □

For w being the Jacobi weight function $w_{\alpha,\beta}(t) = (1-t)^\alpha (1+t)^\beta$, $\alpha,\beta > -1$, the weight function $W^{(\ell)}_{-\frac12}$ becomes

$$W^{(\ell)}_{\alpha,\beta,-\frac12}(\mathbf{x}) = \frac{|T_\ell(x_1) - T_\ell(x_2)|^{2\alpha+1} |T_\ell(x_1) + T_\ell(x_2)|^{2\beta+1}}{\sqrt{1-x_1^2}\sqrt{1-x_2^2}}. \tag{7.27}$$

For $\ell = 1$, this is the weight function in (7.15). For $\ell = 2$, the weight function becomes

$$W^{(\ell)}_{\alpha,\beta,-\frac12}(\mathbf{x}) = 2^{2\alpha+2\beta+2} \frac{|x_1^2 - x_2^2|^{2\alpha+1} |1 - x_1^2 - x_2^2|^{2\beta+1}}{\sqrt{1-x_1^2}\sqrt{1-x_2^2}},$$

so that it coincides, up to the constant multiple, with the weight function (7.1.4) derived via the generalized Gegenbauer weight.

As an example, we consider the cubature rules of degree $2n - 1 = 8m - 1$ for $W^{(2)}_{\alpha,\beta,-\frac12}$, which has $8m^2 + 4m$ number of nodes that are derived from the zeros of the Jacobi polynomials $P_m^{(\alpha,\beta)}$. For $\beta > 0$, the nodes of the cubature rules are propelled away from the circle $x_1^2 + x_2^2 = 1$, and they are also propelled away from diagonals of the square if $\alpha > 0$ as well, as shown in Figure 7.3. The nodes

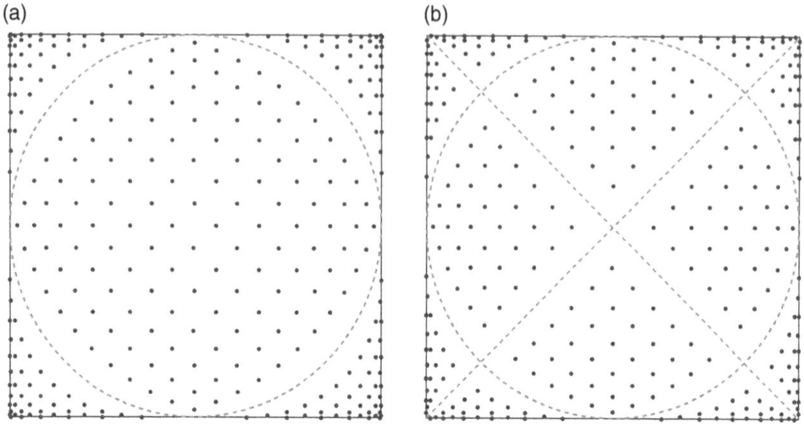

Figure 7.3 The 312-node minimal cubature rule of degree 47 for $W^{(2)}_{\alpha,\beta,-\frac{1}{2}}$ with $\alpha = \frac{1}{2}$ and $\beta = -\frac{1}{2}$ (a) and $\alpha = \frac{1}{2}$ and $\beta = \frac{1}{2}$ (b).

will be further away from the unit circle as α increases and from the diagonals if β increases.

7.6 Notes and Further Results

The orthogonal structure in the first section appears in [178]. The minimal cubature rules of degree $2n-1$ with n even are studied in [179], where the case of the Jacobi weight function $W_{\alpha,\beta,\pm\frac{1}{2}}$ in Subsection 7.2.2 is treated in detail. In particular, a closed-form formula for the kernel $\mathbf{K}_n^*(\mathbf{x},\mathbf{y})$ defined in (6.18) is established and used for an explicit formula for the Lagrange interpolation polynomial $L_n^*(f)$, defined in (6.19), based on the nodes of the minimal cubature rules. Moreover, the Lebesgue constant for $L_n^*(f) = L_n^{(\alpha,\beta)}$ is determined as

$$\left\| L_n^{(\alpha,\beta)} \right\|_\infty = \mathcal{O}(1) \begin{cases} n^{2\max\{\alpha,\beta\}+1}, & \max\{\alpha,\beta\} > -\frac{1}{2} \\ (\log n)^2, & \alpha = \beta = -\frac{1}{2} \end{cases}, \tag{7.28}$$

where the case $\alpha = \beta = -\frac{1}{2}$ is established earlier in [16].

The minimal and near-minimal cubature rules of degree $2n-1$ for odd n are studied in [181] for the Jacobi weight function $W_{\alpha,\beta,\pm\frac{1}{2}}$. The minimal cubature rules in Theorem 7.3.4 are not given explicitly because explicit formulas for the nodes on the diagonal and their cubature weights are unknown. The near-minimal cubature rules are explicit, and a closed-form formula for their corresponding kernel $\mathbf{K}_n^*(x,y)$ is determined and used to show that the Lebesgue constants for the Lagrange interpolation polynomials have the same order of (7.28). Thus, for all practical purposes, the near-minimal cubature rules are more appealing than the minimal ones when n is odd.

By Theorem 6.2.1, the nodes of the near-minimal cubature rules are also zeros of $\mathbb{Q}_{n+1} = \mathbb{P}_{n+1} + \Gamma_1\mathbb{P}_n + \Gamma_2\mathbb{P}_{n-1}$, which contains exactly one polynomial that is not in the ideal $\langle U^{\mathsf{T}}\mathbb{P}_n\rangle$. As shown in Proposition 6.3.4, if $\Gamma_1 = 0$, then $\sigma_0 = \lfloor \frac{n}{2} \rfloor$ or $\sigma_0 = \lfloor \frac{n}{2} \rfloor + 1$, but the converse may not hold when $\sigma_0 = \lfloor \frac{n}{2} \rfloor + 1$. The latter corresponds to the near-minimal rule. For the Jacobi weight $W_{\alpha,\beta,\pm\frac{1}{2}}$, the $\mathbb{Q}_{n+1}$ is determined explicitly in [181] by making use of several relations among Jacobi polynomials of different indices. It follows that $\Gamma_1 = 0$ in this case.

8

Discrete Fourier Transform and Cubature Rules

The orthogonality of the Chebyshev polynomials and Gauss quadrature rules for the Chebyshev weights of the first and the second kind follow from the discrete Fourier transform of exponential functions on the real line. We consider their multivariate extensions, where the periodicity is defined via a lattice, and the lattice translation defines the discrete Fourier transform on the spectral set of the lattice that tiles the Euclidean space by lattice translation. The usual tensor product extension from one variable to several variables corresponds to the $\mathbb{Z}^d$ lattice, and we will discuss several other nontrivial extensions. The discrete Fourier analysis so defined yields cubature rules for exponential functions for the integral over the spectral set of the lattice, which become cubature rules on the fundamental domain of the spectral set for generalized cosine and sine functions, defined as certain symmetric and antisymmetric exponential sums. Furthermore, under appropriate transformation, these generalized trigonometric functions define Chebyshev polynomials that inherit the orthogonality of generalized sine and cosine functions. The integral for the orthogonality is defined on the range of the fundamental domain under the transformation, for which the cubature rules for the generalized trigonometric functions become cubature rules for algebraic polynomials.

In Section 8.1, we define discrete Fourier transforms associated with two lattices; one defines the integral domain, and the other defines the space of exponential functions. Choosing the two lattices as $\mathbb{Z}^2$ and its 45° rotation, we obtain an alternative proof of the near-minimal cubature rules for the Chebyshev weights on the square in Section 8.2. This scheme of a pair of lattices is extended to three variables on a cube in Section 8.3, but the cubature rule so obtained is no longer minimal. The lattices associated with the reflection group $\mathcal{A}_d$ are considered in Section 8.4, which leads to generalized Chebyshev polynomials and a family of Gauss cubature rules on a curved domain.

8.1 Lattice and Discrete Fourier Transforms

A lattice L of the Euclidean space $\mathbb{R}^d$ is a discrete subgroup that contains d linearly independent vectors,

$$L := \{k_1\mathbf{a}_1 + k_2\mathbf{a}_2 + \cdots + k_d\mathbf{a}_d : \ k_i \in \mathbb{Z}, \ i = 1,2,\ldots,d\},$$

where $\mathbf{a}_1,\ldots,\mathbf{a}_d$ are linearly independent column vectors in $\mathbb{R}^d$. Let A be the $d\times d$ matrix that has column vectors $\mathbf{a}_1,\ldots,\mathbf{a}_d$. Then, A is called a *generator matrix* of the lattice L. We can write L as $A\mathbb{Z}^d$,

$$L = A\mathbb{Z}^d := \{A\mathbf{k} : \ \mathbf{k} \in \mathbb{Z}^d\}.$$

The dual lattice $L^\perp$ of L is given by

$$L^\perp := \{\mathbf{x} \in \mathbb{R}^d : \ \mathbf{x}\cdot\mathbf{y} \in \mathbb{Z} \text{ for all } \mathbf{y} \in L\},$$

where $\mathbf{x}\cdot\mathbf{y}$ denotes the usual Euclidean inner product of $\mathbf{x}$ and $\mathbf{y}$. The generator matrix of $L^\perp$, denoted by $A^\perp$, is given by $A^\perp = A(A^\mathsf{T}A)^{-1}$.

A bounded set Ω of $\mathbb{R}^d$ is said to tile $\mathbb{R}^d$ with the lattice L if

$$\sum_{\alpha\in L} \chi_\Omega(\mathbf{x}+\alpha) = 1, \quad \text{for almost all } \mathbf{x} \in \mathbb{R}^d,$$

where χ_Ω denotes the characteristic function of Ω. Tiling and Fourier analysis are closely related as demonstrated by Fuglede's theorem [54]. Let $\langle\cdot,\cdot\rangle_\Omega$ denote the inner product in $L^2(\Omega)$,

$$\langle f,g\rangle_\Omega = \frac{1}{|\Omega|}\int_\Omega f(\mathbf{x})\overline{g(\mathbf{x})}\mathrm{d}\mathbf{x}, \tag{8.1}$$

where $|\Omega|$ denotes the volume of Ω, and the bar denotes the complex conjugation.

Theorem 8.1.1. *Let $\Omega \subset \mathbb{R}^d$ be a bounded domain and L be a lattice of $\mathbb{R}^d$. Then, $\Omega + L = \mathbb{R}^d$ if and only if $\{\mathrm{e}^{2\pi i\alpha\cdot\mathbf{x}} : \alpha \in L^\perp\}$ is an orthonormal basis with respect to the inner product* (8.1).

If $L = L_A$, then $|\Omega| = |\det A|$ and the closure of Ω is the image under A of $[0,1]^d$. Furthermore, since $L_A^\perp = A^{-\mathsf{T}}\mathbb{Z}^d$, we can write $\alpha = A^{-\mathsf{T}}\mathbf{k}$ for $\alpha \in L_A^\perp$ and $\mathbf{k} \in \mathbb{Z}^d$, so that $\alpha\cdot\mathbf{x} = (A^{-\mathsf{T}}\mathbf{k})\cdot\mathbf{x} = \mathbf{k}^\mathsf{T}A^{-1}\mathbf{x}$. Hence, the orthogonality in the theorem is

$$\frac{1}{|\det(A)|}\int_\Omega \mathrm{e}^{2\pi i\mathbf{k}^\mathsf{T}A^{-1}\mathbf{u}}d\mathbf{u} = \delta_{\mathbf{k},0}, \qquad \mathbf{k} \in \mathbb{Z}^d. \tag{8.2}$$

The set Ω is called a *spectral set* for the lattice L.

It is not unique since any affine translation of a spectral set is again a spectral set. We shall fix Ω, so it contains 0 in its interior and assume that the tiling

$\Omega + A\mathbb{Z}^d = \mathbb{R}^d$ holds pointwise and without overlapping. In other words, we require

$$\sum_{\alpha \in L} \chi_\Omega(\mathbf{x}+\alpha) = 1 \qquad \text{for all } \mathbf{x} \in \mathbb{R}^d. \tag{8.3}$$

For example, we can take $\Omega = [-\frac{1}{2}, \frac{1}{2})^d$ for the standard lattice $\mathbb{Z}^d$. Correspondingly, a spectral set satisfying (8.3) is called a *fundamental domain* of the lattice L. If $L = L_A$, we also write $\Omega = \Omega_A$.

A function f defined on $\mathbb{R}^d$ is called a periodic function with respect to the lattice $L = A\mathbb{Z}^d$ if

$$f(\mathbf{x}+A\mathbf{k}) = f(\mathbf{x}) \qquad \text{for all } \mathbf{k} \in \mathbb{Z}^d.$$

If $f \in L^2(\Omega_A)$, then it can be extended periodically so that it has a Fourier series

$$f(\mathbf{x}) = \sum_{\mathbf{k}\in\mathbb{Z}^d} c_{\mathbf{k}} \mathrm{e}^{2\pi i \mathbf{k}^{\mathsf{T}} A^{-1}\mathbf{x}} \quad \text{with} \quad c_{\mathbf{k}} = \frac{1}{|\det(A)|} \int_{\Omega_A} f(\mathbf{x}) \mathrm{e}^{-2\pi i \mathbf{k}^{\mathsf{T}} A^{-1}\mathbf{x}} \mathrm{d}\mathbf{x},$$

which is just a change of variables away from the usual multiple Fourier series.

The Fourier transform of a function $f \in L^1(\mathbb{R}^d)$ and its inverse are defined by, as usual,

$$\hat{f}(\xi) = \int_{\mathbb{R}^d} f(\mathbf{x}) \mathrm{e}^{-2\pi i \xi\cdot\mathbf{x}} \mathrm{d}\mathbf{x} \quad \text{and} \quad f(\mathbf{x}) = \int_{\mathbb{R}^d} \hat{f}(\xi) \mathrm{e}^{2\pi i \xi\cdot\mathbf{x}} \mathrm{d}\xi.$$

One important tool for us is the Poisson summation formula given by

$$\sum_{\mathbf{k}\in\mathbb{Z}^d} f(\mathbf{x}+A\mathbf{k}) = \frac{1}{|\det(A)|} \sum_{\mathbf{k}\in\mathbb{Z}^d} \hat{f}(A^{-\mathsf{T}}\mathbf{k}) \mathrm{e}^{2\pi i \mathbf{k}^{\mathsf{T}} A^{-1}\mathbf{x}}, \tag{8.4}$$

which holds for $f \in L^1(\Omega_A)$ under the usual assumption on the convergence of the series on both sides.

Definition 8.1.2. Let A and B be two nonsingular matrices such that (8.3) holds for both $\Omega = \Omega_A$ and $\Omega = \Omega_B$. Assume that all entries of $N := B^{\mathsf{T}}A$ are integers. Denote

$$\Lambda_N := \{\mathbf{k} \in \mathbb{Z}^d : B^{-\mathsf{T}}\mathbf{k} \in \Omega_A\} \quad \text{and} \quad \Lambda_{N^{\mathsf{T}}} := \{\mathbf{k} \in \mathbb{Z}^d : A^{-\mathsf{T}}\mathbf{k} \in \Omega_B\}.$$

For f, g in $C(\Omega_A)$, the space of continuous functions on Ω_A, we denote

$$\langle f, g\rangle_N := \frac{1}{|\det(N)|} \sum_{\mathbf{k}\in\Lambda_N} f(B^{-\mathsf{T}}\mathbf{k}) \overline{g(B^{-\mathsf{T}}\mathbf{k})} \tag{8.5}$$

and show that exponential functions in the finite subspace $\mathcal{H}_N$, defined by

$$\mathcal{H}_N := \operatorname{span}\left\{\phi_{\mathbf{k}} : \phi_{\mathbf{k}}(\mathbf{x}) := \mathrm{e}^{2\pi i \mathbf{k}^{\mathsf{T}} A^{-1}\mathbf{x}}, \mathbf{k} \in \Lambda_{N^{\mathsf{T}}}\right\}, \tag{8.6}$$

are orthogonal with respect to the discrete inner product $\langle f,g\rangle_N$. We call two points $\mathbf{x},\mathbf{y}\in\mathbb{R}^d$ congruent with respect to the lattice $A\mathbb{Z}^d$ if $\mathbf{x}-\mathbf{y}\in A\mathbb{Z}^d$, which we write as $\mathbf{x}\equiv\mathbf{y} \mod A$.

Theorem 8.1.3. *Let A,B, and N be as in Definition 8.1.2. Then,*

$$\frac{1}{|\det(N)|}\sum_{\mathbf{j}\in\Lambda_N} \mathrm{e}^{2\pi i\mathbf{k}^{\mathsf{T}}N^{-1}\mathbf{j}} = \begin{cases}1, & \text{if } \mathbf{k}\equiv 0 \mod N^{\mathsf{T}},\\ 0, & \text{otherwise}.\end{cases} \tag{8.7}$$

In particular, $\langle\phi_{\mathbf{k}},\phi_{\mathbf{j}}\rangle_N=\delta_{\mathbf{k},\mathbf{j}}$ for all $\phi_{\mathbf{k}}(\mathbf{x})=\mathrm{e}^{2\pi i\mathbf{k}^{\mathsf{T}}A^{-1}\mathbf{x}}\in\mathcal{H}_N$.

Proof Let $f\in L^1(\mathbb{R}^d)$ for which the Poisson summation formula holds. Let

$$\Phi^A f(\mathbf{x}) := \sum_{\mathbf{k}\in\mathbb{Z}^d} f(\mathbf{x}+A\mathbf{k}).$$

Then, by the Poisson summation formula (8.4), we obtain

$$\Phi^A f\left(B^{-\mathsf{T}}\mathrm{j}\right)=\frac{1}{|\det(A)|}\sum_{\mathrm{k}\in\mathbb{Z}^d}\hat f\left(A^{-\mathsf{T}}\mathrm{k}\right)e^{2\pi i\mathbf{k}^{\mathsf{T}}N^{-1}\mathbf{j}},\qquad \mathbf{j}\in\mathbb{Z}^d.$$

Since $\mathbb{R}^d=\Omega_B+B\mathbb{Z}^d$, we can write $A^{-\mathsf{T}}\mathbf{k}=\mathbf{x}_B+B\mathbf{i}$ for every $\mathbf{k}\in\mathbb{Z}^d$ with $\mathbf{x}_B\in\Omega_B$ and $\mathbf{i}\in\mathbb{Z}^d$. The fact that N has integer entries shows that $A^{\mathsf{T}}\mathbf{x}_B=\mathbf{m}\in\mathbb{Z}^d$. That is, every $\mathbf{k}\in\mathbb{Z}^d$ can be written as $\mathbf{k}=\mathbf{m}+N^{\mathsf{T}}\mathbf{i}$ with $\mathbf{m}\in\Lambda_{N^{\mathsf{T}}}$ and $\mathbf{i}\in\mathbb{Z}^d$. It follows that $\mathrm{e}^{2\pi i\mathbf{k}^{\mathsf{T}}N^{-1}\mathbf{j}}=\mathrm{e}^{2\pi i\mathbf{m}^{\mathsf{T}}N^{-1}\mathbf{j}}$ under this decomposition. Consequently,

$$\begin{aligned}\Phi^A f(B^{-\mathsf{T}}\mathbf{j}) &= \frac{1}{|\det(A)|}\sum_{\mathbf{m}\in\Lambda_{N^{\mathsf{T}}}}\sum_{\mathbf{i}\in\mathbb{Z}^d}\hat f(A^{-\mathsf{T}}\mathbf{m}+B\mathbf{i})\mathrm{e}^{2\pi i\mathbf{m}^{\mathsf{T}}N^{-1}\mathbf{j}} \\ &= \frac{1}{|\det(A)|}\sum_{\mathbf{m}\in\Lambda_{N^{\mathsf{T}}}}\Phi^B\hat f(A^{-\mathsf{T}}\mathbf{m})\mathrm{e}^{2\pi i\mathbf{m}^{\mathsf{T}}N^{-1}\mathbf{j}}.\end{aligned} \tag{8.8}$$

Similarly, we have

$$\Phi^B\hat f(A^{-\mathsf{T}}\mathrm{m})=\frac{1}{|\det(B)|}\sum_{j\in\Lambda_N}\Phi^A f(B^{-\mathsf{T}}\mathrm{j})\mathrm{e}^{-2\pi i\mathbf{m}^{\mathsf{T}}N^{-1}\mathbf{j}}. \tag{8.9}$$

Substituting (8.8) into (8.9) leads to the identity

$$\Phi^B\hat f(A^{-\mathsf{T}}\mathrm{m})=\frac{1}{|\det(AB)|}\sum_{\mathrm{k}\in\Lambda_N^{\mathsf{T}}}\Phi^B\widehat{f}(A^{-\mathsf{T}}\mathrm{k})\sum_{\mathrm{j}\in\Lambda_N}\mathrm{e}^{2\pi i(\mathbf{k}-\mathbf{m})^{\mathsf{T}}N^{-1}\mathbf{j}}$$

that holds for f in, say, the Schwartz class of functions, from which (8.7) follows immediately. □

Corollary 8.1.4. *Let A,B, and N be as in Definition 8.1.2. For $f,g\in\mathcal{H}_N$,*

$$\langle f,g\rangle_{\Omega_A}=\langle f,g\rangle_N. \tag{8.10}$$

Proof By the prior theorem and (8.2), the identity (8.7) holds for $f = \phi_{\mathbf{j}}$ and $g = \phi_{\mathbf{k}}$. Since both inner products are sesquilinear, the stated result follows. □

The identity (8.10) can be regarded as a cubature rule for $f\bar{g}$ with $f, g \in \mathcal{H}_N$. In the following, we shall consider generalized trigonometric polynomials that are defined via symmetric or antisymmetric linear combinations of functions in $\mathcal{H}_N$, which are analogs of cosine and sine functions, and they also lead to generalized Chebyshev polynomials, so that the aforementioned cubature rules become cubature rules for algebraic polynomials.

8.2 Discrete Fourier Analysis and Cubature Rules on the Plane

As the first example of the general setup in Section 8.1, we consider an example of $d = 2$ so that Ω_A is the square $[-\frac{1}{2}, \frac{1}{2})^2$ and Ω_B is a rhombus. In Subsection 8.2.1, we specialize the general results for this example and derive cubature rules for trigonometric functions, which are transformed into cubature rules for algebraic polynomials in Subsection 8.2.2.

8.2.1 Cubature Rules for Trigonometric Functions

We choose A as the 2×2 identity matrix so that $\Omega = [-\frac{1}{2}, \frac{1}{2})^2$. Moreover, we choose the matrix B as

$$B = n \begin{bmatrix} 1 & 1 \\ -1 & 1 \end{bmatrix} \qquad \text{so that} \qquad B^{-1} = \frac{1}{2n} \begin{bmatrix} 1 & -1 \\ 1 & 1 \end{bmatrix}.$$

Since B is a rotation, by 45 degrees, of a constant multiple of the diagonal matrix, the domain Ω_B is a rhombus, which we choose as

$$\Omega_B = \{\mathbf{x} \in \mathbb{R}^2 : -n \le x_1 + x_2 < n,\ -n \le x_2 - x_1 < n\},$$

so that it tiles $\mathbb{R}^2$ without overlapping. By our choice, $N = B^{\mathsf{T}} A = B^{\mathsf{T}}$, so that $\Lambda_N = \Lambda_{N^{\mathsf{T}}}$, which we denote by Λ_n, and it is defined by

$$\Lambda_n = \left\{\mathbf{j} \in \mathbb{Z}^2 : -n \le j_1 + j_2 < n,\ -n \le j_2 - j_1 < n\right\}.$$

The cardinality of Λ_n is $|\Lambda_n| = 2n^2$. Furthermore, we denote the space $\mathcal{H}_N$ by $\mathcal{H}_n$, which is given by, since A is the identity matrix,

$$\mathcal{H}_n := \operatorname{span}\left\{\mathrm{e}^{2\pi \mathrm{i}\mathbf{k}\cdot\mathbf{x}} : \mathbf{k} \in \Lambda_n\right\}.$$

Theorem 8.2.1. *Define the set*

$$X_n := \left\{2\mathbf{k} \in \mathbb{Z}^2 : -\tfrac{n}{2} \le k_1, k_2 < \tfrac{n}{2}\right\} \cup \left\{2\mathbf{k}+\mathbf{1} \in \mathbb{Z}^2 : -\tfrac{n+1}{2} \le k_1, k_2 < \tfrac{n-1}{2}\right\}.$$

Then for all $f, g \in \mathcal{H}_n$,

$$\langle f, g\rangle_n := \frac{1}{2n^2} \sum_{\mathbf{k} \in X_n} f\left(\tfrac{\mathbf{k}}{2n}\right)\overline{g\left(\tfrac{\mathbf{k}}{2n}\right)} = \int_{[-\frac{1}{2},\frac{1}{2}]^2} f(\mathbf{x})\overline{g(\mathbf{x})}\mathrm{d}\mathbf{x}.$$

Proof Changing variables from j to $\mathbf{k} = 2nB^{-\mathsf{T}}\mathbf{j}$, or $k_1 = j_1 + j_2$ and $k_2 = j_2 - j_1$, then, as j_1 and j_2 need to be integers and $j_1 = \frac{k_1-k_2}{2}$, $j_2 = \frac{k_1+k_2}{2}$, we see that $\mathbf{j} \in \Lambda_n$ is equivalent to $\mathbf{k} = 2nB^{-\mathsf{T}}\mathbf{j} \in X_n$. Hence, $\langle f, g\rangle_n$ is exactly $\langle f, g\rangle_N$ defined in (8.5), and this theorem follows as a special case of Corollary 8.1.4. □

The set Λ_n lacks symmetry as the inequalities in its definition are half open and half closed. We denote its symmetric counterpart by Λ_n^* and it is defined by

$$\Lambda_n^* := \{\mathbf{j} \in \mathbb{Z}^2 : -n \le j_1 + j_2 \le n,\ -n \le j_1 - j_2 \le n\}.$$

We also denote the counterpart of $\mathcal{H}_n$ by $\mathcal{H}_n^*$ and it is defined by

$$\mathcal{H}_n^* := \operatorname{span}\left\{\mathrm{e}^{2\pi \mathrm{i}\mathbf{k}\cdot\mathbf{x}} : \mathbf{k} \in \Lambda_n^*\right\}.$$

Along the same line, we define the counterpart of X_n as

$$X_n^* := \left\{2\mathbf{k} : -\tfrac{n}{2} \le k_1, k_2 \le \tfrac{n}{2}\right\} \cup \left\{2\mathbf{k}+\mathbf{1} : -\tfrac{n+1}{2} \le k_1, k_2 \le \tfrac{n-1}{2}\right\}.$$

A quick calculation shows $|X_n| = |\Lambda_n| = 2n^2$, whereas $|X_n^*| = 2n^2 + 2n + 1$. We further partition the set X_n^* into three parts,

$$X_n^* = X_n^\circ \cup X_n^e \cup X_n^v,$$

where $X_n^\circ = X_n^* \cap (-n, n)^2$ is the set of interior points of X_n^*, X_n^e consists of those points in X_n^* that are on the edges of the square $[-n, n]^2$ but not on the four vertices, which are corners of the square, while X_n^v consists of those points of X_n^* at the vertices of $[-n, n]^2$.

Theorem 8.2.2. *Define the inner product*

$$\langle f, g\rangle_n^* := \frac{1}{2n^2} \sum_{\mathbf{k} \in X_n^*} c_{\mathbf{k}}^{(n)} f\left(\tfrac{\mathbf{k}}{2n}\right)\overline{g\left(\tfrac{\mathbf{k}}{2n}\right)}, \qquad \text{where} \quad c_{\mathbf{k}}^{(n)} = \begin{cases} 1, & \mathbf{k} \in X_n^\circ \\ \frac{1}{2}, & \mathbf{k} \in X_n^e \\ \frac{1}{4}, & \mathbf{k} \in X_n^v \end{cases}. \tag{8.11}$$

Then, for all $f, g \in \mathcal{H}_n$,

$$\int_{[-\frac{1}{2},\frac{1}{2}]^2} f(x)\overline{g(x)}dx = \langle f, g\rangle_n = \langle f, g\rangle_n^*.$$

Proof It suffices to show that $\langle f,g\rangle_n = \langle f,g\rangle_n^*$. Because of $c_{\mathbf{k}}^{(n)} = 1$ for $\mathbf{k} \in X_n^\circ$, the partial sums over interior points of the two sums agree. The set X_n^e of boundary points can be divided into two parts, $X_n^e = X_n^{e,1} \cup X_n^{e,2}$, where $X_n^{e,1}$ consists of points in X_n that are on the edges of $[-n,n)^2$, but not equal to $(-n,-n)$, and $X_n^{e,2}$ is the complementary of $X_n^{e,1}$ in X_n^e. Evidently, if $\mathbf{x} \in X_n^{e,1}$, then either $\mathbf{x}+(2n,0)$ or $\mathbf{x}+(0,2n)$ belongs to $X_n^{e,2}$. Hence, if f is a periodic function, $f(\mathbf{x}+\mathbf{k}) = f(\mathbf{x})$ for $\mathbf{k} \in \mathbb{Z}^2$, then

$$\sum_{\mathbf{k}\in X_n^e} c_{\mathbf{k}}^{(n)} f\left(\tfrac{\mathbf{k}}{2n}\right) = \frac{1}{2}\sum_{\mathbf{k}\in X_n^e} f\left(\tfrac{\mathbf{k}}{2n}\right) = \sum_{\mathbf{k}\in X_n^{e,1}} f\left(\tfrac{\mathbf{k}}{2n}\right).$$

Furthermore, for $(-n,-n) \in X_n$, X_n^* contains all four vertices $(\pm n, \pm n)$. Since a periodic function takes the same value on all four points, $\sum_{\mathbf{k}\in X_n^v} c_{\mathbf{k}}^{(n)} f\left(\frac{\mathbf{k}}{2n}\right) = f\left(-\frac{1}{2},-\frac{1}{2}\right)$. Consequently, we have proved that $\langle f,g\rangle_n = \langle f,g\rangle_n^*$ if f,g are periodic functions. □

We deduce two cubature rules from the above theorems.

Theorem 8.2.3. *For $n \ge 2$, the cubature rules*

$$\int_{[-\frac{1}{2},\frac{1}{2}]^2} f(\mathbf{x})d\mathbf{x} = \frac{1}{2n^2}\sum_{\mathbf{k}\in X_n^*} c_{\mathbf{k}}^{(n)} f\left(\tfrac{\mathbf{k}}{2n}\right) = \frac{1}{2n^2}\sum_{\mathbf{k}\in X_n} f\left(\tfrac{\mathbf{k}}{2n}\right) \tag{8.12}$$

are exact for $f \in \mathcal{H}_{2n-1}^*$.

Proof It suffices to prove that both cubature formulas in (8.12) are exact for every $\phi_{\mathbf{j}}(\mathbf{x}) = e^{2\pi i \mathbf{j}\cdot\mathbf{x}}$ with $\mathbf{j} \in \Lambda_{2n-1}^*$. For this purpose, we first claim that for any $\mathbf{j} \in \mathbb{Z}^2$, there exist $\nu \in \Lambda_n$ and $\mathbf{m} \in \mathbb{Z}^2$ such that $\mathbf{j} = \nu + B\mathbf{m}$. Indeed, the translations of Ω_B by $B\mathbb{Z}^2$ tile $\mathbb{R}^2$, thus we have $\mathbf{j} = \mathbf{x} + B\mathbf{m}$ for certain $\mathbf{x} \in \Omega_B$ and $\mathbf{m} \in \mathbb{Z}^2$. Since all entries of the matrix B are integers, we further deduce that $\nu := \mathbf{x} = \mathbf{j} - B\mathbf{m} \in \mathbb{Z}^2 \cap \Omega_B = \Lambda_n$.

Let $\mathbf{j} \in \Lambda_{2n-1}^*$ and assume $\mathbf{j} = \nu + B\mathbf{m}$ with $\nu \in \Lambda_n$ and $\mathbf{m} \in \mathbb{Z}^2$. Then, $\phi_{\mathbf{j}}\left(\frac{\mathbf{k}}{2n}\right) = \phi_\nu\left(\frac{\mathbf{k}}{2n}\right)$ for each $\mathbf{k} \in X_n^*$. Consequently, we obtain from Theorem 8.2.2 that

$$\begin{aligned}\sum_{\mathbf{k}\in X_n^*} c_{\mathbf{k}}^{(n)} \phi_{\mathbf{j}}\left(\tfrac{\mathbf{k}}{2n}\right) &= \sum_{\mathbf{k}\in X_n^*} c_{\mathbf{k}}^{(n)} \phi_\nu\left(\tfrac{\mathbf{k}}{2n}\right) = \sum_{\mathbf{k}\in X_n} \phi_\nu\left(\tfrac{\mathbf{k}}{2n}\right)\\ &= \sum_{\mathbf{k}\in X_n} \phi_{\mathbf{j}}\left(\tfrac{\mathbf{k}}{2n}\right) = \int_\Omega \phi_\nu(\mathbf{x})d\mathbf{x} = \delta_{\nu,0}.\end{aligned}$$

Since $\nu = 0$ implies $\mathbf{j} = B\mathbf{m} \in \mathbb{Z}^2$, which gives $\mathbf{j} = \mathbf{m} = 0$, we further obtain that $\delta_{\nu,0} = \delta_{\mathbf{j},0}$. This completes the proof of (8.12). □

8.2.2 Cubature for Algebraic Polynomials

The set Λ_n^* is symmetric with respect to the mappings $(x_1,x_2) \mapsto (-x_1,x_2)$ and $(x_1,x_2) \mapsto (x_1,-x_2)$. It follows that the spaces

$$\mathcal{T}_n := \operatorname{span}\{\cos 2\pi j_1 x_1 \cos 2\pi j_2 x_2 : 0 \le j_1 + j_2 \le n\},$$
$$\mathcal{S}_n := \operatorname{span}\{\sin 2\pi j_1 x_1 \sin 2\pi j_2 x_2 : 1 \le j_1 + j_2 \le n\}$$

are both subspaces of $\mathcal{H}_n^*$ and the space $\mathcal{T}_n$ is, under the changing of variables

$$\mathbf{t} = (t_1,t_2) = (\cos 2\pi x_1, \cos 2\pi x_2), \qquad \mathbf{x} = (x_1,x_2) \in [-\tfrac{1}{2},\tfrac{1}{2}]^2, \tag{8.13}$$

also the space of algebraic polynomials Π_n^2 in $\mathbf{t}$ with a basis given in terms of the Chebyshev polynomials $T_{j_1}(t_1)T_{j_2}(t_2)$. This allows us to translate the cubature rules in Subsection 8.2.1 to those of algebraic polynomials for the product Chebyshev weight function W_0 of the first kind and W_1 of the second kind.

Since $\mathcal{T}_n$ contains functions even in both variables, we only need to consider their values over $X_n^* \cap \{\mathbf{x} : x_1 \ge 0, x_2 \ge 0\}$. Hence, we define

$$\Xi_n := \{(2k_1,2k_2) : 0 \le k_1,k_2 \le \tfrac{n}{2}\} \cup \{(2k_1+1,2k_2+1) : 0 \le k_1,k_2 \le \tfrac{n-1}{2}\}. \tag{8.14}$$

Furthermore, we denote by $\Xi_n^\circ := \Xi_n \cap (0,n)^2$ the subset of interior points of Ξ_n, by Ξ_n^e the set of points in Ξ_n that are on the boundary of $(0,n)^2$ but not on the four corners, and by Ξ_n^v the set of points in Ξ_n that are on the corners of $[0,n]^2$. A quick counting shows that

$$|\Xi_n| = \frac{n(n+1)}{2} + \left\lfloor \frac{n}{2} \right\rfloor + 1 \quad \text{and} \quad |\Xi_n^\circ| = \frac{(n-2)n}{2} + \left\lfloor \frac{n-2}{2} \right\rfloor + 1.$$

Theorem 8.2.4. *Let* $\mathbf{z}_{\mathbf{k},n} = (\cos\frac{k_1}{n}, \cos\frac{k_2}{n})$. *Then, the cubature rule for* W_0

$$\frac{1}{\pi^2}\int_{[-1,1]^2} f(\mathbf{t})W_0(\mathbf{t})d\mathbf{t} = \frac{1}{2n^2}\sum_{\mathbf{k}\in\Xi_n} \lambda_k^{(n)} f(\mathbf{z_k}), \quad \lambda_{\mathbf{k}}^{(n)} := \begin{cases} 4, & \mathbf{k} \in \Xi_n^\circ, \\ 2, & \mathbf{k} \in \Xi_n^e, \\ 1, & \mathbf{k} \in \Xi_n^v, \end{cases} \tag{8.15}$$

and the cubature rule for W_1

$$\frac{1}{\pi^2}\int_{[-1,1]^2} f(\mathbf{t})W_1(\mathbf{t})d\mathbf{t} = \frac{2}{n^2}\sum_{\mathbf{k}\in\Xi_{n+2}^\circ} \sin^2\frac{k_1\pi}{n+2}\sin^2\frac{k_2\pi}{n+2} f(\mathbf{z}_{\mathbf{k},n+2}) \tag{8.16}$$

are degree $2n-1$ *with the number of nodes one more than Möller's lower bound* (5.9).

Proof We note that X_n^* is symmetric in the sense that $\mathbf{k} \in X_n^*$ implies that $(\pm k_1, \pm k_2) \in X_n^*$. Let $g(\mathbf{x}) = f(\cos 2\pi x_1, \cos 2\pi x_2)$. Then, g is even in each of its variables and $g(\frac{\mathbf{k}}{2n}) = f(\mathbf{z}_{\mathbf{k},n})$. Since $f \in \Pi_{2n-1}^2$ implies $g \in \mathcal{T}_{2n-1}^*$, applying the first cubature formula (8.12) to g, we see that (8.15) follows from the following identity,

$$\sum_{k \in X_n^*} c_k^{(n)} g(\tfrac{\mathbf{k}}{2n}) = \sum_{\mathbf{k} \in \Xi_n} \lambda_{\mathbf{k}}^{(n)} f(\mathbf{z}_k b).$$

To prove this identity, let $\mathbf{k}\sigma$ denote the set of distinct elements in $\{(\pm k_1, \pm k_2)\}$; then $g(\frac{k}{2n})$ takes the same value on all points in $\mathbf{k}\sigma$. If $\mathbf{k} \in X_n^*$, $k_1 \neq 0$ and $k_2 \neq 0$, then $\mathbf{k}\sigma$ contains four points; $\sum_{\mathbf{j} \in \mathbf{k}\sigma} c_{\mathbf{k}}^{(n)} g(\frac{\mathbf{j}}{2n}) = 4g(\frac{\mathbf{k}}{2n})$ if $\mathbf{k} \in X_n^\circ$, $\sum_{\mathbf{j} \in \mathbf{k}\sigma} c_{\mathbf{k}}^{(n)} g(\frac{\mathbf{j}}{2n}) = 2g(\frac{\mathbf{k}}{2n})$ if $\mathbf{k} \in X_n^e$, and $\sum_{\mathbf{j} \in \mathbf{k}\sigma} c_{\mathbf{k}}^{(n)} g(\frac{\mathbf{j}}{2n}) = g(\frac{\mathbf{k}}{2n})$ if $\mathbf{k} \in X_n^v$. If $k_1 = 0$ and $k_2 \neq 0$ or $k_2 = 0$ and $k_1 \neq 0$, then $\mathbf{k}\sigma$ contains two points; $\sum_{\mathbf{j} \in \mathbf{k}\sigma} c_{\mathbf{k}}^{(n)} g(\frac{\mathbf{j}}{2n}) = 2g(\frac{\mathbf{k}}{2n})$ if $\mathbf{k} \in X_n^\circ$ and $\sum_{\mathbf{j} \in \mathbf{k}\sigma} c_{\mathbf{k}}^{(n)} g(\frac{\mathbf{j}}{2n}) = g(\frac{\mathbf{k}}{2n})$ if $\mathbf{k} \in X_n^e$. Finally, if $\mathbf{k} = (0,0)$, then $\mathbf{k}\sigma$ contains one point and $g(0,0)$ has coefficient 1. Putting these together proves (8.15).

To prove (8.16), we apply the first cubature formula in (8.12) on the functions

$$\sin(2\pi(k_1+1)x_1)\sin(2\pi(k_2+1)x_2)\sin 2\pi x_1 \sin 2\pi x_2$$

for $0 \le k_1 + k_2 \le 2n-5$. These functions are even in both variables and are elements of $\mathcal{T}_{2n-1}^*$. Furthermore, they are zero when $x_1 = 0$ or $x_2 = 0$ or when (x_1, x_2) are on the boundary of X_n^*. Hence, the change of variables (8.13) shows that the first cubature in (8.12) becomes (8.16) for $U_{k_1}(t_1)U_{k_2}(t_2)$ for $k_1 + k_2 \le 2n-5$. Changing n to $n+2$ proves (8.16). □

It is easy to verify that these two cubature rules are exactly those in Theorem 6.4.4. In other words, using the discrete Fourier transform, we provide an alternative construction for the two near-minimal cubature rules for the product Chebyshev weight functions.

8.3 Cubature Rules on the Cube in Three Variables

We consider the cubature rules on a cube in three variables. As shown in Section 8.2, we choose A as the identity matrix and Ω_A as $\Omega = [-1,1)^3$. If we were to follow the construction for $d = 2$ in Section 8.2, we would choose the spectral set as the octahedron $\{x : -n \le x_1 \pm x_2 \pm x_3 \le n\}$. The octahedron, however, does not tile $\mathbb{R}^3$ by lattice translation. As an alternative, we choose the spectral set as *rhombic dodecahedron* that tiles $\mathbb{R}^3$ by lattice translation with face-centered cubic (fcc) lattice. Thus, we choose the matrix B as the generator matrix of fcc lattice,

$$B = n\begin{pmatrix} 0 & 1 & 1 \\ 1 & 0 & 1 \\ 1 & 1 & 0 \end{pmatrix} \quad \text{and} \quad B^{-1} = \frac{1}{2n}\begin{pmatrix} -1 & 1 & 1 \\ 1 & -1 & 1 \\ 1 & 1 & -1 \end{pmatrix}.$$

The spectral set of the fcc lattice is the rhombic dodecahedron defined by

$$\Omega_B = \{x \in \mathbb{R}^3 : -n \le x_\nu \pm x_\mu < n, 1 \le \nu < \mu \le 3\}.$$

See Figure 8.1. The strict inequality in the definition of Ω_B reflects our require-

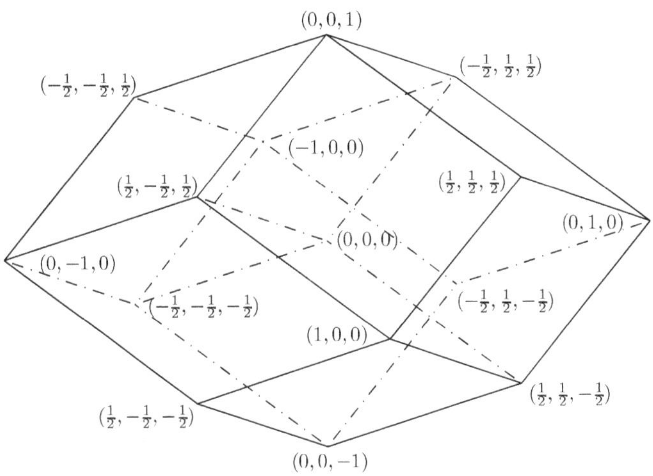

Figure 8.1 Rhombic dodecahedron.

ment that the tiling of the spectral set has no overlapping. Then, $N = B^{\mathsf{T}}$ so that Λ_N, which we denote by Λ_n, is given by

$$\Lambda_n := \{j \in \mathbb{Z}^3 : -n \le -j_1 + j_2 + j_3, j_1 - j_2 + j_3, j_1 + j_2 - j_3 < n\}.$$

Then, $|\Lambda_n| = \det(B) = 2n^3$. Moreover, $\Lambda_{N^{\mathsf{T}}}$, which we denote by $\Lambda_n^\dagger$, is given by

$$\Lambda_n^\dagger = \mathbb{Z}^3 \cap \Omega_B = \{\mathbf{k} \in \mathbb{Z}^3 : -n \le k_\nu \pm k_\mu < n,\ 1 \le \nu < \mu \le 3\}.$$

We again denote the space $\mathcal{H}_N$ by $\mathcal{H}_n$, which is given by

$$\mathcal{H}_n := \operatorname{span}\{\mathrm{e}^{2\pi i k\cdot x} :\ k \in \Lambda_n^\dagger\}.$$

Then, $\dim \mathcal{H}_n = |\Lambda_n^\dagger| = \det(B) = 2n^3$. Furthermore, as in the case of $d = 2$, we define

$$X_n := \{2\mathbf{k} : -\tfrac{n}{2} \le k_1, k_2, k_3 < \tfrac{n}{2}\} \cup \{2\mathbf{k} + 1 : -\tfrac{n+1}{2} \le k_1, k_2, k_3 < \tfrac{n-1}{2}\}$$

and defines its symmetric counterpart by X_n^* so that

$$X_n^* := \{2\mathbf{k} : -\tfrac{n}{2} \le k_1, k_2, k_3 \le \tfrac{n}{2}\} \cup \{2\mathbf{k}+1 : -\tfrac{n+1}{2} \le k_1, k_2, k_3 \le \tfrac{n-1}{2}\}.$$

Then, $|X_n^*| = n^3 + (n+1)^3$. The set X_n^* is further partitioned into four parts,

$$X_n^* = X_n^\circ \cup X_n^f \cup X_n^e \cup X_n^v,$$

where $X_n^\circ = X_n^* \cap (-n,n)^2$ is the set of interior points, X_n^f contains the points in X_n^* that are on the faces of $[-n,n]^3$ but not on the edges or vertices, X_n^e contains the points in X_n^* that are on the edges of $[-n,n]^3$ but not on the corners or vertices, while X_n^v denotes the points of X_n^* at the vertices of $[-n,n]^3$.

Theorem 8.3.1. *Define the inner product*

$$\langle f,g\rangle_n := \frac{1}{2n^3} \sum_{\mathbf{k}\in X_n} f\left(\tfrac{\mathbf{k}}{2n}\right)\overline{g\left(\tfrac{\mathbf{k}}{2n}\right)},$$

and

$$\langle f,g\rangle_n^* := \frac{1}{2n^3} \sum_{\mathbf{k}\in X_n^*} c_{\mathbf{k}}^{(n)} f\left(\tfrac{\mathbf{k}}{2n}\right)\overline{g\left(\tfrac{k}{2n}\right)}, \qquad \textit{where} \quad c_k^{(n)} = \begin{cases} 1, & \mathbf{k}\in X_n^\circ, \\ \frac{1}{2}, & \mathbf{k}\in X_n^f, \\ \frac{1}{4}, & \mathbf{k}\in X_n^e, \\ \frac{1}{8}, & \mathbf{k}\in X_n^v. \end{cases} \tag{8.17}$$

Then, for all $f,g \in \mathcal{H}_n$,

$$\langle f,g\rangle = \int_{[-\frac{1}{2},\frac{1}{2}]^3} f(\mathbf{x})\overline{g(\mathbf{x})}\mathrm{d}\mathbf{x} = \langle f,g\rangle_n = \langle f,g\rangle_n^*.$$

Proof Changing variables from $\mathbf{j}$ to $\mathbf{k} = 2nB^{-\mathsf{T}}\mathbf{j}$, or $\mathbf{j} = B^{\mathsf{T}}\mathbf{k}/(2n)$, then, as j_1, j_2, j_3 are integers and $j_1 = \frac{k_2+k_3}{2}$, $j_2 = \frac{k_1+k_3}{2}$, $j_3 = \frac{k_1+k_2}{2}$, we see that $\mathbf{j} \in \Lambda_n$ is equivalent to $2nB^{-\mathsf{T}}\mathbf{j} \in X_n$, so that $\langle f,g\rangle_n$ is $\langle f,g\rangle_N$ in (8.5) and is equal to $\langle f,g\rangle$ by Corollary 8.1.4. Next, we show $\langle f,g\rangle_n = \langle f,g\rangle^*$ if $f\bar{g}$ is periodic. The interior points of X_n and X_n^* are the same, so that $c_{\mathbf{k}}^{(n)} = 1$ for $\mathbf{k} \in X_n^\circ$. Each point $\mathbf{k}$ in X_n^f has exactly one opposite point $\mathbf{k}^*$ in X_n^f under translation by $\pm n\varepsilon_i$, where $\varepsilon_1 = (1,0,0)$, $\varepsilon_2 = (0,1,0)$, and $\varepsilon_3 = (0,0,1)$, and only one of them is in X_n, so that $f(\mathbf{x_k}) = \frac{1}{2}[f(\mathbf{x_k}) + f(\mathbf{x_k^*})]$ if f is periodic, which explains why we define $c_{\mathbf{k}}^{(n)} = \frac{1}{2}$ for $\mathbf{k} \in X_n^f$. Moreover, only three edges of X_n^* are in $X_n^* \setminus X_n$, and each point in X_n^e corresponds to exactly four points in X_n^e under integer translations $\pm n\varepsilon_i$ and only one among the four is in X_n, so we define $c_{\mathbf{k}}^{(n)} = \frac{1}{4}$ for $\mathbf{k} \in X_n^e$. Finally, all eight corner points can be derived from translations $n\varepsilon_i$ points, used repeatedly, and exactly one, $(-n,-n,-n)$, is in $X_n^* \setminus X_n$, so that we define $c_{\mathbf{k}}^{(n)} = \frac{1}{8}$ for $\mathbf{k} \in X_n^v$. □

We also denote the symmetric counterpart of $\Lambda_n^\dagger$ by $\Lambda_n^{\dagger*}$,

$$\Lambda_n^{\dagger*} := \{\mathbf{j} \in \mathbb{Z}^3 : -n \le j_\nu \pm j_\mu \le n, 1 \le \nu < \mu \le 3\} \tag{8.18}$$

and denote the counterpart of $\mathcal{H}_n$ by $\mathcal{H}_n^*$, which is defined accordingly by

$$\mathcal{H}_n^* := \operatorname{span}\left\{e^{2\pi i \mathbf{k}\cdot\mathbf{x}} : \mathbf{k} \in \Lambda_n^{\dagger*}\right\}.$$

Theorem 8.3.2. *Let $\mathcal{I}f = \int_{[-\frac{1}{2},\frac{1}{2}]^3} f(\mathbf{x})d\mathbf{x}$. For $n \ge 2$, the cubature rules*

$$\mathcal{I}f = \frac{1}{2n^3} \sum_{\mathbf{k}\in X_n^*} c_k^{(n)} f\left(\tfrac{\mathbf{k}}{2n}\right) \quad \text{and} \quad \mathcal{I}f = \frac{1}{2n^3} \sum_{\mathbf{k}\in X_n} f\left(\tfrac{\mathbf{k}}{2n}\right) \tag{8.19}$$

are exact for $f \in \mathcal{H}_{2n-1}^*$.

Proof Let $\phi_\mathbf{k}(\mathbf{x}) = e^{2\pi i \mathbf{k}\cdot\mathbf{x}}$. As in the proof of Theorem 8.2.3, for any $\mathbf{i} \in \mathbb{Z}^3$, there exist $\nu \in \Lambda_n^\dagger$ and $\mathbf{j} \in \mathbb{Z}^3$ such that $\mathbf{i} = \nu + B\mathbf{j}$. Assume now $\mathbf{i} \in \Lambda_{2n-1}^{\dagger*}$. Then, the integral of $\phi_\mathbf{i}$ over $[-1,1]^3$ is $\delta_{\mathbf{i},0}$. Let us suppose $\mathbf{i} = \nu + B\mathbf{j}$ with $\nu \in \Lambda_n$ and $\mathbf{j} \in \mathbb{Z}^3$. Then, $\phi_\mathbf{i}(\frac{\mathbf{k}}{2n}) = \phi_\nu(\frac{\mathbf{k}}{2n})$ for each $\mathbf{k} \in X_n^*$. Consequently, we obtain by Theorem 8.3.1 that

$$\begin{aligned}\sum_{\mathbf{k}\in X_n^*} c_\mathbf{k}^{(n)} \phi_\mathbf{j}\left(\tfrac{\mathbf{k}}{2n}\right) &= \sum_{\mathbf{k}\in X_n^*} c_\mathbf{k}^{(n)} \phi_\nu\left(\tfrac{\mathbf{k}}{2n}\right) = \sum_{\mathbf{k}\in X_n} \phi_\nu\left(\tfrac{\mathbf{k}}{2n}\right)\\ &= \sum_{\mathbf{k}\in X_n} \phi_\mathbf{j}\left(\tfrac{\mathbf{k}}{2n}\right) = \mathcal{I}\phi_\nu = \delta_{\nu,0}.\end{aligned}$$

Since $\nu = 0$ implies $\mathbf{i} = \mathbf{j} = 0$, we further obtain that $\delta_{\nu,0} = \delta_{\mathbf{i},0}$. Hence, the cubature rules (8.19) are exact for each $\phi_\mathbf{i}$ with $\mathbf{i} \in \Lambda_{2n-1}^{\dagger*}$. The proof is completed. □

We can also translate the cubature rules in the above theorem into those for the Chebyshev weight functions of the first and the second kind under the change of variables

$$t_1 = \cos 2\pi x_1, \quad t_2 = \cos 2\pi x_2, \quad t_3 = \cos 2\pi x_3, \quad x \in \left[-\tfrac{1}{2},\tfrac{1}{2}\right]^3,$$

which maps the subspace of $\mathcal{H}_n^*$ that consists of functions even in each of its variables to the space of algebraic polynomials defined by

$$\Pi_n^* = \operatorname{span}\{T_{k_1}(x_1)T_{k_2}(x_2)T_{k_3}(x_3) : k_1,k_2,k_3 \ge 0, k_\nu + k_\mu \le n, 1 \le \nu < \mu \le 3\}.$$

The index set of the space Π_n^* is depicted in Figure 8.2. To evaluate functions that are even in each of its variables on X_n^*, we only need to consider $X_n^* \cap \{x : x_1,x_2,x_3 \ge 0\}$. Hence, we define,

$$\Xi_n := \{2k : 0 \le k_1,k_2,k_3 \le \tfrac{n}{2}\} \cup \{2k+1 : 0 \le k_1,k_2,k_3 \le \tfrac{n-1}{2}\}. \tag{8.20}$$

Moreover, we denote by Ξ_n°, Ξ_n^f, Ξ_n^e and Ξ_n^v the subsets of Ξ_n that consist of

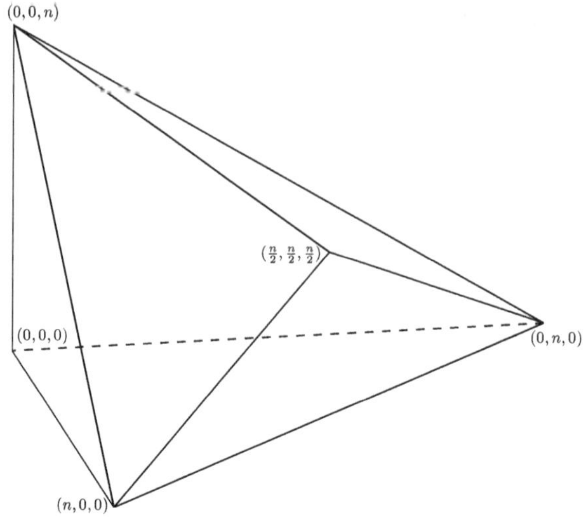

Figure 8.2 Index set of Π_n^*.

interior points, points on the faces but not on the edges, points on the edges but not on the vertices, and points on the vertices, of $[0,n]^3$, respectively. Then,

$$|\Xi_n| = \left(\lfloor \tfrac{n}{2} \rfloor + 1\right)^3 + \left(\lfloor \tfrac{n-1}{2} \rfloor + 1\right)^3 = \begin{cases} \frac{(n+1)^3}{4} + \frac{3(n+1)}{4}, & n \text{ is even,} \\ \frac{(n+1)^3}{4}, & n \text{ is odd.} \end{cases} \tag{8.21}$$

Theorem 8.3.3. *Let* $\mathbf{z}_{\mathbf{k},n} = \left(\cos \frac{k_1}{n}, \cos \frac{k_2}{n}, \cos \frac{k_3}{n}\right)$. *The cubature rule for* W_0

$$\frac{1}{\pi^3} \int_{[-1,1]^3} f(\mathbf{t}) W_0(\mathbf{t}) d\mathbf{t} = \frac{1}{2n^2} \sum_{\mathbf{k} \in \Xi_n} \lambda_{\mathbf{k}}^{(n)} f(\mathbf{z}_{\mathbf{k}}), \quad \lambda_{\mathbf{k}}^{(n)} := \begin{cases} 8, & k \in \Xi_n^\circ, \\ 4, & k \in \Xi_n^f, \\ 2, & k \in \Xi_n^e, \\ 1, & k \in \Xi_n^v, \end{cases} \tag{8.22}$$

and the cubature rule for W_1

$$\frac{1}{\pi^3} \int_{[-1,1]^3} f(\mathbf{t}) W_1(\mathbf{t}) d\mathbf{t} = \frac{4}{n^3} \sum_{\mathbf{k} \in \Xi_{n+2}^\circ} \sin^2 \tfrac{k_1 \pi}{n+2} \sin^2 \tfrac{k_2 \pi}{n+2} \sin^2 \tfrac{k_3 \pi}{n+2} f(\mathbf{z}_{\mathbf{k}}) \tag{8.23}$$

are exact for Π_{2n-1}^*. *In particular, they are exact for* Π_{2n-1}^3.

The proof of this theorem is parallel to that of Theorem 8.2.4, and we shall skip the details.

We observe that Π_{2n-1}^* contains Π_{2n-1}^3 as a subspace as the index set for the latter is the integers in the simplex. but it does not contain Π_{2n}^3 since, for example, $T_n(x_1)T_n(x_2)$ is in Π_{2n}^3 but not in Π_{2n-1}^3. Hence, the cubature (8.22) is of degree $2n-1$. The number of nodes of these cubature rules, as shown in (8.23), is

$$N = \frac{1}{4}n^3 + \mathcal{O}(n^2),$$

which is substantially greater than Möller's lower bound $\frac{1}{6}n^3 + \mathcal{O}(n^2)$, whereas it is much better than the number of nodes, n^3, of the product Gauss quadrature rules of degree $2n-1$.

Remark 8.3.4. *The cubature rule evaluates polynomials in Π_{2n-1}^* exactly, which is more than Π_{2n-1}^3 (see Figure 8.2). Our degree of precision of a cubature rule is measured in terms of the highest total degree of polynomials evaluated exactly by the cubature rule. We observe that if the precision were measured by the maximal degree of product polynomials, then the product Gauss cubature rule of n^3 nodes would be the minimal one since it evaluates all polynomials in $\Pi_{2n-1} \times \Pi_{2n-1} \times \Pi_{2n-1}$ exactly.*

8.4 $\mathcal{A}_d$ Lattice and Gauss Cubature Rule

For the Euclidean plane $\mathbb{R}^2$, there are essentially two tilings by translations of convex fundamental sets: tiling by rectangular and hexagonal domains. In this section, we consider the hexagonal tiling for the plane and its extension to $\mathbb{R}^d$, which are tilling by the fundamental domain of the $\mathcal{A}_d$ lattice. In Subsection 8.4.1, we discuss the $\mathcal{A}_d$ lattice and define the generalized cosine function, invariant under $\mathcal{A}_d$, and sine function, skew-invariant under $\mathcal{A}_d$. Under a change of variables via symmetric polynomials in Subsection 8.4.2, these functions become the algebraic polynomials, called generalized Chebyshev polynomials of the first and the second kind, that are orthogonal on a curved domain. In Subsection 8.4.3, we show that the common zeros of generalized Chebyshev polynomials of the second kind are the nodes of Gauss cubature rules.

8.4.1 Generalized Sine and Cosine Functions Associated with $\mathcal{A}_d$ Lattice

It is more convenient to adopt homogeneous coordinates. For $d \ge 2$, we identify $\mathbb{R}^d$ with the hyperplane $\mathbb{R}_{\mathrm{H}}^{d+1}$ in $\mathbb{R}^{d+1}$, defined by

$$\mathbb{R}_{\mathrm{H}}^{d+1} := \left\{(t_1, t_2, \ldots, t_{d+1}) \in \mathbb{R}^{d+1} : t_1 + t_2 + \cdots + t_{d+1} = 0\right\}.$$

We use *homogeneous coordinates* $\mathbf{t} \in \mathbb{R}_{\mathrm{H}}^{d+1}$ to describe our lattice in d-variables. In $\mathbb{R}_{\mathrm{H}}^{d+1}$, the lattice is given by

$$\mathbb{Z}_{\mathrm{H}}^{d+1} := \mathbb{Z}^{d+1} \cap \mathbb{R}_{\mathrm{H}}^{d+1} = \{(k_1, k_2, \ldots, k_{d+1}) \in \mathbb{Z}^{d+1} : k_1 + k_2 + \cdots + k_{d+1} = 0\}$$

and the fundamental domain for the lattice is

$$\Omega_{\mathrm{H}} := \{\mathbf{t} \in \mathbb{R}_{\mathrm{H}}^{d+1} \mid -1 \le t_i - t_j \le 1,\ 1 \le i < j \le d+1\},$$

which tiles $\mathbb{R}_{\mathrm{H}}^{d+1}$ in the sense that the characteristic function Ξ_Ω of Ω_{H} satisfies

$$\sum_{\mathbf{k} \in \mathbb{Z}_{\mathrm{H}}^{d+1}} \Xi_\Omega(\mathbf{t} + k) = 1 \quad \text{for almost all } \mathbf{t} \in \mathbb{R}_{\mathrm{H}}^{d+1}.$$

For $d = 2$, this is equivalent to the hexagonal tiling of $\mathbb{R}^2$, as shown in Figure 8.3. Up to an affine transformation, the fundamental domain is a rhombic

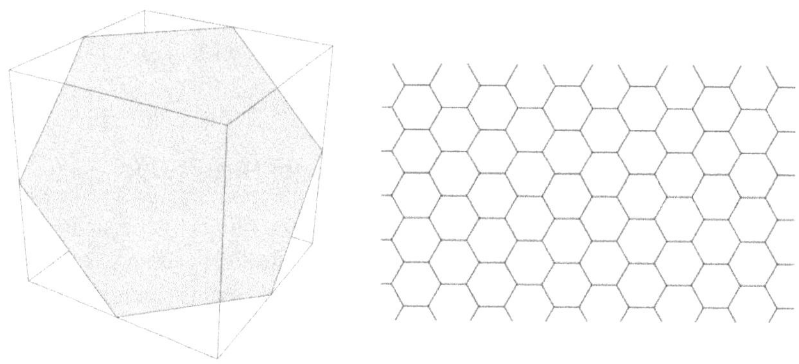

Figure 8.3 Ω_{H} in $\mathbb{R}_{\mathrm{H}}^3$ and hexagonal tilling.

dodecahedron.

The lattice is the root lattice of the reflection group $\mathcal{A}_d$. Under homogeneous coordinates, the group $\mathcal{A}_d$ is generated by the reflections $\{\sigma_{ij} : 1 \le i < j \le d+1\}$, where σ_{ij} is defined by

$$\mathbf{t}\sigma_{ij} := \mathbf{t} - 2\frac{\langle \mathbf{t}, \mathbf{e}_{i,j} \rangle}{\langle \mathbf{e}_{i,j}, \mathbf{e}_{i,j} \rangle}\mathbf{e}_{i,j} = \mathbf{t} - (t_i - t_j)\mathbf{e}_{i,j}, \quad \text{where } \mathbf{e}_{i,j} := e_i - e_j.$$

In particular, $\sigma_{i,j}$ is a transposition, so that the group $\mathcal{A}_d$ is exactly the permutation group of $d+1$ elements. For $\mathbf{t} \in \mathbb{R}_{\mathrm{H}}^{d+1}$ and $\sigma \in \mathcal{A}_d$, the action of σ on $\mathbf{t}$ is denoted by $\mathbf{t}\sigma$, which means permutation of the element in $\mathbf{t}$ by σ. In homogeneous coordinates, the generator matrix A of the lattice $\mathcal{A}_d$ is of size

$(d+1)\times d$. The matrix A and the generator matrix $A^\perp = A(A^\mathsf{T}A)^{-1}$ of the dual lattice are given by

$$A := \begin{pmatrix} 1 & 0 & \dots & 0 & 0 \\ 0 & 1 & \dots & 0 & 0 \\ \vdots & \vdots & \ddots & \vdots & \vdots \\ 0 & 0 & \dots & 1 & 0 \\ 0 & 0 & \dots & 0 & 1 \\ -1 & -1 & \dots & -1 & -1 \end{pmatrix} \quad \text{and} \quad A^\perp = \frac{1}{d+1}\begin{pmatrix} d & -1 & \dots & -1 & -1 \\ -1 & d & \dots & -1 & -1 \\ \vdots & \vdots & \ddots & \vdots & \vdots \\ -1 & -1 & \dots & d & -1 \\ -1 & -1 & \dots & -1 & d \\ -1 & -1 & \dots & -1 & -1 \end{pmatrix}.$$

To describe the dual lattice of $\mathbb{Z}_\mathrm{H}^{d+1}$ explicitly, we introduce the notation

$$\mathbb{H} := \{\mathbf{k}\in\mathbb{Z}_\mathrm{H}^{d+1} \mid k_1 \equiv \cdots \equiv k_{d+1} \mod (d+1)\}.$$

For $\mathbf{j}\in\mathbb{Z}^d$, let $\mathbf{k} = (d+1)A^\perp\mathbf{j}$. Then, $\mathbf{k}\in\mathbb{Z}_\mathrm{H}^{d+1}$ and, furthermore, $\mathbf{k}\in\mathbb{H}$ as a quick computation shows. Moreover, the definition of $A^\perp$ leads to $\mathbf{j} = A^\mathsf{T}\mathbf{k}/(d+1)$, which shows $\mathbf{j}\in\mathbb{Z}^d$ if $\mathbf{k}\in\mathbb{H}$. Consequently, the dual lattice of $\mathbb{Z}_\mathrm{H}^{d+1}$ is given by

$$\left(\mathbb{Z}_\mathrm{H}^{d+1}\right)^\perp = \left\{\frac{\mathbf{k}}{d+1} : \mathbf{k}\in\mathbb{H}\right\}.$$

A function f on $\mathbb{R}_\mathrm{H}^{d+1}$ is called H-periodic if $f(\mathbf{t}) = f(\mathbf{t}+\mathbf{j})$ for all $\mathbf{j}\in\mathbb{Z}_\mathrm{H}^{d+1}$. The exponential functions defined by

$$\phi_\mathbf{k}(\mathbf{t}) := \mathrm{e}^{\frac{2\pi}{d+1}\mathrm{i}\mathbf{k}\cdot\mathbf{t}}, \qquad \mathbf{k}\in\mathbb{H},$$

are periodic since $\mathbf{j}\cdot\mathbf{k} = j_1(k_1-k_{d+1})+\cdots+j_d(k_d-k_{d+1}) \equiv 0 \mod d+1$ by $\mathbf{k}\in\mathbb{H}$. Define the inner product

$$\langle f,g\rangle_{\Omega_\mathrm{H}} = \frac{1}{|\Omega_\mathrm{H}|}\int_{\Omega_\mathrm{H}} f(\mathbf{t})\overline{g(\mathbf{t})}\mathrm{d}\mathbf{t} = \frac{1}{\sqrt{d+1}}\int_{\Omega_\mathrm{H}} f(\mathbf{t})\overline{g(\mathbf{t})}\mathrm{d}\mathbf{t}.$$

By Theorem 8.1.1, $\{\phi_\mathbf{k} : \mathbf{k}\in\mathbb{H}\}$ consists of an orthonormal basis in $L^2(\Omega_\mathrm{H})$,

$$\langle\phi_\mathbf{j},\phi_\mathbf{k}\rangle_{\Omega_\mathrm{H}} = \delta_{\mathbf{j},\mathbf{k}}, \qquad \mathbf{j},\mathbf{k}\in\mathbb{H}.$$

We denote Λ_N in Definition 8.1.2 by $\mathbb{H}_n$, which is given by

$$\mathbb{H}_n := \left\{\mathbf{k}\in\mathbb{H} : \tfrac{\mathbf{k}}{(d+1)n}\in\Omega_H\right\} = \left\{\mathbf{k}\in\mathbb{H} : -n < \frac{k_i-k_j}{d+1} \le n,\ 1\le i<j\le d+1\right\}.$$

The finite-dimensional space $\mathcal{H}_N$ of (8.6) becomes

$$\mathcal{H}_n := \operatorname{span}\{\phi_\mathbf{k} : \mathbf{k}\in\mathbb{H}_n\} \qquad \text{with} \quad \dim\mathcal{H}_n = \det(N) = (d+1)n^d.$$

Then, Corollary (8.1.4) for $\mathcal{A}_d$ becomes the following proposition.

Theorem 8.4.1. *For $n \geq 0$, define*

$$\langle f, g \rangle_n := \frac{1}{(d+1)n^d} \sum_{j \in \mathbb{H}_n} f\left(\tfrac{\mathbf{j}}{(d+1)n}\right) \overline{g\left(\tfrac{\mathbf{j}}{(d+1)n}\right)}, \quad f, g \in C(\overline{\Omega}_{\mathrm{H}}).$$

Then, it satisfies

$$\langle f, g \rangle = \langle f, g \rangle_n, \qquad f, g \in \mathcal{H}_n. \tag{8.24}$$

Remark 8.4.2. *As in the case of* (8.12) *for the square domain, we could define another inner product on the symmetrized domain $\mathbb{H}_n^*$ and show that it is equal to those defined in* (8.24)*. Such an inner product, however, is a weighted sum, and its weights depend on the decomposition of the set of nodes on the boundary of $\mathbb{H}_n^*$, which can be rather tedious; see [94] for $d = 2$ and [93] for $d > 2$.*

The group $\mathcal{A}_d$ is divided into two parts, $\mathcal{A}_d^+$ (or $\mathcal{A}_d^-$) consists of elements with an even (or odd) number of products of reflections σ_{ij}. The action of $\sigma \in \mathcal{A}_d$ on the function $f : \mathbb{R}_{\mathrm{H}}^{d+1} \mapsto \mathbb{R}$ is defined by $\sigma f(\mathbf{t}) = f(\mathbf{t}\sigma)$. A function f in homogeneous coordinates is called *invariant* if $\sigma f = f$ for all $\sigma \in \mathcal{A}_d$ and skew-symmetric if $\sigma f = (-1)^{\pm} f$ for $f \in \mathcal{A}_d^{\pm}$. We consider $\mathcal{A}_d$ invariant and skew-invariant functions derived from $\phi_{\mathbf{k}}$, which are analogs of cosine and sine functions. For $d = 1$, these functions are $\cos(\pi t)$ and $\sin(\pi t)$ as $\mathcal{A}_1$ is the group of sign changes. For $d > 1$, we define analogs of cosine and sine functions as follows.

Definition 8.4.3. For $\mathcal{A}_d$, we define the operators $\mathcal{P}^+$ and $\mathcal{P}^-$ by

$$\mathcal{P}^{\pm} f(\mathbf{t}) := \frac{1}{(d+1)!} \left(\sum_{\sigma \in \mathcal{A}^+} f(\mathbf{t}\sigma) \pm \sum_{\sigma \in \mathcal{A}^-} f(\mathbf{t}\sigma) \right),$$

where $\mathcal{A}^+$ (or $\mathcal{A}^-$) contains an even (or odd) number of products of reflections σ_{ij}. Then, the functions defined by

$$\mathbf{C}_{\mathbf{k}}(\mathbf{t}) := \mathcal{P}^+ \phi_{\mathbf{k}}(\mathbf{t}) \qquad \text{and} \qquad \mathbf{S}_{\mathbf{k}}(\mathbf{t}) := \mathrm{i}^{-1} \mathcal{P}^- \phi_{\mathbf{k}}(\mathbf{t}) \tag{8.25}$$

are called $\mathcal{A}_d$ *cosine* function and $\mathcal{A}_d$ *sine* function.

By definition, $\mathcal{C}_{\mathbf{k}}$ is invariant, and $\mathcal{S}_{\mathbf{k}}$ is skew-invariant. The spectral set Ω_{H} is the union of congruent simplexes under the action of the group $\mathcal{A}_d$. We define its *fundamental simplex* $\triangle_{\mathrm{H}}$ by

$$\triangle_{\mathrm{H}} := \left\{ \mathbf{t} \in \mathbb{R}_{\mathrm{H}}^{d+1} \mid 0 \leq t_i - t_j \leq 1, \quad 1 \leq i < j \leq d+1 \right\}.$$

For $d = 2$, $\triangle_{\mathrm{H}}$ is the triangle in $\mathbb{R}_{\mathrm{H}}^3$ with vertices

$$\mathbf{v}_0 = (0,0,0), \quad \mathbf{v}_1 = \frac{1}{3}(1,1,-2), \quad \mathbf{v}_2 = \frac{1}{3}(2,-1,-1). \tag{8.26}$$

By the symmetry in their definition, $\mathcal{A}_d$ cosine and sine functions are determined by their restrictions on $\triangle_{\mathrm{H}}$. Moreover, $\phi_{\mathbf{k}}$ and $\phi_{\mathbf{k}\sigma}$ with $\sigma \in \mathcal{A}_d$ lead to the same $\mathsf{C}_{\mathbf{k}}$ since $\mathsf{C}_{\mathbf{k}\sigma}(\mathbf{t}) = \mathsf{C}_{\mathbf{k}}(\mathbf{t}\sigma) = \mathsf{C}_{\mathbf{k}}(\mathbf{t})$ for $\mathbf{t} \in \triangle_{\mathrm{H}}$ and $\sigma \in \mathcal{A}_d$. Consequently, we only need to consider $\mathcal{A}_d$ cosine functions in the space

$$\mathcal{C}(\mathcal{A}) := \{\mathsf{C}_{\mathbf{k}} : \mathbf{k} \in \Lambda\} \quad \text{with} \quad \Lambda := \{\mathbf{k} \in \mathbb{H} : k_1 \ge k_2 \ge \cdots \ge k_{d+1}\}. \tag{8.27}$$

Similarly, $\mathcal{A}_d$ sine functions satisfy $\mathsf{S}_{\mathbf{k}\sigma}(\mathbf{t}) = \mathsf{S}_{\mathbf{k}}(\mathbf{t}\sigma) = \mathsf{S}_{\mathbf{k}}(\mathbf{t})$, $\sigma \in \mathcal{A}^+$ and $\mathsf{S}_{\mathbf{k}\sigma}(\mathbf{t}) = \mathsf{S}_{\mathbf{k}}(\mathbf{t}\sigma) = -\mathsf{S}_{\mathbf{k}}(\mathbf{t})$, $\sigma \in \mathcal{A}^-$. In particular, it follows that $\mathsf{S}_{\mathbf{k}}(\mathbf{t}) = 0$ whenever two or more components of $\mathbf{k}$ are equal. Thus, we only need to consider $\mathcal{A}_d$ sine functions in the space

$$\mathcal{S}(\mathcal{A}) := \{\mathsf{S}_{\mathbf{k}} : \mathbf{k} \in \Lambda^\circ\} \quad \text{with} \quad \Lambda^\circ := \{\mathbf{k} \in \mathbb{H} : k_1 > k_2 > \cdots > k_{d+1}\}. \tag{8.28}$$

The index set Λ° consists of the interior points of Λ. The boundary of Λ depends on the integer decomposition of $d+1$. A *composition* $\mathbf{p}$ of $d+1$ is a decomposition of $d+1$ such that

$$\mathbf{p} = (p_1, \ldots, p_\ell) \in \mathbb{Z}^\ell, \quad p_i > 0, \quad 1 \le i \le \ell \le d+1, \quad |\mathbf{p}| := p_1 + \cdots + p_\ell = d+1.$$

The order of p_i in $\mathbf{p}$ matters so that a composition is different from a partition. Let $\{t\}^p = (t, t, \ldots, t) \in \mathbb{R}^p$. For each composition $\mathbf{p}$ of $d+1$, define

$$\Lambda^{\mathbf{p}} := \left\{\mathbf{k} = \left(\{k_1\}^{p_1}, \{k_2\}^{p_2}, \cdots, \{k_\ell\}^{p_\ell}\right) \in \mathbb{H} : k_i > k_j, 1 \le i < j \le \ell\right\}.$$

Then, $\Lambda^\circ = \Lambda^{\{1\}^{d+1}}$ and $\Lambda = \bigcup_{\mathbf{p} \in \mathcal{C}_{d+1}} \Lambda^{\mathbf{p}}$, where $\mathcal{C}_{d+1}$ denotes the set of compositions of $d+1$. Let $\mathbf{k}\mathcal{A}_d = \{\mathbf{k}\sigma : \ \sigma \in \mathcal{A}_d\}$ be the orbit of $\mathbf{k}$ under $\mathcal{A}_d$. If $\mathbf{k}, \mathbf{j} \in \Lambda$ and $\mathbf{k} \ne \mathbf{j}$, then $\mathbf{k}\mathcal{A}_d \cap \mathbf{j}\mathcal{A}_d = \varnothing$. Hence,

$$|\mathbf{k}\mathcal{A}_d| = \binom{d+1}{\mathbf{p}} = \frac{(d+1)!}{p_1! p_2! \cdots p_\ell!} \qquad \text{if } \mathbf{k} \in \Lambda^{\mathbf{p}}.$$

Let $\mathcal{A}_d(\mathbf{k})$ be the stabilizer of $\mathbf{k}$ in $\mathcal{A}_d$. Then, $\mathcal{A}_d / \mathcal{A}_d(\mathbf{k})$ is isomorphic to $\mathbf{k}\mathcal{A}_d$, so that

$$\mathsf{C}_{\mathbf{k}}(\mathbf{t}) = \frac{1}{|\mathcal{A}|} \sum_{\mathbf{j} \in \mathbf{k}\mathcal{A}_d} \sum_{\sigma \in \mathcal{A}_d(\mathbf{j})} \phi_{\mathbf{j}\sigma}(\mathbf{t}) = \frac{1}{|\mathbf{k}\mathcal{A}_d|} \sum_{\mathbf{j} \in \mathbf{k}\mathcal{A}_d} \phi_{\mathbf{j}}(\mathbf{t}). \tag{8.29}$$

More generally, we show that these functions satisfy several recurrence relations.

Lemma 8.4.4. *The $\mathcal{A}_d$ sine and cosine satisfy the recurrence relations,*

$$\mathsf{C}_{\mathbf{j}}(\mathbf{t})\mathsf{C}_{\mathbf{k}}(\mathbf{t}) = \frac{1}{|\mathcal{A}_d|}\sum_{\sigma\in\mathcal{A}_d}\mathsf{C}_{\mathbf{k}+\mathbf{j}\sigma}(\mathbf{t}), \qquad \mathbf{j},\mathbf{k}\in\Lambda, \tag{8.30}$$

$$\mathsf{C}_{\mathbf{j}}(\mathbf{t})\mathsf{S}_{\mathbf{k}}(\mathbf{t}) = \frac{1}{|\mathcal{A}|}\sum_{\sigma\in\mathcal{A}_d}\mathsf{S}_{\mathbf{k}+\mathbf{j}\sigma}(\mathbf{t}), \qquad \mathbf{j},\mathbf{k}\in\Lambda, \tag{8.31}$$

$$\mathsf{S}_{\mathbf{j}}(\mathbf{t})\mathsf{S}_{\mathbf{k}}(\mathbf{t}) = \frac{1}{|\mathcal{A}|}\sum_{\sigma\in\mathcal{A}_d}(-1)^{|\sigma|}\mathsf{C}_{\mathbf{k}+\mathbf{j}\sigma}(\mathbf{t}), \quad \mathbf{j},\mathbf{k}\in\Lambda. \tag{8.32}$$

Proof From the definition of $\mathsf{C}_{\mathbf{k}}$, we obtain that

$$\begin{aligned}\mathsf{C}_{\mathbf{k}}(\mathbf{t})\mathsf{C}_{\mathbf{k}}(\mathbf{t}) &= \frac{1}{|\mathcal{A}_d|}\sum_{\sigma\in\mathcal{A}_d}\phi_{\mathbf{k}\sigma}(\mathbf{t})\times\frac{1}{|\mathcal{G}|}\sum_{\sigma\in\mathcal{A}_d}\phi_{\mathbf{k}\sigma}(\mathbf{t}) = \frac{1}{|\mathcal{A}_d|^2}\sum_{\tau\in\mathcal{A}_d}\sum_{\sigma\in\mathcal{A}_d}\phi_{\mathbf{k}\tau+\mathbf{j}\sigma}(\mathbf{t})\\ &= \frac{1}{|\mathcal{A}_d|^2}\sum_{\tau\in\mathcal{A}_d}\sum_{\sigma\in\mathcal{A}_d}\phi_{(\mathbf{k}+\mathbf{j}\sigma\tau^{-1})\tau}(\mathbf{t}) = \frac{1}{|\mathcal{A}_d|^2}\sum_{\tau\in\mathcal{A}_d}\sum_{\sigma\in\mathcal{A}_d}\phi_{(\mathbf{k}+\mathbf{j}\sigma)\tau}(\mathbf{t})\\ &= \frac{1}{|\mathcal{A}_d}\sum_{\sigma\in\mathcal{A}_d}\mathsf{C}_{\mathbf{k}+\mathbf{j}\sigma}(\mathbf{t}),\end{aligned}$$

which proves (8.30). The other two relations, (8.31) and (8.32), can be established similarly. □

The orthogonality of $\phi_{\mathbf{k}}$ passes down to that of generalized sine and cosine functions. We define

$$\langle f,g\rangle_{\triangle_{\mathrm{H}}} = \frac{1}{|\triangle_{\mathrm{H}}|}\int_{\triangle_{\mathrm{H}}} f(\mathbf{t})\overline{g(\mathbf{t})}\mathrm{d}\mathbf{t} = \frac{(d+1)!}{\sqrt{d+1}}\int_{\triangle_{\mathrm{H}}} f(\mathbf{t})\overline{g(\mathbf{t})}\mathrm{d}\mathbf{t}.$$

Theorem 8.4.5. *For* $\mathbf{k},\mathbf{j}\in\Lambda$,

$$\langle\mathsf{C}_{\mathbf{k}},\mathsf{C}_{\mathbf{j}}\rangle_{\triangle_{\mathrm{H}}} = \frac{\delta_{\mathbf{k},\mathbf{j}}}{|\mathbf{k}\mathcal{A}_d|} = \frac{p_1!p_2!\cdots p_\ell!}{(d+1)!}\delta_{\mathbf{k},\mathbf{j}}, \qquad \mathbf{k}\in\Lambda^{\mathbf{p}}, \tag{8.33}$$

and for $\mathbf{k},\mathbf{j}\in\Lambda^\circ$,

$$\langle\mathsf{S}_{\mathbf{k}},\mathsf{S}_{\mathbf{j}}\rangle_{\triangle_{\mathrm{H}}} = \frac{1}{(d+1)!}\delta_{\mathbf{k},\mathbf{j}}. \tag{8.34}$$

Proof By definition, if $f\bar{g}$ is invariant under $\mathcal{A}_d$, then $\langle f,g\rangle_{\Omega_{\mathrm{H}}} = \langle f,g\rangle_{\triangle_{\mathrm{H}}}$, from which both statements follow. For (8.33), the invariance is evident, and we only have to use the orthogonality of $\phi_{\mathbf{k}}$ over Ω_{H} and (8.29). For (8.34), we use the invariance of $\mathsf{S}_{\mathbf{k}}(\mathbf{t})\overline{\mathsf{S}_{\mathbf{j}}(\mathbf{t})}$ under $\mathcal{A}_d$ and the orthogonality of $\phi_{\mathbf{k}}$ on Ω_H. □

For $n\in\mathbb{N}_0$, we define the finite spaces $\mathcal{C}_n\subset\mathcal{C}(\mathcal{A}_d)$ and $\mathcal{S}_n\subset\mathcal{S}(\mathcal{A}_d)$ of cosine and sine functions by

$$\mathcal{C}_n = \operatorname{span}\{\mathsf{C}_{\mathbf{k}}:\ \mathbf{k}\in\Lambda_n\} \quad\text{and}\quad \mathcal{S}_n = \operatorname{span}\{\mathsf{S}_{\mathbf{k}}:\ \mathbf{k}\in\Lambda_n^\circ\},$$

where the index sets $\Lambda_n^\circ \subset \Lambda_n \subset \mathbb{H}_n$ are defined by

$$\begin{aligned} \Lambda_n &= \{\mathbf{k} \in \mathbb{H} : k_{d+1} \le k_d \le \cdots \le k_1 \le k_{d+1} + (d+1)n\}, \\ \Lambda_n^\circ &= \{\mathbf{k} \in \mathbb{H} : k_{d+1} \le k_d < \cdots < k_1 < k_{d+1} + (d+1)n\}. \end{aligned} \tag{8.35}$$

Since Λ_n contains integer points in a regular simplex, it is easy to see that

$$\dim \mathcal{C}_n = |\Lambda_n| = \binom{n+d}{d} \quad \text{and} \quad \dim \mathcal{S}_n = |\Lambda_n^\circ| = \binom{n-1}{d}. \tag{8.36}$$

Since the congruent of $\{\frac{\mathbf{k}}{(d+1)n} : \mathbf{k} \in \Lambda_n\}$ under $\mathcal{A}_d$ is the interior $\mathbb{H}_n^\circ$ of $\mathbb{H}_n$ and $\mathsf{S}_\mathbf{k}(\frac{\mathbf{j}}{(d+1)n}) = 0$ when $\mathbf{j} \in \mathbb{H}_n \setminus \mathbb{H}_n^\circ$, Theorem 8.4.6 is equivalent to Theorem 8.4.1.

Theorem 8.4.6. *Let the discrete inner product $\langle \cdot,\cdot \rangle_{\triangle^\circ,n}$ be defined by*

$$\langle f,g \rangle_{\triangle^\circ,n} = \frac{d!}{n^d} \sum_{\mathbf{j} \in \Lambda_n^\circ} f\left(\tfrac{\mathbf{j}}{(d+1)n}\right) \overline{g\left(\tfrac{\mathbf{j}}{(d+1)n}\right)}.$$

Then, it satisfies

$$\langle f,g \rangle_{\triangle_\mathrm{H}} = \langle f,g \rangle_{\triangle^\circ,n}, \qquad f,g \in \mathcal{S}_n. \tag{8.37}$$

One can also define an inner product so that an analog of (8.37) holds for $\mathcal{C}_n$, which is equivalent to the one defined as a weighted sum as described in Remark 8.4.2.

8.4.2 Chebyshev Polynomials Associated with $\mathcal{A}_d$

In the case of $d = 1$, the classical Chebyshev polynomials $T_n(t) = \cos n\theta$ and $U_n(t) = \sin(n+1)\theta/\sin\theta$ are defined by setting $t = \cos\theta$. For $d > 1$, we define their analogs associated with $\mathcal{A}_d$, for which we need $\mathcal{A}_d$ sine and cosine functions indexed by the vertices of the simplex $\triangle_\mathrm{H}$.

The simplex $\triangle_\mathrm{H}$ has $d+1$ vertices. We start with one element in the simplex, which is proportion to

$$\mathbf{v}^\circ := \frac{d+1}{2}\big(d, d-2, \ldots, -(d-2), -d\big),$$

and satisfies $\mathbf{v}^\circ \in \mathbb{H}$. For $d = 2$, this is the element $(\mathbf{v}_1 + \mathbf{v}_2)/2$ on the boundary of the simplex, where $\mathbf{v}_1$ and $\mathbf{v}_2$ are given in (8.26). We need the $\mathcal{A}_d$ sine function indexed by $\mathbf{v}^\circ$.

Lemma 8.4.7. *For $\mathbf{t} \in \mathbb{H}$,*

$$\mathsf{S}_{\mathbf{v}^\circ}(\mathbf{t}) = \frac{(2i)^{\frac{d(d+1)}{2}}}{(d+1)!} \prod_{1 \le \mu < \nu \le d+1} \sin \pi(t_\mu - t_\nu). \tag{8.38}$$

Proof Let $\beta = (d, d-1, \ldots, 1, 0) \in \mathbb{N}_0^{d+1}$. Then, a quick computation shows

$$e^{\frac{2\pi i}{d+1}\mathbf{v}^\circ \cdot \mathbf{t}} = e^{-\pi i d(t_1+\cdots+t_{d+1})} e^{2\pi i \beta \cdot \mathbf{t}} = e^{2\pi i \beta \cdot \mathbf{t}}.$$

Since $\mathcal{A}_d$ is equivalent to the permutation group $\mathcal{S}_{d+1}$, the well-known Vandermonde determinant states

$$\sum_{\sigma \in \mathcal{S}_{d+1}} \operatorname{sign}(\sigma) \mathbf{x}^{\beta\sigma} = \det\left[x_i^{d+1-j}\right]_{1 \le i,j \le d+1} = \prod_{1 \le i < j \le d+1} (x_i - x_j),$$

which gives, when setting $x_j = e^{2\pi i t_j}$, that

$$\mathbf{S}_{\mathbf{v}^\circ}(\mathbf{t}) = \frac{1}{|\mathcal{A}_d|} \sum_{\sigma \in \mathcal{A}_d} \operatorname{sign}(\sigma) e^{\frac{2\pi i}{d+1}\mathbf{v}^\circ \cdot \mathbf{t}\sigma} = \frac{1}{(d+1)!} \prod_{1 \le \mu < \nu \le d+1} \left(e^{2i\pi t_\mu} - e^{2i\pi t_\nu}\right),$$

from which the identity (8.38) follows immediately. □

For $\mathcal{A}_d$ cosine, we need d nonzero vertices of the simplex $\triangle_{\mathrm{H}}$. Let

$$\mathbf{v}^k = \left(\{d+1-k\}^k, \{-k\}^{d+1-k}\right), \qquad k = 1, 2, \ldots, d.$$

Then, $\frac{\mathbf{v}^k}{d+1}$ are vertices of the simplex $\triangle_{\mathrm{H}}$. For $d = 2$, they are the vertices $\mathbf{v}_1$ and $\mathbf{v}_2$ in (8.26). We define a mapping $\mathbf{t} \in \mathbb{R}_{\mathrm{H}}^{d+1} \mapsto \mathbf{z} \in \mathbb{C}^d$ by

$$z_k = z_k(\mathbf{t}) := \mathsf{C}_{\mathbf{v}^k}(\mathbf{t}) = \frac{1}{|\mathbf{v}^k \mathcal{A}_d|} \sum_{\mathbf{j} \in \mathbf{v}^k \mathcal{A}_d} e^{\frac{2\pi i}{d+1}\mathbf{j} \cdot \mathbf{t}}, \quad k = 1, 2, \ldots, d. \tag{8.39}$$

Lemma 8.4.8. *The coordinates $z_1, \ldots, z_d$ are the first d elementary symmetric polynomials of $\{e^{2\pi i t_1}, \ldots, e^{2\pi i t_{d+1}}\}$ and they satisfy*

$$z_k = \frac{1}{\binom{d+1}{d}} \frac{\mathsf{S}_{\mathbf{v}^\circ + \mathbf{v}^k}(\mathbf{t})}{\mathsf{S}_{\mathbf{v}^\circ}(\mathbf{t})}, \qquad 1 \le k \le d. \tag{8.40}$$

Moreover,

$$\overline{z}_k = z_{d+1-k}, \qquad 0 \le k \le d+1. \tag{8.41}$$

Proof By the homogeneity of $\mathbf{t}$, $\mathbf{v}^k \cdot \mathbf{t} = (d+1)(t_1 + \cdots + t_k)$ and, consequently,

$$z_k = \frac{1}{\binom{d+1}{k}} \sum_{J \subset \mathbb{N}_{d+1}, |J| = k} e^{2\pi i \sum_{j \in J} t_j},$$

where $\mathbb{N}_{d+1} = \{1, 2, \ldots, d+1\}$, so that z_k is the kth elementary symmetric polynomials of $\{e^{2\pi i t_1}, \ldots, e^{2\pi i t_{d+1}}\}$. Moreover, since $\mathbf{v}^k \cdot \mathbf{t} = -(d+1)(t_{k+1} + \cdots + t_{d+1})$, the relation (8.41) follows. Next, we prove (8.40). By (8.31),

$$z_k \mathsf{S}_{\mathbf{v}^\circ}(\mathbf{t}) = \mathsf{C}_{\mathbf{v}^k}(\mathbf{t}) \mathsf{S}_{\mathbf{v}^\circ}(\mathbf{t}) = \frac{1}{(d+1)!} \sum_{\sigma \in \mathcal{A}_d} \mathsf{S}_{\mathbf{v}^\circ + \mathbf{v}^k \sigma}(\mathbf{t}).$$

Assume $\mathbf{l} = \mathbf{v}^k\sigma \neq \mathbf{v}^k$ for some $\sigma \in \mathcal{A}_d$. By the definition of $\mathbf{v}^k$, there exists an integer $1 \le j \le d$ such that $l_j = -k$ and $l_{j+1} = d+1-k$. Consequently,

$$\frac{(d+2-2j)(d+1)}{2} - k = \frac{(d-2j)(d+1)}{2} + (d+1-k) = v^\circ_{j+1} + l_{j+1},$$

which shows that $\mathbf{v}^\circ + \mathbf{v}^k\sigma$ is in $\Lambda \setminus \Lambda^\circ$ so that $\mathsf{S}_{\mathbf{v}^\circ+\mathbf{v}^k\sigma}(\mathbf{t}) = 0$. Thus, the sum is over the stabilizer $\mathcal{A}_d(\mathbf{v}^k)$ and

$$z_k \mathsf{S}_{\mathbf{v}^\circ}(\mathbf{t}) = \frac{1}{(d+1)!} \sum_{\sigma \in \mathcal{A}_d(\mathbf{v}^k)} \mathsf{S}_{\mathbf{v}^\circ+\mathbf{v}^k\sigma}(\mathbf{t}) = \frac{k!(d+1-k)!}{(d+1)!} \mathsf{S}_{\mathbf{v}^\circ+\mathbf{v}^k}(\mathbf{t}),$$

as the stabilizer of $\mathbf{v}^k$ has cardinality $k!(d+1-k)!$. □

Definition 8.4.9. Define the index mapping $\alpha : \Lambda \to \mathbb{N}_0^d$,

$$\alpha_i = \alpha_i(\mathbf{k}) := \frac{k_i - k_{i+1}}{d+1}, \qquad 1 \le i \le d. \tag{8.42}$$

Under the change of variables (8.39) and this index map, define

$$T_\alpha(\mathbf{z}) := \mathsf{C}_\mathbf{k}(\mathbf{t}) \quad \text{and} \quad U_\alpha(\mathbf{z}) := \frac{\mathsf{S}_{\mathbf{k}+\mathbf{v}^\circ}(\mathbf{t})}{\mathsf{S}_{\mathbf{v}^\circ}(\mathbf{t})}, \qquad \alpha \in \mathbb{N}_0^d. \tag{8.43}$$

The mapping (8.42) is an isomorphism. Indeed, since $k_1 + \cdots + k_{d+1} = 0$ for $\mathbf{k} \in \Lambda$, it is easy to verify that the inverse of α is given by

$$k_i = \sum_{\mu=1}^{d} \sum_{\nu=1}^{\mu} \alpha_\nu - (d+1) \sum_{\nu=1}^{i-1} \alpha_\nu, \qquad 1 \le i \le d+1.$$

In particular, $\alpha(\mathbf{v}^k) = \left(\{0\}^{k-1}, 1, \{0\}^{d-k}\right)$ is the kth coordinate vector of $\mathbb{R}^d$.

Proposition 8.4.10. *Let P_α denote either T_α or U_α. Then $\{P_\alpha : |\alpha| = n\}$ consists of linearly independent polynomials of degree n.*

Proof Since $\mathsf{C}_\mathbf{k}$ is symmetric and $z_1, \ldots, z_d$ are the first d elementary symmetric polynomials in of $e^{2\pi i t_1}, \ldots, e^{2\pi i t_{d+1}}$, $T_\alpha(\mathbf{z})$ is a polynomial of $\mathbf{z}$ by the fundamental theorem of symmetric functions and its highest degree term is a single monomial $\prod_{i=1}^d z_i^{\frac{k_i - k_{i+1}}{d+1}}$, so that it is of degree $(k_1 - k_{d+1})/(d+1) = |\alpha|$ under (8.42). For $\mathsf{S}_\mathbf{k}$, we observe that the proof of Lemma 8.4.7 also shows that, for each $\mathbf{k} \in \Lambda$,

$$\mathsf{S}_{\mathbf{k}+\mathbf{v}^\circ}(\mathbf{t}) = \det\left(x_i^{\lambda_j+\beta}\right)_{1 \le i,j \le d+1} \quad \text{with } \lambda = (k_1 - k_{d+1}, k_2 - k_{d+1}, \ldots, k_d - k_{d+1}, 0),$$

where $x_i = e^{2\pi i t_i}$. Since $\lambda \in \Lambda$ is a partition, $\mathsf{S}_{\mathbf{k}+\mathbf{v}^\circ}(\mathbf{t})$ is divisible by $\mathsf{S}_{\mathbf{v}^\circ}$ in the ring $\mathbb{Z}[x_1, \ldots, x_d]$, so that the quotient

$$s_\lambda(x_1, \ldots, x_d) = \mathsf{S}_{\mathbf{k}+\mathbf{v}^\circ}(\mathbf{t}) / \mathsf{S}_{\mathbf{v}^\circ}(\mathbf{t})$$

is a symmetric polynomial in $x_1,\ldots,x_d$, which is the Schur function in the variables $x_1,\ldots,x_d$ corresponding to the partition λ [99, p. 40]. Hence, as in the case $\mathsf{C}_{\mathbf{k}}$, this shows that $\mathsf{S}_{\mathbf{k}}$ is a polynomial in $\mathbf{z}$ and its highest degree term is again a single monomial $\mathbf{z}^{\alpha}$ [99, (3.5)]. Finally, the relation (8.44) follows form $-(k_i-k_j)=k_j-k_i$ and (8.42). □

We call T_α and U_α Chebyshev polynomials of the first and the second kind associated with $\mathcal{A}_d$. Using the index sets in (8.35), $\{T_\alpha : |\alpha| = n\}$ corresponds to $\{\mathsf{C}_{\mathbf{k}} : \mathbf{k} \in \Lambda_n\}$ and $\{U_\alpha : |\alpha| = n\}$ corresponds to $\{\mathsf{S}_{\mathbf{k}+\mathbf{v}^\circ} : \mathbf{k} \in \Lambda^\circ_{n+d}\}$.

Theorem 8.4.11. *Let P_α denote either T_α or U_α. Then,*

$$\overline{P_\alpha(\mathbf{z})} = P_{\alpha_d,\alpha_{d-1},\ldots,\alpha_1}(\mathbf{z}), \quad \alpha \in \mathbb{N}_0^d, \tag{8.44}$$

and they satisfy the recursion relation, for $1 \le i \le d$,

$$\binom{d+1}{i} z_i P_\alpha(\mathbf{z}) = \sum_{\mathbf{j}\in\mathbf{v}^i\mathcal{A}_d} P_{\alpha+\alpha(\mathbf{j})}(\mathbf{z}), \quad \alpha \in \mathbb{N}_0^d, \tag{8.45}$$

in which the components of $\alpha(\mathbf{j})$, $\mathbf{j} \in \mathbf{v}^i\mathcal{A}_d$, have values in $\{-1,0,1\}$, $U_\alpha(z) = 0$ whenever α has a component $\alpha_i = -1$, and for the coordinates vector $\varepsilon_1,\ldots\varepsilon_d$,

$$T_0(z) = 1, \quad T_{\varepsilon_k}(z) = z_k, \qquad 1 \le k \le d,$$
$$U_0(z) = 1, \quad U_{\varepsilon_k}(z) = \binom{d+1}{k} z_k, \quad 1 \le k \le d.$$

Proof The relation (8.44) follows from $-(k_i-k_j) = k_j - k_i$ and (8.42). The relation (8.45) follows immediately from (8.30) and (8.31). The values of P_0 and P_{ε_k} follow from definitions and (8.40). If α has a component $\alpha_i = -1$ then, by (8.42), $k_i(\alpha) = k_{i+1}(\alpha) - (d+1)$. A quick computation shows then $k_i(\alpha) + v_i^\circ = k_{i+1}(\alpha) + v_{i+1}^\circ$, which implies that $\mathbf{k}+\mathbf{v}^\circ \in \Lambda\setminus\Lambda^\circ$, so that $\mathsf{S}_{\mathbf{k}+\mathbf{v}^\circ}(\mathbf{t}) = 0$ and $U_\alpha(z) = 0$. □

It is easy to see that $|\alpha(\mathbf{j})|$, $\mathbf{j} \in \mathbf{v}^i$, also takes value in $\{-1,0,1\}$. As a result, in terms of the total degree $|\alpha|$ of P_α, the right-hand side of (8.45) contains only polynomials of degree $n-1, n$ and $n+1$, so that it is a three-term relation. For $d = 2$, the relation (8.45) can be rewritten as a recursive relation,

$$P_{\alpha+\varepsilon_1}(z) = 3z_1P_\alpha(z) - P_{\alpha+(-1,1)}(z) - P_{\alpha-\varepsilon_2}(z),$$
$$P_{\alpha+\varepsilon_2}(z) = 3z_2P_\alpha(z) - P_{\alpha+(1,-1)}(z) - P_{\alpha-\varepsilon_1}(z),$$

where $\varepsilon_1 = (1,0)$ and $\varepsilon_2 = (0,1)$, which can be used to compute a polynomial P_α from polynomials of lower degree recursively.

Both Chebyshev polynomials inherit orthogonality from those of $\mathsf{C}_{\mathbf{k}}$ and

$\mathsf{S}_\mathbf{k}$. To be more precise, by (8.44), we define real coordinates of $\mathbf{x} = (x_1, \ldots, x_d)$ by

$$x_k = \frac{z_k + z_{d+1-k}}{2}, \quad x_{d+1-k} = \frac{z_k - z_{d+1-k}}{2i} \quad \text{for} \quad 1 \le k \le \lfloor \tfrac{d}{2} \rfloor, \tag{8.46}$$

as well as $x_{\frac{d+1}{2}} = \frac{1}{\sqrt{2}} z_{\frac{d+1}{2}}$ when d is odd. Instead of $\mathbf{t} \mapsto \mathbf{z}$ from $\mathbb{R}_{\mathrm{H}}^{d+1}$ to $\mathbb{C}^d$ defined in (8.39) we can make a change of variable $\mathbf{t} \mapsto \mathbf{x}$ from $\mathbb{R}_{\mathrm{H}}^{d+1}$ to $\mathbb{R}^d$. The Jacobian of the changing variable is given in terms of

$$\mathsf{w}(\mathbf{x}) = \mathsf{w}(\mathbf{x}(\mathbf{t})) := \prod_{1 \le \mu < \nu \le d+1} \sin^2 \pi (t_\mu - t_\nu).$$

Lemma 8.4.12. *The Jacobian of the changing variable* $\mathbf{t} \mapsto \mathbf{x}$ *is given by*

$$\mathrm{J}(\mathbf{x}) = \left| \det \left[\frac{\partial (x_1, x_2, \ldots, x_d)}{\partial (t_1, t_2, \ldots, t_d)} \right] \right| = 2^{\frac{d(d+2)}{2}} \frac{\pi^d}{\prod_{k=1}^d \binom{d+1}{k}} [\mathsf{w}(\mathbf{x})]^{\frac{1}{2}}. \tag{8.47}$$

Proof Let $\partial_i = \frac{\partial}{\partial t_i}$, $1 \le i \le d$. We first prove by induction that

$$\det \left[\frac{\partial (z_1, z_2, \ldots, z_d)}{\partial (t_1, t_2, \ldots, t_d)} \right] = \prod_{k=1}^d \frac{2\pi \mathrm{i}}{\binom{d+1}{k}} \prod_{1 \le \mu < \nu \le d+1} |\sin \pi (t_\mu - t_\nu)|. \tag{8.48}$$

Regarding $t_1, t_2, \ldots, t_{d+1}$ as independent variables, we obtain

$$\frac{\partial z_k}{\partial t_j} = \frac{2\pi i}{\binom{d+1}{k}} \sum_{j \in I \subseteq \mathbb{N}_{d+1}, |I| = k} \mathrm{e}^{2\pi \mathrm{i} \sum_{\nu \in I} t_\nu}.$$

For each fixed j, let $\mathbb{N}^{(j)} = \mathbb{N}_m \setminus \{j\}$ and split $I \subset \mathbb{N}_{d+1}$ as two parts, one contains $\{j, d+1\}$ and one does not. We then obtain, after canceling the common factor,

$$\begin{aligned}
\frac{\partial z_k}{\partial t_j} - \frac{\partial z_k}{\partial t_{d+1}} &= \frac{2\pi \mathrm{i}}{\binom{d+1}{k}} \left(\sum_{j \in I \subseteq \mathbb{N}_{d+1}^{\{d+1\}}, |I| - k} \mathrm{e}^{2\pi \mathrm{i} \sum_{\nu \in I} t_\nu} - \sum_{d+1 \in I \subseteq \mathbb{N}_{d+1}^{\{j\}}, |I| = k} \mathrm{e}^{2\pi \mathrm{i} \sum_{\nu \in I} t_\nu} \right) \\
&= \frac{2\pi \mathrm{i}}{\binom{d+1}{k}} \left(\mathrm{e}^{2\pi \mathrm{i} t_j} - \mathrm{e}^{2\pi \mathrm{i} t_{d+1}} \right) \sum_{I \subseteq \mathbb{N}_d^{\{j\}}, |I| = k-1} \mathrm{e}^{2\pi \mathrm{i} \sum_{\nu \in I} t_\nu}.
\end{aligned}$$

Hence, setting $f_{j,k}^d := \sum_{I \subseteq \mathbb{N}_d^{\{j\}}, |I| = k-1} \mathrm{e}^{2\pi \mathrm{i} \sum_{\nu \in I} t_\nu}$ and defining $F_d := \left[f_{j,k}^d \right]_{1 \le j,k \le d}$, we have shown that

$$\det \left(\frac{\partial (z_1, z_2, \ldots, z_d)}{\partial (t_1, t_2, \ldots, t_d)} \right) = \prod_{k=1}^d \frac{2\pi i}{\binom{d+1}{k}} \prod_{j=1}^d \left(\mathrm{e}^{2\pi \mathrm{i} t_j} - \mathrm{e}^{2\pi \mathrm{i} t_{d+1}} \right) \det(F_d).$$

By definition, $f^d_{j,1} = 1$. Let $\mathbb{N}^{j,d} = \mathbb{N} \setminus \{j,d\}$. Splitting $\mathbb{N}^{\{j\}}$ into two parts, one contains d and one does not, it follows that

$$f^d_{j,k} - f^d_{d,k} = \left(\mathrm{e}^{2\pi i t_d} - \mathrm{e}^{2\pi i t_j}\right) \sum_{I \subseteq \mathbb{N}_d^{\{j,d\}}, |I|=k-2} \mathrm{e}^{2\pi \mathrm{i} \sum_{\nu\in I} t_\nu} = \left(\mathrm{e}^{2\pi i t_d} - \mathrm{e}^{2\pi i t_j}\right) f^{d-1}_{j,k-1},$$

which permits us to apply induction and show

$$\det(F_d) = \prod_{1\le\mu<d-1} \left(\mathrm{e}^{2\pi i t_\mu} - \mathrm{e}^{2\pi i t_d}\right) \det(F_{d-1}) = \cdots = \prod_{1\le\mu<\nu\le d} \left(\mathrm{e}^{2\pi i t_\mu} - \mathrm{e}^{2\pi i t_\nu}\right).$$

This proves (8.48) upon using (8.38). Now, a quick computation using (8.46) shows that

$$\begin{aligned}
\mathrm{J}(\mathbf{x}) &= \left|\det\left[\frac{\partial(x_1,x_2,\ldots,x_d)}{\partial(z_1,z_2,\ldots,z_d)}\right]\right| \times \left|\det\left[\frac{\partial(z_1,z_2,\ldots,z_d)}{\partial(x_1,x_2,\ldots,x_d)}\right]\right| \\
&= 2^{\frac{d(d+2)}{2}} \frac{\pi^d}{\prod_{k=1}^d \binom{d+1}{k}} \prod_{1\le\mu<\nu\le d+1} |\sin\pi(t_\mu - t_\nu)|,
\end{aligned}$$

which is (8.47). This completes the proof. □

The image of the simplex $\triangle_{\mathrm{H}}$ under the mapping $\mathbf{t} \mapsto \mathbf{x}$ is the domain

$$\triangle^*_{\mathrm{H}} := \left\{ \mathbf{x} = \mathbf{x}(\mathbf{t}) \in \mathbb{R}^d : \mathbf{t} \in \mathbb{R}^{d+1}_H, \prod_{1\le i<j\le d+1} \sin\pi(t_i - t_j) \ge 0 \right\}.$$

For $d = 2$, the domain is bounded by a deltoid (also called Steiner's hypocycloid). We depict the domain for $d = 2$ and $d = 3$ in Figure 8.4. The change of

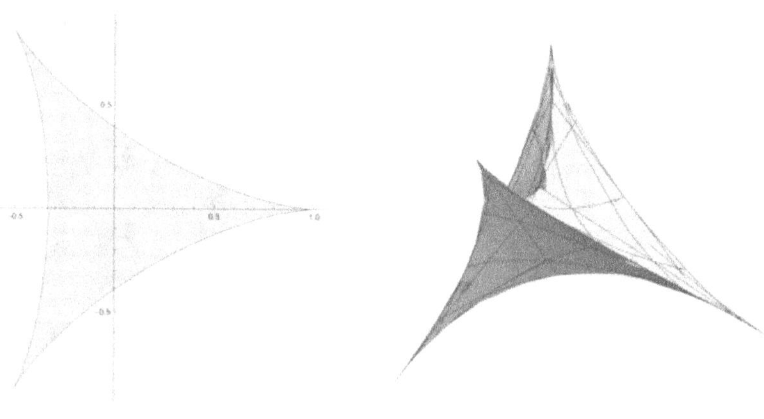

Figure 8.4 The domain $\triangle^*_{\mathrm{H}}$ for $d = 2$ and $d = 3$.

variables $\mathbf{t} \mapsto \mathbf{x}$ shows immediately that

$$\frac{1}{|\triangle_H|}\int_{\triangle_H} f(\mathbf{t})d\mathbf{t} - c_{-\frac{1}{2}}\int_{\triangle_H^*} f(\mathbf{x})\mathrm{w}^{-\frac{1}{2}}(\mathbf{x})d\mathbf{x}, \tag{8.49}$$

where the constant $c_{-\frac{1}{2}}$ is, by (8.47),

$$c_{-\frac{1}{2}} = \frac{1}{|\triangle_H|}\frac{\mathrm{w}^{\frac{1}{2}}(\mathbf{x})}{\mathrm{J}(\mathbf{x})} = 2^{-\frac{d(d+2)}{2}}\frac{(d+1)!}{\pi^d\sqrt{d+1}}\prod_{k=1}^{d}\binom{d+1}{k}.$$

Moreover, by (8.38) and (8.32),

$$\mathrm{w}(\mathbf{x}) = \frac{[(d+1)!]^2}{2^{d(d+1)}}[\mathrm{S}_{\mathbf{v}^\circ}(\mathbf{t})]^2 = \frac{[(d+1)!]^2}{2^{d(d+1)}}\sum_{\sigma\in\mathcal{A}_d}(-1)^{|\sigma|}\mathrm{C}_{\mathbf{v}^\circ+\mathbf{v}^\circ\sigma}(\mathbf{t}), \tag{8.50}$$

so that w is a polynomial in $\mathbf{z}$ with a total degree of $2d$ and, it follows by (8.34),

$$c_{-\frac{1}{2}}\int_{\triangle_H^*}\mathbf{w}^{\frac{1}{2}}(\mathrm{x})d\mathbf{x} = \frac{1}{|\triangle_H|}\int_{\triangle_H}\mathrm{w}(\mathbf{x}(\mathbf{t}))d\mathbf{t} = \frac{(d+1)!}{2^{d(d+1)}}.$$

Then, the normalization constant of $\mathrm{w}^{\frac{1}{2}}$ is given by

$$c_{\frac{1}{2}} = \left[\int_{\triangle_H^*}\mathbf{w}^{\frac{1}{2}}(\mathbf{x})d\mathbf{x}\right]^{-1} = \frac{2^{d(d+1)}}{(d+1)!}c_{-\frac{1}{2}}.$$

The Chebyshev polynomials T_α and U_α are orthogonal on $\triangle_H^*$. Let

$$\langle f,g\rangle_{\mathbf{w}^{\pm\frac{1}{2}}} = c_{\pm\frac{1}{2}}\int_{\triangle_H^*} f(\mathbf{z})\overline{g(\mathbf{z})}\mathbf{w}^{\pm\frac{1}{2}}(\mathbf{x})d\mathbf{x}.$$

Theorem 8.4.13. *The polynomials U_α and T_α are orthogonal and satisfy*

$$\langle U_\alpha, U_\beta\rangle_{\mathrm{w}^{\frac{1}{2}}} = \frac{\delta_{\alpha,\beta}}{(d+1)!},\quad \alpha,\beta\in\mathbb{N}_0^d, \tag{8.51}$$

$$\langle T_\alpha, T_\beta\rangle_{\mathrm{w}^{-\frac{1}{2}}} = d_\alpha\delta_{\alpha,\beta},\qquad d_\alpha = \frac{1}{\binom{d+1}{\mathbf{p}}}\ \textit{if}\ \alpha\in\wedge^{\mathbf{p}}, \tag{8.52}$$

where $\wedge^{\mathbf{p}}$ is the image of $\Lambda^{\mathbf{p}}$ under the index mapping (8.42),

$$\wedge^{\mathbf{p}} - \left\{\alpha\in\mathbb{N}_0^d : \alpha = \alpha(\mathbf{k}),\quad \mathbf{k}\in\Lambda^{\mathbf{p}}\right\}.$$

And $\{T_\alpha : |\alpha| - n\}$ is a basis for $\mathcal{V}_n^d(\mathrm{w}^{-\frac{1}{2}})$ and $\{U_\alpha : |\alpha| = n\}$ is a basis for $\mathcal{V}_n^d(\mathrm{w}^{\frac{1}{2}})$.

This is an immediate consequence of Theorem 8.4.5 and the correspondence.

8.4.3 Zeros of Chebyshev Polynomials and Gauss Cubature Rule

We start with the zeros of Chebyshev polynomials associated with $\mathcal{A}_d$. Let Λ_n and Λ_n° be defined as in (8.35). Let Y_n and Y_n° be the image of $\left\{\frac{\mathbf{j}}{(d+1)n} : \mathbf{j} \in \Lambda_n\right\}$ and $\left\{\frac{\mathbf{j}}{(d+1)n} : \mathbf{j} \in \Lambda_n^\circ\right\}$ under the mapping $\mathbf{t} \mapsto \mathbf{x}$, respectively,

$$Y_n := \left\{\mathbf{x}\left(\tfrac{\mathbf{j}}{(d+1)n}\right) : \mathbf{j} \in \Lambda_n\right\} \quad \text{and} \quad Y_n^\circ := \left\{\mathbf{x}\left(\tfrac{\mathbf{j}}{(d+1)n}\right) : \mathbf{j} \in \Lambda_{n+d}^\circ\right\}.$$

Theorem 8.4.14. *The set Y_{n+d}° consists of common zeros of $\{U_\alpha(x) : |\alpha| = n\}$ and $|Y_{n+d}^\circ| = \binom{n+d-1}{n} = \dim \Pi_{n-1}^d$.*

Proof By the definition of U_α and (8.42), if $|\alpha| = n$, then the corresponding index $\mathbf{k} \in \Lambda^\circ$ of $\mathsf{S}_{\mathbf{v}^\circ+\mathbf{k}}/\mathsf{S}_{\mathbf{v}^\circ}$ satisfies $k_1 - k_{d+1} = (d+1)n$. Hence, to show that U_α vanishes on Y_{n+d}°, it suffices to show, by the definition of $\mathbf{v}^\circ$, that

$$\mathsf{S}_{\mathbf{k}}\left(\tfrac{\mathbf{j}}{(d+1)n}\right) = 0 \quad \text{for } \mathbf{k} \in \Lambda^\circ \text{ and } k_1 - k_{d+1} = (d+1)n.$$

Making use of the transposition $\sigma_{i,j}$ that interchanges i and j, a simple computation shows

$$\begin{aligned}\phi_{\mathbf{k}}&\left(\tfrac{\mathbf{j}}{(d+1)n}\right) - \phi_{\mathbf{k}}\left(\tfrac{\mathbf{j}\sigma_{1,d+1}}{(d+1)n}\right)\\ &= 2\mathrm{i}\sin\frac{\pi(j_1 - j_{d+1})(k_1 - k_{d+1})}{(d+1)^2 n}\, \mathrm{e}^{\frac{\pi \mathrm{i}}{(d+1)^2 n}\left((j_1+j_{d+1})(k_1+k_{d+1}) + 2\sum_{\nu=2}^{d} j_\nu k_\nu\right)},\end{aligned}$$

which is 0 whenever $\mathbf{j} \in \Lambda_n$ and $\mathbf{k} \in \Lambda_n$ satisfying $k_1 - k_{d+1} = (d+1)n$. Hence,

$$\mathsf{S}_{\mathbf{k}}\left(\tfrac{\mathbf{j}}{(d+1)n}\right) = \frac{1}{2|\mathcal{A}_d|} \sum_{\sigma \in \mathcal{A}_d} \mathrm{sign}(\sigma)\left[\phi_{\mathbf{k}}\left(\tfrac{\mathbf{j}}{(d+1)n}\right) - \phi_{\mathbf{k}}\left(\tfrac{\mathbf{j}\sigma_{1,d+1}}{(d+1)n}\right)\right] = 0.$$

This shows U_α vanishes on Y_{n+d}°. □

The above theorem shows that $\mathbb{U}_n$ has $|X_n^\circ| = \binom{n+d-1}{d} = \dim \Pi_{n-1}^d$ real, common zeros. By Theorem 4.1.1, the weight function $W_{\frac{1}{2}}$ admits the Gauss cubature rule of degree $2n - 1$. More precisely, we obtain the following:

Theorem 8.4.15. *The Gauss cubature rules of degree $2n - 1$ for $\mathrm{w}_{\frac{1}{2}}$ on $\triangle_{\mathrm{H}}^*$ are given by*

$$c_{\frac{1}{2}} \int_{\triangle_{\mathrm{H}}^*} f(x)\mathbf{w}_{\frac{1}{2}}(x)\mathrm{d}x = \frac{2^{d(d+1)}}{(d+1)(n+d)^d} \sum_{\mathbf{y} \in Y_{n+d}^\circ} \lambda_{\mathbf{y}} f(\mathbf{y}), \quad \forall f \in \Pi_{2n-1}^d,$$

where $\lambda_{\mathbf{y}} = \prod_{1 \le \nu < \mu \le d+1} \sin^2 \pi(t_\mu - t_\nu)$ under $\mathbf{y} = \mathbf{x}(\mathbf{t})$.

Proof The existence of the cubature rule is already established. The normalization constants satisfy $c_{\frac{1}{2}}/c_{-\frac{1}{2}} = 2^{d(d+1)}/(d+1)!$, by the change of variables

$\mathbf{t} \mapsto \mathbf{x}$, (8.49) and (8.50), we obtain

$$c_{\frac{1}{2}} \int_{\triangle_{\mathrm{H}}^*} f(\mathbf{x}) \mathsf{w}^{\frac{1}{2}}(\mathbf{x}) d\mathbf{x} = (d+1)! \frac{1}{|\triangle_{\mathrm{H}}|} \int_{\triangle_{\mathrm{H}}} f(\mathbf{x}(\mathbf{t})) |\mathsf{S}_{\mathbf{v}^\circ}(\mathbf{t})|^2 d\mathbf{t}$$
$$= \frac{d!(d+1)!}{(n+d)^d} \sum_{j \in \Lambda_{n+d}^\circ} \left| \mathbf{S}_{\mathbf{v}^\circ}\left(\tfrac{\mathbf{j}}{(n+d)(d+1)}\right)\right|^2 f\left(\mathbf{z}\left(\tfrac{\mathbf{j}}{(n+d)(d+1)}\right)\right),$$

the second step follows from applying Theorem 8.4.6 on $f(z(\mathbf{t}))|\mathsf{S}_{\mathbf{v}^\circ}(\mathbf{t})|^2$, which, by (8.31) and the second identity in (8.50), are in $\mathcal{S}_{n+d}$ if f is of degree $n-d$, and $g(\mathbf{t}) = 1$. This determines the weights of the Gauss cubature rule and gives the stated formula by (8.38). □

The nodes of this cubature rule for $n = 50$ were depicted in Figure 8.5.

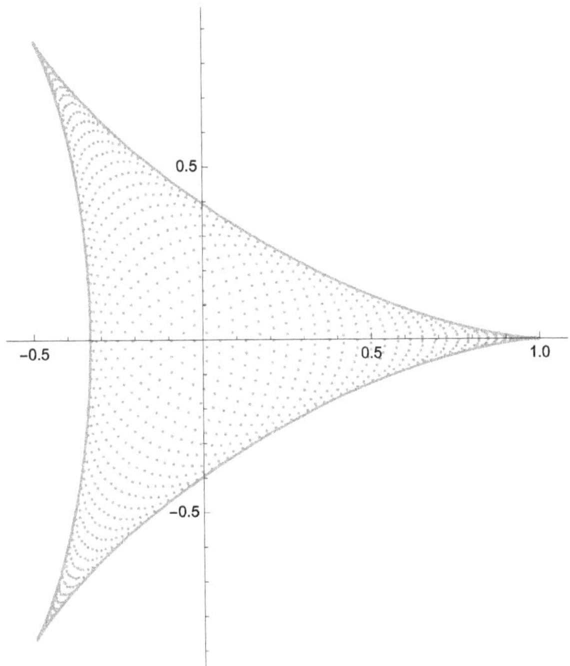

Figure 8.5 The points X_n° with $n = 50$.

The orthogonal polynomials $\{T_\alpha : |\alpha| = n\}$, however, do not have $\dim \Pi_{n-1}^d$ common zeros. Hence, $\mathsf{w}_{-\frac{1}{2}}$ does not admit the Gauss cubature rules. However, a cubature rule of degree $2n-1$ with the nodes

$$X_n = \left\{\mathbf{x}\left(\frac{\mathbf{k}}{(d+1)n}\right) \Big| \mathbf{k} \in \Lambda_n\right\},$$

exists for $\mathsf{w}_{-\frac{1}{2}}$, which has $|X_n| = \binom{n+d}{n} = \dim \Pi_n^d$ points, substantially more than that of a Gauss cubature rule.

8.5 Notes and Further Results

The translation tiling and Fourier analysis have been studied extensively (see [78] for further background). Fuglede proved Theorem 8.1.1 in [54], and he also conjectured in the same paper that a set $\Omega \subset \mathbb{R}^d$ is spectral if, and only if, it can tile the space by translations. The conjecture is false, as shown by a counterexample in [150], but it is true if Ω is assumed to be a convex body, defined as a compact convex set with a nonempty interior, as proved in [72] for $d = 2$ and in [89] for $d \geq 2$.

A convex body that tiles by translations must be a polytope, and all such polytopes are characterized in [154] and, later rediscovered, in [102]. Let Ω be a convex polytope in $\mathbb{R}^d$. A facet of Ω is a $(d-1)$-dimensional face and a belt of Ω is a collection of its facets that contain a translation of a given $(d-2)$-dimensional face of Ω. Then, a convex body $\Omega \subset \mathbb{R}^d$ can tile the space by translations if, and only if, the following four conditions are satisfied

1. Ω is a convex polytope.
2. Ω is centrally symmetric.
3. All facets of Ω are centrally symmetric.
4. Each belt of Ω consists of either four or six facets.

Moreover, a convex body Ω that satisfies the four conditions admits a face-to-face tiling by translations along a certain lattice. For $d = 2$, this shows that Ω is spectral if and only if Ω is either a quadrilateral or a centrally symmetric hexagon. For $d = 3$, the convex polytopes that can tile by translations were classified in [52] in 1885 as five types: parallelepiped, the hexagonal prism, the rhombic dodecahedron, the elongated dodecahedron, and the truncated octahedron.

The discrete Fourier analysis with lattice (cf. [45, 64]) has been studied extensively in signal processing and applications. The content of Section 8.1 is developed in [94], where it is applied to the hexagonal lattice in two variables. The setup of the discrete Fourier analysis in Section 8.1 permits two lattices: one determines the integral domain, and the other determines the space of exponentials orthogonal under the discrete inner product. For the corresponding cubature rule for algebraic polynomials, the first determines the integral for which the cubature rule holds, and the second determines the subspace of polynomials for which the cubature rule is exact. Possible choices of the two lattices and fundamental domains in two variables are explored in [96]. From the point of view of cubature rules with fewer nodes, the most interesting cases are the near minimal cubature rules in Subsection 8.2, first discussed in [95], and the Gauss cubature rule on the deltoid, which is the case of $\mathcal{A}_2$ in Section 8.4.

The study of $\mathcal{A}_d$ tiling in Section 8.4 follows that of [93]. The discrete inner product in (8.24) is based on the finite set $\mathbb{H}_n$, where $\mathbb{H}_n$ is not symmetric as it misses half of the boundary points. Its restriction on the generalized sine functions leads to (8.37) on $\triangle_{\mathrm{H}}^{\circ}$ since the sine functions vanish on the boundary points. By taking an average, we can define another inner product based on the symmetrized point set $\mathbb{H}^*$, which retains property (8.24) and leads to an inner product on $\triangle_{\mathrm{H}}$, when restricted to cosine functions, that satisfies an analog of (8.37) for cosine functions. This last inner product leads to the cubature rule of degree $2n-1$ for the weight function $\mathrm{w}^{-\frac{1}{2}}$ since the cosine functions correspond to the Chebyshev polynomials of the first kind associated with $\mathcal{A}_d$. While this outline can be carried out fairly straightforwardly for $d = 2$, the geometry for $d = 3$ is not as lucid, and it takes considerable effort, relying on 3D geometry, in [92] to work out all details for $d = 3$, which paves the way for the general $\mathcal{A}_d$ case in [93]. For readers interested in the tango between discrete Fourier analysis, interpolation, and cubature rules, we refer to [93] for further discussions.

The Chebyshev polynomials associated with $\mathcal{A}_2$ in two variables were first studied in [80] (see also [81]). They were further studied and extended to several variables by several authors [7, 10, 48, 49]. They can also be defined for other Weyl groups of a crystallographic root system (see [81, 76, 77] and, more recently, [66]). The full extension of the Chebyshev polynomials to $\mathcal{A}_2$ is carried out in [10], where these polynomials appear as eigenfunctions of the radial part of the Laplacian–Beltrami operator on certain symmetric spaces, extending the study in [80] in two variables. There have been many other symmetric and antisymmetric orthogonal trigonometric functions that come from discrete Fourier transform [76, 77, 70], which lead to generalized sine and cosine functions, with symmetry defined as under various finite groups, and analogs of Chebyshev polynomials. The discrete Fourier transforms under groups other than $\mathcal{A}_2$ also lead to cubature rules [104, 105, 71] that are exact for subspaces of trigonometric functions defined according to an order associated with the underlying group, which, however, do not correspond to the subspaces of algebraic polynomials of total degree. As a result, they do not lead to cubature rules of high precision in our setting other than those we already encountered.

The Gauss cubature rules in Theorem 8.4.15 are the only other examples than those given in Section 4.2. The nodes of the cubature rules admit a unique Lagrange interpolation polynomial of total degree. Moreover, the Lagrange interpolation polynomials based on the symmetrized points in $\mathcal{H}_n^*$ have the Lebesgue constant in the order of $(\log n)^d$. Similar sets of nodes on the domain bounded by the deltoid are also studied in [111, 112]. The polynomial interpolation based on nodes of cubature rules is another aspect of the approach via discrete Fourier analysis, which holds in each of the cases mentioned earlier.

The polynomials $U_\alpha(x)$ on the deltoid are observed in [3] to be the generalized characteristic polynomials of a special $n \times (n+1)$ Toeplitz matrix A. Let

$$A_{n,n+1}(z,\bar{z}) := \begin{bmatrix} z & \bar{z} & \bar{a} & & & \bigcirc \\ c & z & \bar{z} & \bar{c} & & \\ & \ddots & \ddots & \ddots & \ddots & \\ & & c & z & \bar{z} & \bar{c} \\ \bigcirc & & & a & z & \bar{z}, \end{bmatrix},$$

where $n+1$ columns of vectors $\mathbb{C}^n$. For $0 \le k \le n$, let $A_k(z,\bar{z})$ be a submatrix of $A_{n,n+1}(z,\bar{z})$ with kth column removed. The generalized characteristic polynomials of A are defined as $P_k^n(z,\bar{z}) = \det A_k(z,\bar{z})$ with $1 \le k \le n+1$. It was observed in [3] that this family of generalized characteristic polynomials has $\binom{n+1}{2}$ real common zeros. In [180], it is shown that $a^{-n+k}\bar{a}^{-k}P_k^n(3az, 3\bar{a}\bar{z})$ is a dilation of the Chebyshev polynomials $U_{k,n-k}$ of the second kind in Section 8.4. Moreover, the polynomials P_k^n have $\binom{n+1}{2}$ real common zeros if $|c| \ge 2|c-a|$ and $c(c-a) \in \mathbb{R}$, and they are orthogonal polynomials for a positive definite linear functional $\mathcal{I}$. This shows that $\mathcal{I}$ admits the Gauss cubature rule of degree $2n-1$, but it is unknown if it has an explicit integral representation. The definition of generalized characteristic polynomials is given for a general matrix of size $n \times n+d$ [141]. Several conjectures about their eigenvalues remain open [3, 141, 180].

9
Cubature Rules and Polynomial Ideals

The existence of cubature rules of degree m can be characterized by the set of m-orthogonal polynomials that vanish on its nodes. Let V be the set of nodes of a cubature rule. If a polynomial f vanishes on V, then so is the multiple of f by any polynomial h. Thus, V is the zero set, called variety, of a polynomial ideal I generated by m-orthogonal polynomials. In this chapter, we explore the general structure of cubature rules in the language of ideals and varieties. Section 9.1 reviews results on polynomial ideals and varieties, which are used in Section 9.2 to discuss the structure of cubature rules.

9.1 Polynomial Ideals and Varieties

We consider the space of polynomials in d variables with coefficients in a field $\mathbb{K}$. For short, we denote $\Pi^d = \mathbb{K}[x_1,\ldots,x_d]$. In most algebraic geometry, $\mathbb{K} = \mathbb{C}$. For studying cubature rules, we need $\mathbb{K} = \mathbb{R}$, so that $\Pi^d = \mathbb{R}[x_1,\ldots,x_d]$.

If $f \in \Pi^d$, we write f as a finite sum of monomials, $f = f_\alpha \mathbf{x}^\alpha + \sum_{\alpha \succ \gamma} f_\gamma \mathbf{x}^\gamma$, under a fixed monomial order, which we assume to be the graded lexicographical order. Let $\mathrm{LT}(f)$ denote the leading term of f under the monomial, so that $\mathrm{LT}(f) = f_\alpha \mathbf{x}^\alpha$ in the abovementioned expansion of f. The division algorithm for polynomials in several variables is given as follows.

Theorem 9.1.1. *Let $F = (f_1,\ldots,f_s)$ be an ordered tuple of polynomials in Π^d. Then, every $f \in \Pi^d$ can be written as*

$$f = h_1 f_1 + \cdots + h_s f_s + r,$$

where h_i, $r \in \Pi^d$ and either $r = 0$ or r is a linear combination of monomials, none of which is divisible by any of $\mathrm{LT}(f_1),\ldots,\mathrm{LT}(f_s)$. Furthermore, if $h_i f_i \neq 0$, then the degree of $\mathrm{LT}(h_i f_i)$ is no more than the degree of $\mathrm{LT}(f)$.

A subset I of Π^d is called a polynomial ideal if $0 \in I$, $f+g \in I$ if $f,g \in I$, and $fh \in I$ if $f \in I$ and $h \in \Pi^d$. Let $f_1,\dots,f_r$ be polynomials in Π^d. Then, the set

$$\langle f_1,f_2,\dots,f_r\rangle = \left\{h_1 f_1 + \cdots h_r f_r : h_1,\dots,h_r \in \Pi^d\right\}$$

is a polynomial ideal, generated by the polynomials $f_1,f_2,\dots,f_r$. By the Hilbert basis theorem, every ideal $I \subset \Pi^d$ has a finite generating set $\langle f_1,\dots,f_r\rangle$ for some $f_1,\dots,f_r \in I$. The generating set is also called the basis of the ideal. A basis is said to be a Gröbner basis if

$$\langle \mathrm{LT}(f_1),\dots,\mathrm{LT}(f_s)\rangle = \langle \mathrm{LT}(I)\rangle,$$

where $\mathrm{LT}(I) = \{\mathrm{LT}(f) : f \in I\}$ denotes the set of leading terms of elements of the ideal I, and $\langle \mathrm{LT}(I)\rangle$ denotes the ideal generated by the elements of $\mathrm{LT}(I)$. Every ideal has a Gröbner basis.

A finite set $\{f_1,\dots,f_r\}$ is called an H-basis of the ideal I if it is a basis of I and, for all $f = h_1 f_1 + \cdots + h_r f_r$ in I, the degrees of $h_j f_j$ for $1 \le j \le r$ are no greater than the degree of f. Let $\mathrm{LT_H}(f)$ denote the homogeneous polynomial of the highest degree in f, so that $f = \mathrm{LT_H}(f) + $lower degree terms. Then, $\{f_1,\dots,f_r\}$ is an H-basis if

$$\langle \mathrm{LT_H}(f_1),\dots,\mathrm{LT_H}(f_s)\rangle = \langle \mathrm{LT_H}(f)(I) : f \in I\rangle,$$

where $\mathrm{LT_H}(I) = \{\mathrm{LT_H}(f) : f \in I\}$ and $\langle \mathrm{LT_H}(I)\rangle$ is the ideal generated by the elements of $\mathrm{LT_H}(I)$. Every ideal has an H-basis. Moreover, if $\{f_1,\dots,f_m\}$ is an H-basis of I, then every $f \in \Pi^d$ can be written as $f = \sum_{i=1}^m f_i g_i$ for some polynomials $g_i \in \Pi^d$ that satisfy $\deg g_i + \deg f_i \le \deg f$.

Let $I \subset \Pi^d$ be an ideal. Two polynomials f and g in Π^d are called congruent modulus I if $f - g \in I$, which we write as

$$f \equiv g \mod I.$$

This defines an equivalence relation on Π^d, which allows us to define the quotient of Π^d modulo I, denoted by Π^d/I, as

$$\Pi^d/I = \{[f] : f \in \Pi_n^d\} \quad \text{with} \quad [f] = \{g \in \Pi^d : g \equiv f \mod I\}.$$

The dimension of the quotient Π^d/I, as a vector space, is called the codimension of the ideal,

$$\operatorname{codim} I := \dim \Pi^d/I.$$

Applying the division algorithm on a Gröbner basis shows that every $f \in I$ is congruent modulo I to a unique polynomial r, which is a linear combination of the monomials in the complement of $\langle \mathrm{LT}(I)\rangle$. Moreover, let

$$\Lambda_I = \{\alpha \in \mathbb{N}_0^d : \mathbf{x}^\alpha \notin \langle \mathrm{LT}(I)\rangle\}.$$

Then, the elements of $\{\mathbf{x}^\alpha : \alpha \in \Lambda_I\}$ are linearly independent modulo I in the sense that if

$$\sum_{\alpha\in\Lambda_I} c_\alpha \mathbf{x}^\alpha \equiv 0 \mod I,$$

then $c_\alpha = 0$ for all $\alpha \in \Lambda_I$. As a result, we have the following correspondence for the quotient ring.

Proposition 9.1.2. *Let I be an ideal. Then, Π^d/I is isomorphic as a vector space to $\mathcal{S}_I := \operatorname{span}\{x^\alpha : \alpha \in \Lambda_I\}$.*

Let $f_1,\ldots,f_r$ be polynomials in Π^d. The set of zeros of $f_1,\ldots,f_r$, denoted by

$$\mathsf{V}(f_1,\ldots,f_r) = \left\{\mathbf{x} \in \mathbb{R}^d : f_i(\mathbf{x}) = 0, \quad 1 \le i \le r\right\},$$

is called the affine variety, or variety in short, of $f_1,\ldots,f_r$. If $I = \langle f_1,\ldots,f_r\rangle$, we write $\mathsf{V} = \mathsf{V}(I)$. An affine variety can be an algebraic curve if $d = 2$ and an algebraic surface of dimension $d-1$ or less if $d > 2$, and it could be a finite set of points, which we call dimension 0. For studying cubature rules, we are particularly interested in the latter case.

Theorem 9.1.3. *Let I be an ideal and assume that $\mathsf{V} = \mathsf{V}(I)$ is a finite set. Let $|\mathsf{V}|$ denote the cardinality of V. Then*

$$|\mathsf{V}| \le \operatorname{codim} I.$$

Proof Let $r = |\mathsf{V}|$ and $\mathsf{V} = \{\mathbf{x}_1,\ldots,\mathbf{x}_r\}$, where $\mathbf{x}_j = (x_{j,1},\ldots,x_{j,d})$. Since $\mathbf{x}_1 \ne \mathbf{x}_j$ for $j \ge 1$, $\mathbf{x}_1$ and $\mathbf{x}_j$ differ in at least one component, say $x_{1,k} \ne x_{j,k}$. Then, $g_j(\mathbf{x}) = (x_k - x_{j,k})/(x_{1,k} - x_{j,k})$ is a linear polynomial that satisfies $g_j(\mathbf{x}_1) = 1$ and $g_j(\mathbf{x}_j) = 0$. In particular, the polynomial $f_1(\mathbf{x}) = g_2(\mathbf{x})\cdots g_r(\mathbf{x})$ is a polynomial of degree r that satisfies $f_1(\mathbf{x}_1) = 1$ and $f_1(\mathbf{x}_j) = 0$ for $2 \le j \le r$. Likewise, we can define polynomials f_j for $j = 2,\ldots,r$, so that $f_j(\mathbf{x}_k) = \delta_{j,k}$, $1 \le j,k \le r$. To prove $r \le \operatorname{codim} I$, it is sufficient to show that $[f_1],\ldots,[f_r] \in \Pi^d/I$ are linearly independent. Suppose $\sum_{i=1}^r a_i[f_i] = 0$ for scalars a_i, or equivalently, $\sum_{i=1}^r a_i f_i = g \mod I$. Then, $g \in I$ and g vanishes on all points of V. By the property of f_j, we obtain $0 = g(\mathbf{x}_j) = a_j f_j(\mathbf{x}_j) = a_j$ for $1 \le j \le r$, which proves the desired linear independence. □

Let V be a set in $\mathbb{C}^d$ or $\mathbb{R}^d$. We denote by $I(\mathsf{V})$ the ideal of polynomials that vanish on V; that is,

$$I(\mathsf{V}) = \left\{p \in \Pi^d \mid p(\mathbf{x}) = 0, \mathbf{x} \in \mathsf{V}\right\}.$$

Theorem 9.1.4. *Let V be a finite set. Then, $|\mathsf{V}| = \operatorname{codim} I(\mathsf{V})$.*

Proof Let $V = \{\mathbf{x}_1, \ldots, \mathbf{x}_r\}$. Following the proof of the previous theorem, it suffices to show that $[f_1], \ldots, [f_r]$ are a basis of Π^d/I. Furthermore, it suffices to show that they span Π^d/I since their linear independence has already been established. Let $[g] \in \Pi^d/I$. Consider $h = g - \sum_{k=1}^N a_k f_k$ with $a_k = g(\mathbf{x}_k)$. Since $f_j(\mathbf{x}_k) = \delta_{j,k}$, then $h(\mathbf{x}_k) = 0$ for $1 \le k \le N$, so that $h \in I(V)$. Thus, $h \equiv 0 \mod I$. Consequently, $[g] = \sum_{k=1}^N a_k[f_k]$ in Π^d/I. The proof is completed. □

Let $\mathcal{S}_I$ be the space defined in Proposition 9.1.2. We obtain a result on polynomial interpolation based on the points in a finite variety.

Theorem 9.1.5. *Let I be a polynomial ideal in Π^d and V be its affine variety. Assume V is a finite set. Then, there is a unique polynomial $p \in \mathcal{S}_I$ such that $p(\mathbf{x}) = f(\mathbf{x})$, $\mathbf{x} \in V$ for a generic function f on V.*

Proof By the isomorphism between Π^d/I and $\mathcal{S}_I$, the dimension of $\mathcal{S}_I$ is equal to the codimension of I. Since V is a finite set, it follows that $|V| = \dim \mathcal{S}_I$. Any $p \in \mathcal{S}_I$ can be written as $p(\mathbf{x}) = \sum_{\alpha \in \Lambda_I} a_\alpha \mathbf{x}^\alpha$. For each $\alpha \in \Lambda_I$, we define a column vector $\Phi(x) = [\mathbf{x}^\alpha : \alpha \in \Lambda_I]$, where the elements are arranged according to a fixed monomial order. Let $V = \{\mathbf{x}_1, \ldots, \mathbf{x}_N\}$, $N = |V|$. It is sufficient to prove that the invertibility of the square matrix

$$M = [\Phi(\mathbf{x}_1) \cdots \Phi(\mathbf{x}_N)]$$

that has $\Phi(\mathbf{x}_i)$ as columns. Suppose it is not and $\operatorname{rank} M = r - 1 < N$. Without loss of generality, we may assume that the first $r-1$ columns of the matrix are linearly independent, and there exist scalars $a_1, a_2, \ldots, a_r$, not all zero, such that $\sum_{k=1}^r a_k \Phi(\mathbf{x}_k) = 0$. In terms of the components of the vector Φ, this leads to

$$L(p) := \sum_{k=1}^{r} a_k p(\mathbf{x}_k) = 0 \quad \text{for } p(\mathbf{x}) = \mathbf{x}^\alpha,\ \alpha \in \Lambda_I. \tag{9.1}$$

There are at least two a_is that are not zero. We may assume a_1 and a_r are nonzero. Since V is the variety of I, $L(p) = 0$ for $p \in I$. By the division algorithm, or the isomorphism, any polynomial $P \in \Pi_n^d$ can be written as $P = p + q$, where $p \in I$ and $q \in \mathcal{S}_I$. Consequently, $L(P) = 0$ for all $P \in \Pi^d$. In particular, writing $\mathbf{x}_k = (x_{k,1}, \ldots, x_{k,d})$, we conclude

$$\begin{aligned}\sum_{k=1}^{r-1} a_k (x_{r,i} - x_{k,i}) \mathbf{x}_k^\alpha &= x_{r,i} \sum_{k=1}^{r} a_k \mathbf{x}_k^\alpha - \sum_{k=1}^{r} a_k x_{k,i} \mathbf{x}_k^\alpha \\ &= x_{r,i} L(\mathbf{x}^\alpha) - L(x_i \mathbf{x}^\alpha) = 0\end{aligned}$$

for $\alpha \in \Lambda_I$ and $1 \le i \le d$. In terms of $\Phi(\mathbf{x}_k)$, this shows $\sum_{k=1}^{r-1} a_k (x_{k,i} - x_{r,i}) \Phi(\mathbf{x}_k) = 0$. Since $\Phi(\mathbf{x}_1), \ldots, \Phi(\mathbf{x}_{r-1})$ are linearly independent, we conclude that $a_k(x_{k,i} - x_{r,i}) = 0$ for $1 \le k \le r-1$ and $1 \le i \le d$. Since $a_1 \ne 0$, it follows that $x_{1,i} = x_{r,i}$ for

$1 \le i \le d$, that is, $\mathbf{x}_1 = \mathbf{x}_r$, which contradicts the assumption that $\mathbf{x}_k$s are distinct. Thus, the matrix M must be invertible, and the polynomial interpolation on V has a unique solution in $\mathcal{S}_I$. □

9.2 Structure of Cubature Rules

Let $\mathcal{I}(f)$ be a nonnegative integral or a positive definite linear functional. Let $\mathsf{C}_m(f)$ be a positive cubature rule of degree m for $\mathcal{I}(f)$ with N nodes and let $\mathsf{V} = \{x_1, \ldots, x_N\}$ be the set of nodes of the cubature rule. We define the matrix $\mathsf{M}_k = \left[x_k^\alpha\right]_{1 \le k \le N, |\alpha| \le k}$ of size $\dim \Pi_k^d \times N$ as in Section 5.1. For $m = 2n-1$ and $m = 2n-2$, we again define

$$\tau_k = \mathsf{M}_{n-1+k} \quad \text{and} \quad \sigma_k = \tau_k - \tau_{k-1}, \qquad k = 0, 1, 2, \ldots.$$

We consider the polynomial ideal $I(\mathsf{V})$, which consists of polynomials that vanish on V, or in other words, polynomials that have V as common zeros, and reformulate some of the results in Section 5.1.

Theorem 9.2.1. *Let $m = 2n-1$ or $2n-2$. Let* V *be the set of nodes of a positive cubature rule of degree m. Then,*

(i) $I(\mathsf{V})$ *consists of m-orthogonal polynomials of degree at least n and*

$$\dim I(\mathsf{V}) \cap \Pi_{n+\ell}^d = r_{n+\ell}^d - \tau_\ell, \qquad \ell = 0, 1, \ldots m-n.$$

Moreover, the $r_{n+\ell}^d - \tau_\ell$ linearly independent polynomials in $I(\mathsf{V}) \cap \Pi_{n+\ell}^d$ are of degree $n+\ell$.

(ii) *If $\sigma_{s+1} = 0$ for some $s < m-n$, then*

$$N = |\mathsf{V}| = \dim \Pi_{n-1}^d + \sigma_0 + \cdots + \sigma_s.$$

Proof The item (i) is a reformulation of Proposition 5.1.5. We now prove (ii). If $\sigma_{s+1} = 0$, then $I(\mathsf{V}) \cap \Pi_{n+s+1}^d$ contains r_{n+s+1}^d linearly independent m-orthogonal polynomials of degree $n+s+1$. Since x_i maps $I(\mathsf{V}) \cap \Pi_{n+s+1}^d$ to $I(\mathsf{V}) \cap \Pi_{n+s+2}^d$, we have $\Pi^d / I(\mathsf{V}) = \Pi_{n+s}^d / (I(\mathsf{V}) \cap \Pi_{n+s}^d)$. Hence, it follows that $\dim(I(\mathsf{V}) \cap \Pi_{n+s}^d) = r_{n+s}^d - \tau_s$. Consequently,

$$\operatorname{codim} I(V) = \tau_s = \dim \Pi_{n-1}^d + \sigma_0 + \cdots + \sigma_s.$$

Since $|V| = \operatorname{codim} I(V)$, this completes the proof. □

For $m = 2n-1$ or $2n-2$ and $\ell \ge 0$, an m-orthogonal polynomial of degree

$n+\ell$ can be written as $\mathbf{a}^{\mathsf{T}}\mathbb{Q}_{n+\ell,\ell}$, where $\mathbf{a} \in \mathbb{R}^{r^d_{n+\ell}}$ and $\mathbb{Q}_{n+\ell,\ell}$ denotes the column vector of polynomials, which we also regard as a set.

$$\mathbb{Q}_{n+\ell,\ell} = \mathbb{P}_{n+\ell} + \Gamma_{1,\ell}\mathbb{P}_{n+\ell-1} + \cdots + \Gamma_{2n+2\ell-m-1,\ell}\mathbb{P}_{m-n-\ell+1},$$

where $\Gamma_{j,\ell} : r^d_{n+\ell} \times r^d_{n+\ell-j}$. We also assume $\Gamma_{0,0} = 0$, so that $\mathbb{Q}_{n,0} = \mathbb{P}_n$ if $m = 2n-1$ and $\mathbb{Q}_{n,0} = \mathbb{P}_n + \Gamma_{1,0}\mathbb{P}_{n-1}$ if $m - 2n - 2$. Thus, the $r^d_{n+\ell} - \sigma_\ell$ linearly independent polynomial in $I(\mathsf{V}) \cap \Pi^d_{n+\ell}$, stated in (i) of Theorem 9.2.1, can be written as $U_\ell^{\mathsf{T}}\mathbb{Q}_{n+\ell,\ell}$, where U_ℓ is a matrix of $r_{n+\ell} \times (r^d_{n+\ell} - \sigma_\ell)$ with $\operatorname{rank} U_\ell = r^d_{n+\ell} - \sigma_\ell$. Moreover, since $\sigma_{\sigma+1} = 0$, U_{s+1} is a square matrix and invertible, which we can choose as the identity matrix. As a consequence of the theorem, we see that the ideal $\mathcal{I}(\mathsf{V})$ has a basis

$$\left\langle U_0^{\mathsf{T}}\mathbb{Q}_{n,0}, U_1^{\mathsf{T}}\mathbb{Q}_{n+1,1}, \ldots, U_s^{\mathsf{T}}\mathbb{Q}_{n+s,s}, \mathbb{Q}_{n+s+1,s+1}\right\rangle. \tag{9.2}$$

This basis is larger than it is necessary. For example, for $1 \le i \le d$, $x_i U_0^{\mathsf{T}}\mathbb{Q}_{n,0}(x)$ belong to $I(\mathsf{V}) \cap \Pi^d_{n+1}$, so that a part of $U_1^{\mathsf{T}}\mathbb{Q}_{n+1,1}$ comes from $U_0^{\mathsf{T}}\mathbb{P}_n$. As it is given, the basis (9.2) is a basis for the vector space in $I(\mathsf{V}) \cap \Pi^d_{n+s+1}$; moreover, every polynomial of degree $n+s+2$ in $I(\mathsf{V})$ can be written in terms of $x_i\mathbb{Q}_{n+s+1,s+1}$ by the three-term relations (3.15) and (3.12), so are polynomials of degree $> n+s+2$ in $I(\mathsf{V})$. Consequently, the basis (9.2) is an H-basis.

Theorem 9.2.1 provides the necessary conditions for the existence of a cubature rule of degree m. We now give sufficient conditions stated in terms of the polynomial ideal.

Theorem 9.2.2. *Let $\mathcal{I}$ be a square positive linear functional. Let I be an ideal generated by an H-basis of m-orthogonal polynomials and let V be the variety of I. If V is a finite set of real points and $|\mathsf{V}| = \operatorname{codim} I$, then there is a cubature rule of degree m that has V as its set of nodes.*

Proof Since the basis is an H-basis, enlarging the basis by applying $p \mapsto x_i p$ if necessary, we can assume that the basis is of the form (9.2), where all U_j have full rank, which implies in particular that

$$I \cap \Pi^d_{n+s+1} = \operatorname{span}\left\{U_0^{\mathsf{T}}\mathbb{Q}_{n,0}, U_1^{\mathsf{T}}\mathbb{Q}_{n+1,1}, \ldots, U_s^{\mathsf{T}}\mathbb{Q}_{n+s,s}, \mathbb{Q}_{n+s+1,s+1}\right\}.$$

Let V_k be a matrix of size $r^d_{n+k} \times \sigma_k$ with rank σ_k such that the matrix $(U_k|V_k)$, whose columns are columns of U_k and V_k, is invertible. Define the space

$$\mathcal{V}_n(I) = \Pi^d_{n-1} \cup \operatorname{span}\{V_0^T\mathbb{Q}_{n,0}, V_1^T\mathbb{Q}_{n+1,1}, \ldots, V_m^T\mathbb{Q}_{n+s,s}\},$$

which is the complement of the space spanned by the basis. Then, $\mathcal{V}_n(I) \cap I = \{0\}$ since $I \cap \Pi^d_{n-1} = \{0\}$ due to m-orthogonality. Thus, $\Pi^d_{n+s} = \mathcal{V}_n(I) \oplus (I \cap \Pi^d_{n+s})$. Since $\Pi^d/I = \Pi^d_{n+s}/(I \cap \Pi^d_{n+s})$, it follows that $\dim \mathcal{V}_n(I) = \operatorname{codim} I$. Moreover, by Proposition 9.1.2, the space $\mathcal{V}_n(I)$ is isomorphic to $\mathcal{S}_I$. Hence, by Theorem 9.1.5, the interpolation process $f \mapsto Lf$ defined by $Lf(\mathbf{x}) = f(\mathbf{x})$, $\mathbf{x} \in \mathsf{V}$, has a unique solution $Lf \in \mathcal{V}_n(I)$. The polynomial Lf is a linear functional that satisfies $Lf(\mathbf{x}_k) = f(\mathbf{x}_k)$, $1 \le k \le N$, so that we can write Lf as

$$Lf(\mathbf{x}) = \sum_{k=1}^{N} f(\mathbf{x}_k)\ell_k(\mathbf{x}),$$

where ℓ_k are unique polynomials in $\mathcal{V}_n(I)$ determined by $\ell_k(\mathbf{x}_j) = \delta_{k,j}$, $1 \le k, j \le N$. Applying the linear functional $\mathcal{L}$ on Lf leads to a cubature formula

$$\mathsf{C}(f) = \sum_{k=1}^{N} \lambda_k f(\mathbf{x}_k) \qquad \text{with} \quad \lambda_k = \mathcal{L}(\ell_k),$$

which satisfies, by the uniqueness of Lf, that $\mathsf{C}(f) = \mathcal{I}(f)$ for $f \in \mathcal{V}_n(I)$. We only need to prove that $\mathsf{C}(f) = \mathcal{I}(f)$ holds for all $f \in \Pi_m^d$. If $f \in I$, then f vanishes on V so that $\mathsf{C}(f) = 0$. Moreover, since the basis (9.2) is an H-basis that consists of m-orthogonal polynomials, we must have $\mathcal{I}(f) = 0$. Thus, $\mathsf{C}(f) = \mathcal{I}(f)$ for $f \in I$. Now, using the invertibility of $(U_0|V_0)$, every $f \in \Pi_n^d$ can be written as $f = g + h$, where $g \in \operatorname{span}\{U_0^{\mathsf{T}}\mathbb{Q}_{n,0}\}$ and $h \in \Pi_{n-1}^d \cup \operatorname{span}\{V_0^T\mathbb{Q}_{n,0}\}$. Hence, it follows that

$$\mathsf{C}(f) = \mathsf{C}(g) + \mathsf{C}(h) = 0 + \mathsf{C}(h) = 0 + \mathcal{I}(h) = \mathcal{I}(g) + \mathcal{I}(h) = \mathcal{I}(f)$$

for every $f \in \Pi_n^d$. Likewise, using the invertibility of $(U_k|V_k)$, we can also show that $\mathsf{C}(f) = \mathcal{I}(f)$ for all $f \in \Pi_{n+s+1}^d$. Since the basis (9.2) contains $\mathbb{Q}_{n+s+1,s+1}$, each $f \in \Pi_k^d$ with $k > n+s+1$ can be written as $f = g + h$, where $g \in I$ is a linear combination of the components of $\mathbf{x}^\alpha\mathbb{Q}_{n+s+1,s+1}$, where $|\alpha| = k - n - s - 1$, and h is a polynomial of degree at most $k-1$. Hence, we can use induction on k to show that $\mathsf{C}(f) = \mathcal{I}(f)$ using the m-orthogonality of $\mathbb{Q}_{n+s+1,s+1}$. This verifies that $\mathsf{C}(f)$ is a cubature rule of degree $2n-1$ and completes the proof. □

By these results, the essential condition for establishing a cubature rule is $|\mathsf{V}| = \operatorname{codim} I$. It is known that $|\mathsf{V}| \le \operatorname{codim} I$ in general, but the equality is rare. For polynomial ideals in $\mathbb{C}[\mathbf{x}]$, $|\mathsf{V}| = \operatorname{codim} I$ if I is a radical ideal. The fact that we are dealing with $\mathbb{R}[\mathbf{x}]$ makes the problem much more complicated. According to Theorem 9.2.2, the existence of a cubature rule of degree m amounts to identifying a large set of zeros of a family of m-orthogonal polynomials. If $m = 2n-1$ or $m = 2n-2$, the lowest degree of m-orthogonal polynomials is n. In this regard, the following proposition is useful.

Proposition 9.2.3. *Assume that the cubature formula of degree $2n-1$ is positive and the polynomials $U^{\mathsf{T}}\mathbb{P}_n$ vanish on the nodes of the cubature rule, then there is a nonnegative definite matrix M such that*

$$A_{n-1,i}(I-M)A_{n-1,j}^{\mathsf{T}} = A_{n-1,j}(I-M)A_{n-1,i}^{\mathsf{T}}, \qquad 1 \le i, j \le d, \qquad (9.3)$$

so that U satisfies $MU = 0$.

Proof As a necessary condition, (9.3) is the same as (6.8). The proof follows exactly as in Theorem 6.2.5, which depends only on $U^{\mathsf{T}}\mathbb{P}_n$ and is independent of $\mathbb{Q}_{n+1}$. □

The theorem applies to all cubature rules, not just minimal or minimal ones. As an example, let us verify that the condition $|\mathrm{V}| = \operatorname{codim} I$ holds for product cubature rules on a cube for the weight function

$$W(\mathbf{x}) = w_1(x_1)\dots w_d(x_d),$$

where each weight function w_i is defined on an interval in $\mathbb{R}$. Up to a change of variables, we can assume that all w_i are defined on $[-1,1]$ so that W is defined on $[-1,1]^d$. Thus, we consider the product cubature rule of degree $2n-1$ given by

$$\int_{[-1,1]^d} f(\mathbf{x})W(\mathbf{x})\mathrm{d}\mathbf{x} = \sum_{k_1=1}^{n}\cdots\sum_{k_d=1}^{n}\lambda_{k_1,1}\cdots\lambda_{k_d,d}f(x_{k_1,1},\dots,x_{k_d,d}),$$

where for each i, the number $\lambda_{k_i,i}$ are the weights and $x_{k_i,i}$, $1\le k_i\le n$, are zeros of orthogonal polynomial $p_n(w_i)$ associated with w_i, and they are the nodes of the Gauss quadrature rule of degree $2n-1$ for w_i. A basis of orthogonal polynomials for W on $[-1,1]^d$ is given by

$$P_{\mathbf{k}}(\mathbf{x}) = p_{k_1}(w_1;x_1)\cdots p_{k_d}(w_d;x_d), \qquad \mathbf{k} = (k_1,\dots,k_d),$$

where the degree of $P_{\mathbf{k}}$ is $|\mathbf{k}|$. In particular, let $e_1,\dots,\dots,e_d$ be the standard Euclidean coordinate vectors. Then, $P_{ne_i}(\mathbf{x}) = p_n(x_i)$, $1\le i\le d$, are orthogonal polynomials of degree n with respect to W. Let $I = \langle P_{ne_1},\dots,P_{ne_d}\rangle$. Then, the variety V of I consists of $|\mathrm{V}| = n^d$ points of the product cubature rule. Since the leading term of $P_{ne_i}(\mathbf{x})$ is $\gamma_i x_i^n$, $\langle \mathrm{LT}(\mathcal{I})\rangle$ contains $x_1^n,\dots,x_d^n$. Multiplying I by $\mathbf{x}^\alpha$, it follows easily that $\mathrm{LT}(\mathcal{I})$ contains $\{\mathbf{x}^\alpha : |\alpha| = 2n-1\}$. Consequently, $\Pi^d/I = \Pi^d_{2n-1}/I$. Moreover, it follows readily that $\mathcal{S}_I = \{x^\alpha : 0\le\alpha_i<n,\ 1\le i\le d\}$. Consequently, by Proposition 9.1.2,

$$\operatorname{codim} I = \#\{\alpha : 0\le\alpha_i<n, 1\le i\le d\} = n^d = |\mathrm{V}|.$$

9.3 Notes and Further Results

Our main reference for the first section is [33]. In particular, Proposition 9.1.2 is stated in [33, Proposition 4, p. 229]. The H-basis was introduced in [101] and studied recently in [109, 132]. The connection between cubature rules and ideal and variety has long been observed; see, for example, [146]. Using algebraic geometry tools in the subject appeared in [106, 107] and in [137], we follow [174].

We do not know, however, if the cubature rule in the statement of Theorem 9.2.2 is positive or not. In the case that $\mathcal{I} = \langle U^{\mathsf{T}}\mathbb{P}_n, \mathbb{Q}_{n+1}\rangle$, the positivity is established in Theorem 6.1.2, where $|\mathsf{V}| = \dim \Pi_{n-1}^d + \sigma_0 = \operatorname{codim} I$, and the positivity of the cubature rule is established via the generalized Christoffel–Darboux formula in (6.18), which relies on the conditions (6.8)–(6.10). It is possible to extend these results to the ideal in (9.2) for $s > 1$, including the positivity, with complicated notations and formulas. However, this requires an extension of (6.8)–(6.10) that characterizes the common zeros, which are given by nonlinear matrix equations (cf. [167]) that we choose not to discuss since they are complex and no longer practical.

10
Epilogue: Two Open Problems

Many questions around cubature rules remain open. We discuss two problems in this chapter; both are fundamental for further study. The first concerns better lower bounds of the number of nodes, and the second discusses cubature rules of more than two variables.

10.1 Better Lower Bound for the Number of Nodes

For $d = 2$, we have discussed a large collection of minimal cubature rules on the square. However, it does not include the integral against the constant weight function. Indeed, numerical calculation indicates that Möller's lower bound is not sharp for cubature rules of degree >11 on the square. The number of nodes of cubature rules on a domain can differ for different weight functions.

Let W_0 be the product Chebyshev weight on the cube $[-1,1]^d$ defined by

$$W_0(\mathbf{x}) := \frac{1}{\pi^d} \prod_{i=1}^{d} \frac{1}{\sqrt{1-x_i^2}}, \qquad \mathbf{x} \in [-1,1]^d.$$

Then, $W(\mathbf{x})\mathrm{d}\mathbf{x}$ is known as the equilibrium measure on $[-1,1]^d$. More generally, the equilibrium measure $\mathrm{d}\mu$ on a compact domain Ω in $\mathbb{R}^d$ is defined as the one that minimizes the energy functional

$$I(\nu) := \int\int \log \frac{1}{|z-w|} \mathrm{d}\nu(z)\mathrm{d}\nu(w)$$

over Borel probability measures ν supported on Ω. In particular, the equilibrium measures for the unit ball $\mathbb{B}^d$ and the simplex Δ^d are associated with the Chebyshev weight on the respective domain. For $\mathbb{B}^d$, it is

$$W_0^{\mathbb{B}}(\mathbf{x}) = \frac{1}{\sqrt{1-\|x\|^2}}, \qquad \mathbf{x} \in \mathbb{B}^d,$$

and for the simplex Δ^d, it is

$$W_0^{\Delta}(x) = \frac{1}{\sqrt{x_1 \dots x_d(1-|x|)}}, \qquad \mathbf{x} \in \Delta^d.$$

Conjecture 10.1.1. *Let Ω be a compact convex domain in $\mathbb{R}^d$. A cubature rule of degree $2n-1$ for the integral $\int_\Omega f(\mathbf{x})W(\mathbf{x})\mathrm{d}\mathbf{x}$ has the smallest number of nodes when $W(\mathbf{x})\mathrm{d}\mathbf{x}$ is the equilibrium measure.*

The conjecture holds for the square $[-1,1]^2$. For cubature rules of degree $2n-2$, however, the conjecture does not hold since the Gauss cubature rule of even degree exists for the product Chebyshev weight W_1 of the second kind but does not exist for W_0. The requirement that Ω is compact is likely necessary because of the example of the minimal cubature rule of degree $2n-1$ on the domain $\{x : \|x\| \ge 1\}$ in Section 5.5, which holds for a weight function depending on n.

If the conjecture were true, the lower bound in Section 5.4 would imply a lower bound stronger than Möller's lower bound for weight functions on the unit ball and the simplex. However, the lower bounds for the ball and the simplex in Section 5.4 are improvable and, hence, not sharp, so the immediate question is to find a sharp lower bound on these domains. As we have seen, the nodes of a minimal cubature rule of degree m should be the variety of a polynomial ideal generated by m-orthogonal polynomials. Working with specific cases of lower degrees suggests the following conjecture.

Conjecture 10.1.2. *The number of nodes N of a cubature rule of degree $2n-1$ or $2n-2$ for the integral $\int_\Omega f(\mathbf{x})W(\mathbf{x})\mathrm{d}\mathbf{x}$ satisfies*

$$N \ge \frac{2}{3}n^2 + \mathcal{O}(n) \tag{10.1}$$

if W and its domain are given by

(i) $W_\mu(\mathbf{x}) = (1-\|\mathbf{x}\|^2)^\mu$ on the disk $\Omega = \mathbb{B}^2$;
(ii) $W_{\boldsymbol{\kappa}}(\mathbf{x}) = x_1^{\kappa_1}x_2^{\kappa_2}(1-x_1-x_2)^{\kappa_3}$ on the triangle $\Omega = \Delta^2$.

If this conjecture were true, it would provide reassurance to finding cubature rules by solving the system of moment equations (2.3) numerically since the latter requires matching the number of equations and the number of variables, as discussed in 2.6, which satisfies (10.1) when $d = 2$. Admittedly, the conjecture is a wild guess, and there is little positive empirical evidence to support it. Cubature rules found numerically are of degrees too low to provide convincing support.

As seen from centrally symmetric integrals, cubature rules of the same degree need more nodes if the integral is symmetric, and the weight function W_μ is invariant under the large group, the rotation group, so we should expect a stronger lower bound for W_μ. The conjecture, however, could not hold for cubature rules for all rotationally symmetric integrals because of the artificial example in Theorem 5.5.2.

Pushing even further, we conjecture that the lower bound (10.1) is normal and the cubature rules that attain Möller's lower bound are exceptional. In other words, the lower bound (10.1) for the number of nodes of cubature rules holds for almost all integrals.

10.2 A Family of Cubature Rules on the Cube

For $d > 2$, little is known about minimal cubature rules. In particular, for centrally symmetric integrals, we do not know any cubature rules of degree $2n-1$ that attain Möller's lower bound (5.11). Moreover, the landscape for $d > 3$ appears even more obscure, as illustrated by the example in this section.

We consider cubature rules for the product Chebyshev weight function on the cube $[-1,1]^d$. For $d = 2$, a family of minimal cubature rules is given in Section 6.4. We are interested in the case for higher dimensions.

We start with the Gauss–Lobatto formula for the Chebyshev weight w_0 of the first kind on $[-1,1]$ that takes the form

$$\int_{-1}^{1} f(x)w_0(x)dx = \frac{1}{n}\left(\frac{1}{2}f(-1) + \sum_{j=1}^{n-1} f\left(\cos\tfrac{j\pi}{n}\right) + \frac{1}{2}f(1)\right) := \mathsf{Q}_n f, \quad (10.2)$$

which again holds for all $f \in \Pi_{2n-1}$, and we factor $\mathsf{Q}_n f$ into a sum of two. The factorization depends on whether n is even or n is odd. Define

$$n = 2m: \quad \begin{aligned} \mathsf{Q}_n^{\mathsf{E}} f &= \frac{1}{n}\left(\frac{1}{2}f(-1) + \sum_{j=1}^{m-1} f\left(\cos\tfrac{2j\pi}{n}\right) + \frac{1}{2}f(1)\right), \\ \mathsf{Q}_n^{\mathsf{O}} f &= \frac{1}{n}\sum_{j=1}^{m} f\left(\cos\tfrac{(2j-1)\pi}{n}\right) \end{aligned}$$

and define

$$n = 2m-1: \quad \begin{aligned} \mathsf{Q}_n^{\mathsf{E}} f &= \frac{1}{n}\left(\sum_{j=1}^{m-1} f\left(\cos\tfrac{2j\pi}{n}\right) + \frac{1}{2}f(1)\right), \\ \mathsf{Q}_n^{\mathsf{O}} f &= \frac{1}{n}\left(\frac{1}{2}f(-1) + \sum_{j=1}^{m-1} f\left(\cos\tfrac{(2j-1)\pi}{n}\right)\right), \end{aligned}$$

where we use the superscripts E and O to signify that the sum is taken over even indices or odd indices, respectively. Evidently, the quadrature (10.2) becomes

$$\int_{-1}^{1} f(x) w_0(x) dx = \mathsf{Q}_n^{\mathsf{E}} f + \mathsf{Q}_n^{\mathsf{O}} f, \qquad \forall f \in \Pi_{2n-1}.$$

Lemma 10.2.1. *For $n \geq 0$ and $k \in \mathbb{N}_0$,*

$$\mathsf{Q}_n^{\mathsf{E}} T_k = \begin{cases} 0, & k \neq 0 \mod n, \\ \frac{1}{2}, & k = 0 \mod n, \end{cases} \quad \textit{and} \quad \mathsf{Q}_n^{\mathsf{O}} T_k = \begin{cases} 0, & k \neq 0 \mod n, \\ \frac{1}{2}, & k = 0, 2n, 4n, \ldots \\ -\frac{1}{2}, & k = 0, n, 3n, \ldots. \end{cases}$$

Proof The proof follows from elementary trigonometric identities. For example, for $n = 2m$, an elementary calculation shows that

$$\mathsf{Q}_n^{\mathsf{O}} T_k = \frac{1}{n} \sum_{j=1}^{m} \cos k \frac{(2j-1)\pi}{2m} = \frac{\sin k\pi}{4m \sin \frac{k\pi}{2m}} = \frac{\sin k\pi}{2n \sin \frac{k\pi}{n}}$$

from which $\mathsf{Q}_n^{\mathsf{O}} T_k = 0$ for $k \neq 0 \mod n$ follows immediately. The case when k is a multiple of n follows from the first equal sign of the abovementioned equation without summing it up. Similarly,

$$\mathsf{Q}_n^{\mathsf{E}} T_k = \frac{1}{n} \left(\frac{1}{2} \cos k\pi + \sum_{j=1}^{m-1} \cos k \frac{j\pi}{m} + \frac{1}{2} \right) = \frac{\sin k\pi \cos \frac{k\pi}{n}}{2n \sin \frac{k\pi}{n}},$$

from which the stated result follows. The proof for $n = 2m - 1$ is similar and is omitted. □

We write $\sigma \in \{\mathsf{E}, \mathsf{O}\}^d$ if $\sigma = (\sigma_1, \ldots, \sigma_d)$ with $\sigma_i = \mathsf{E}$ or $\sigma_i = \mathsf{O}$. For $\sigma \in \{\mathsf{E}, \mathsf{O}\}^d$ and $f : \mathbb{R}^d \mapsto \mathbb{R}$, we define the sum

$$\mathsf{Q}_n^{\sigma} f = \mathsf{Q}_n^{\sigma_1} \cdots \mathsf{Q}_n^{\sigma_d} f, \qquad \sigma \in \{\mathsf{E}, \mathsf{O}\}^d,$$

as a d-fold multiple sum in which Q^{σ_j} is applied to the jth variable of f. We further define

$$\tilde{\sigma}_i = \begin{cases} \mathsf{E} & \sigma_i = \mathsf{O} \\ \mathsf{O} & \sigma_i = \mathsf{E}. \end{cases}$$

For each $\sigma \in \{E, O\}^d$, we then define

$$\mathsf{C}_{n,d}^{\sigma} f := \mathsf{Q}_n^{\sigma} f + \mathsf{Q}_n^{\tilde{\sigma}} f = \mathsf{Q}_n^{\sigma_1} \ldots \mathsf{Q}_n^{\sigma_d} f + \mathsf{Q}_n^{\tilde{\sigma}_1} \ldots \mathsf{Q}_n^{\tilde{\sigma}_d} f$$

and show that each of $\mathsf{C}_{n,d}^{\sigma} f$ is a cubature rule for the Chebyshev weight on the cube.

Theorem 10.2.2. *For $d \ge 2$ and each $\sigma \in \{\mathsf{E},\mathsf{O}\}^d$, the cubature formula*

$$\int_{[-1,1]^d} f(\mathbf{x})W_0(\mathbf{x})d\mathbf{x} = 2^{d-1}\mathsf{C}_{n,d}^{\sigma} f,$$

is exact for $f \in \Pi_{2n-1}^d$, and its number of nodes, N, satisfies

$$N = \frac{n^d}{2^{d-1}}\left(1 + o(1)n^{-1}\right).$$

Proof For $\mathbf{k} \in \mathbb{N}_0^d$ let $T_{\mathbf{k}}(x) := T_{k_1}(x_1)\cdots T_{k_d}(x_d)$. It suffices to establish the cubature rule for $f \in \{T_{\mathbf{k}} : |\mathbf{k}| \le 2n-1\}$ since this set is an orthogonal basis of Π_{2n-1}^d. We have

$$\int_{[-1,1]^d} T_{\mathbf{k}}(\mathbf{x})W_0(\mathbf{x})d\mathbf{x} = 2^{d-1}\left[\mathsf{Q}_n^{\sigma_1}T_{k_1}\cdots\mathsf{Q}_n^{\sigma_d}T_{k_d} + I_n^{\widetilde{\sigma}_1}T_{k_1}\cdots I_n^{\widetilde{\sigma}_d}T_{k_d}\right].$$

By the orthogonality of T_k, the left-hand side equals 1 if $\mathbf{k} = \mathbf{0} = (0,\ldots,0)$ and 0 if $\mathbf{k} \ne \mathbf{0}$. From the definition of $\mathsf{Q}_n^{\mathsf{E}}$ and $\mathsf{Q}_n^{\mathsf{O}}$, it is evident that $\mathsf{Q}_n^{\mathsf{E}}1 = \mathsf{Q}_n^{\mathsf{O}}1 = 1/2$. Hence, for $\mathbf{k} = \mathbf{0}$, the right-hand side is equal to $2^{d-1}(2^{-d} + 2^{-d}) = 1$, verifying the equation for $\mathbf{k} = \mathbf{0}$. Assume now $0 < |\mathbf{k}| \le 2n-1$. If one of $k_i \ne 0 \mod n$, then $\mathsf{C}_{n,d}^{\sigma}T_{\mathbf{k}} = 0$ by Lemma 10.2.1. We are left with the case that $k_i = 0 \mod n$ for all i. Since $|\mathbf{k}| \le 2n-1$, there can be at most one $k_i = n$. Furthermore, $|\mathbf{k}| > 0$ shows that there is exactly one $k_i = n$. Thus, the right-hand side becomes $\mathsf{Q}_n^{\sigma_i}T_n + \mathsf{Q}_n^{\widetilde{\sigma}_i}T_n = \mathsf{Q}_n^{\mathsf{E}}T_n + \mathsf{Q}_n^{\mathsf{O}}T_n$, which is zero as $\mathsf{Q}_n^{\mathsf{E}}T_n = 1/2$ and $\mathsf{Q}_n^{\mathsf{O}}T_n = -1/2$ according to Lemma 10.2.1. □

For $d = 2$, Theorem 10.2.2 contains two distinct cubature rules of degree $2n-1$, $\sigma = (\mathsf{E},\mathsf{E})$ and $\sigma = (\mathsf{E},\mathsf{O})$. If n is even, then the number of nodes for the case (E,O) is $\frac{1}{2}n(n+2)$, which agrees with Möller's lower bound (5.9), whereas if n is odd, the number of nodes is $\frac{1}{2}(n+1)^2$, which is one more than Möller's lower bound. The number of nodes for the case (E,E) is one more than the lower bound for both n even and odd. Thus, the theorem provides an alternative proof for the minimal and near-minimal cubature rules in Section 6.4.

For $d = 3$, there are four distinct cubature rules of degree $2n-1$ that correspond to $\sigma = (\mathsf{E},\mathsf{E},\mathsf{E}), (\mathsf{E},\mathsf{E},\mathsf{O}), (\mathsf{E},\mathsf{O},\mathsf{E}), (\mathsf{O},\mathsf{E},\mathsf{E})$, respectively. For $n = 2m$, the number of nodes for the case $\sigma = (\mathsf{E},\mathsf{E},\mathsf{E})$ is

$$\frac{1}{4}\left[(n+1)^3 + (n+1)\right],$$

and the number of nodes for $\sigma = (\mathsf{E},\mathsf{E},\mathsf{O}), (\mathsf{E},\mathsf{O},\mathsf{E}), (\mathsf{O},\mathsf{E},\mathsf{E})$ is

$$\frac{1}{4}\left[(n+1)^3 - (n+1)\right].$$

Thus, these cubature rules of degree $2n-1$ have the number of nodes

$$N = \frac{n^3}{4}\left(1+o(1)n^{-1}\right). \tag{10.3}$$

Comparing with the lower bound (5.11) for $d = 3$, which is given by

$$N \ge \dim \Pi_{n-1}^3 + \mathscr{P}_{n-1}^2 - \delta_n = \frac{n^3}{6} + \frac{3n^2}{4} + \frac{5n}{6} - \delta_n = \frac{n^3}{6}\left(1+o(1)n^{-1}\right),$$

the number of nodes for the cubature rules in Theorem 10.2.2 is substantially larger than the lower bound. However, the number of nodes (10.3) is the smallest among known cubature rules of degree $2n-1$ for $d = 3$. These results suggest the following conjecture.

Conjecture 10.2.3. *Let $d \ge 3$. The number of nodes N for the minimal cubature rule of degree $2n-1$ for W_0 on $[-1,1]^d$ satisfies*

$$N \ge \frac{n^d}{2^{d-1}}\left(1+o(1)n^{-1}\right). \tag{10.4}$$

If this were true, one might entertain an even broader conjecture that this lower bound holds for all centrally symmetric weight integrals on the cube, which, as we have seen, holds for $d = 2$.

The likewise conjecture, however, is unlikely to hold for other domains. Indeed, the lower bound (10.4) holds for all centrally symmetric integrals for $d = 2$, as shown by Möller's lower bound, but it fails to hold for the unit disk and the triangle, as shown in Section 5.4. This indicates a strong dependence of the problem on the geometry of the domain.

10.3 Notes and Further Results

The equilibrium measure in the multivariate setting is studied via pluripotential theory in $\mathbb{C}^d$, see [90, 131]. If Ω is convex and centrally symmetric, an explicit expression is given in [9]. Moreover, several examples of sets Ω with their equilibrium measure are given explicitly in [6].

For $d = 2$, Theorem 10.2.2 was first established in [20], which provided an alternative proof of the minimal and near-minimal cubature rules in the results in [110], which appeared in Section 6.4. The construction is utilized in [39] to establish Theorem 10.2.2 for $d > 2$. The number of nodes (10.3) also holds for the cubature rules derived via the discrete Fourier transform, as shown in Theorem 8.3.3, and for the lattice cubature rules derived in [29], which provide some support for the validity of Conjecture 10.2.3.

It is worth pointing out that the system of moment equations for cubature rules of degree $2n-1$ is of dimension $(d+1)N \times \dim \Pi_{2n-1}^d$. Hence, the presumption of more variables than equations for $d = 3$ becomes

$$N \geq \frac{n^3}{3}\left(1+o(1)n^{-1}\right), \tag{10.5}$$

which is stronger than the conjectured lower bound of (10.4). It is unclear, however, if the conjecture could hold even for (10.5). After all, only cubature rules of lower degrees are known for $d \geq 3$, and those are mostly symmetric ones, which utilize more nodes.

The classical lower bound (2.2), or (5.11) for the centrally symmetric integrals, gives

$$N \geq \frac{n^d}{d!}(1+o(1)n^{-1}),$$

which is a function of both n and d. For n small, there have been constructions of cubature rules with the number of nodes close to the lower bound when d is large [65, 119, 156]. For example, the cubature rule of degree 5 in [65] contains cubature rules of degree 5 with d^2+7d+1 nodes, whereas the lower bound is d^2+d+1, but the cubature rules are not positive. A method proposed in [156] can reduce the size of invariant cubature rules on the simplex and the sphere, which utilizes two combinatorial objects, namely t-designs and orthogonal arrays. This is used in [135], where several families of positive cubature rules are constructed on the simplex and, as a result, on the sphere.

11

Addendum: Cubature Rules on the Triangle and Simplex

Cubature rules on triangles and simplexes are of special interest for the finite element method (FEM) and hp-FEM, which often employ triangulation of the domain. They also serve as a good specimen for constructing cubature rules on regular domains by solving moment equations numerically. In this addendum, we discuss the structure of symmetric cubature rules for integrals on triangles and simplexes.

As shown in Corollary 5.2.4, Möller's lower bound holds for the triangle, but, if Conjecture 10.1.2 is true, then Möller's lower bound is far from sharp by Theorem 5.4.9, which indicates that we do not have a viable lower bound for cubature rules on the triangle. Consequently, apart from a handful of cubature rules of lower degrees, we do not know how to tell if a rule is minimal nor a method to construct a minimal cubature in general. Most of the known cubature rules on these domains are found numerically by solving nonlinear systems of equations. If Conjecture 10.1.2 were true, then cubature rules constructed numerically would be good candidates. However, due to difficulties in solving nonlinear systems of equations, it is often more practical to consider invariant cubature rules under the symmetric group for triangles and even more so for simplexes, as this approach can significantly reduce the size of the system as indicated by the Sobolev theorem, although the invariant cubature rules require more nodes.

In Section 11.1, we describe the method of constructing symmetric cubature rules on the triangle and discuss, in particular, the structure of the moment equations and invariant polynomials. In Section 11.2, we consider cubature rules on the triangle with many nodes on the boundary of the triangle, called Lobatto cubature rules, which could be of advantage for FEM. In Section 11.3, we consider the structure of invariant cubature rules on the simplex, clarifying notations and formulations and paying particular attention to cubature rules on the three-dimensional tetrahedra.

11.1 Constructing Cubature Rules via Moment Equations

The existence of minimal cubature rules is equivalent to solving the system of nonlinear equations. Let $\{\phi_{\mathbf{k}} : |\mathbf{k}| \le m\}$ be a basis of Π_m^d. Then, the existence of the cubature rule $\mathsf{C}f = \sum_{i=1}^N \lambda_i f(\mathbf{x}_i)$ of degree m for the integral $\mathcal{I}f$ is equivalent to solving

$$\sum_{i=1}^{N} \lambda_i \phi_{\mathbf{k}}(\mathbf{x}_i) = \mathcal{I}(\phi_{\mathbf{k}}), \qquad |\mathbf{k}| \le m,$$

with both λ_i and $\mathbf{x}_i$ as variables, which is a nonlinear system of $\dim \Pi_m^d$ equations and $(d+1)N$ variables. If $\phi_{\mathbf{k}} = \mathbf{x}^{\mathbf{k}}$, this is the system of moment equations, but other choices of bases are often more convenient.

For cubature rules of lower degrees, it is possible to solve the nonlinear system of equations numerically, and many known cubature rules of lower degrees are found this way in the literature. As discussed in Section 2.6, it is often necessary to set up the nonlinear system with the number of equations equal to the number of variables so that Newton's root-finding method is applicable. To reduce the size of the nonlinear system, it is advantageous to consider cubature rules that possess the same symmetry as the integral, so the Sobolev theorem is applicable, which restricts the polynomials in Π_m^d to polynomials invariant under the group that defines the symmetry.

We illustrate how this can be carried out for the triangle in $\mathbb{R}^2$ by considering invariant cubature rules on the triangle for the integral against a symmetric weight function W,

$$\int_{\Delta^2} f(\mathbf{x}) W(\mathbf{x}) d\mathbf{x} = \mathsf{C}(f), \qquad f \in \Pi_m^2,$$

where $\Delta^2 = \{\mathbf{y} \in \mathbb{R}^2 : y_1 \ge 0, y_2 \ge 0, y_1 + y_2 \le 1\}$. Such a cubature rule also leads to cubature rules on the unit disk $\mathbb{B}^2$ and the unit sphere $\mathbb{S}^2$ that are invariant under the octahedral group by Corollary 2.5.6. In particular, if W is the Chebyshev weight

$$W_0(\mathbf{x}) = \frac{1}{\sqrt{x_1 x_2 (1 - x_1 - x_2)}},$$

then the corresponding invariant cubature rule over the unit sphere $\mathbb{S}^2$ is for integral against the surface measure $d\sigma$. For applications in the FEM method, the interesting case is $W(x) = 1$.

Instead of the triangle Δ^2, it is convenient to work with the equilateral triangle $\triangle$ defined by

$$\triangle = \left\{(x_1, x_2) : x_1 \le 1/2, \quad \sqrt{3}x_2 - x_1 \le 1, \quad -\sqrt{3}x_2 - x_1 \le 1\right\},$$

which is mapped from Δ^2 by an affine transformation

$$\Psi : (y_1, y_2) \in \Delta^2 \mapsto (x_1, x_2) \in \triangle, \quad x_1 = \sqrt{3}(y_1 + y_2)/2 - 1, \quad x_2 = \sqrt{3}(y_2 - y_1)/2,$$

as depicted in Figure 11.1. The symmetric group $\mathcal{S}_\triangle$ of $\triangle$ is generated by a

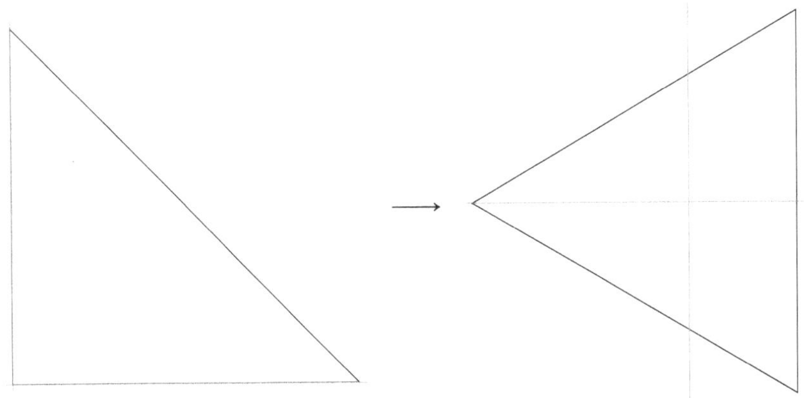

Figure 11.1 $\Psi : \Delta^2 \mapsto \triangle$.

rotation through an angle $\frac{2\pi}{3}$ and a reflection about the x-axis. Let $\Pi_n^\triangle$ be the space of polynomials of degree at most n that are invariant under the group $\mathcal{S}_\triangle$. A convenient basis for $\Pi_n^\triangle$ can be given in terms of the polar coordinates $\mathbf{x} = (r\cos\theta, r\sin\theta)$,

$$\Pi_n^\triangle = \operatorname{span}\left\{r^{2k}(r^3\cos 3\theta)^j,\ 0 \le 2k + 3j \le n\right\}. \tag{11.1}$$

The dimension of this space is given by

$$\dim \Pi_n^\triangle = \left\lfloor \tfrac{1}{12}(n^2 + 6n + 12) \right\rfloor.$$

Working with functions $g(r, \theta) = f(r\cos\theta, r\sin\theta)$ in polar coordinates, an invariant cubature rule is built upon the basic sum

$$\mathsf{C}(r, \theta) g = \frac{1}{6}\sum_{j=1}^{3}\left\{g\left(r, \theta + \frac{2\pi j}{3}\right) + g\left(r, -\theta + \frac{2\pi j}{3}\right)\right\},$$

which is just a sum over the $\mathcal{S}_\triangle$-orbit of the point (r, θ). To describe the symmetric cubature rule on $\triangle$, we let $\Lambda_\triangle$ denote the triangle

$$\Lambda_\triangle = \left\{(x_1, x_2) : 0 \le x_1 \le 1/2, \quad 0 \le x_2 \le \sqrt{3}x_1\right\},$$

which is one of the six congruent triangles under $\mathcal{S}_\triangle$ that makes up $\triangle$. An invariant cubature rule on $\triangle$ is determined by the nodes on $\Lambda_\triangle$ and their associated weights.

Definition 11.1.1. An invariant cubature rule on the triangle $\triangle$ is called type $[m_0; m_1, m_2, m_3; m_4, m_5]$ if it has m_0 node at the origin, m_1 node at the vertex $(1/2, \sqrt{3}/2)$, m_2 node at $(1/2, 0)$, m_3 nodes on the two sides $\theta = 0$ and $\theta = \pi/3$ (not at the vertices) of $\Lambda_\triangle$, m_4 nodes on the side $x_1 = 1/2$ (not at the vertices) of $\Lambda_\triangle$, and m_5 nodes in the interior of $\Lambda_\triangle$.

By definition, m_0, m_1, and m_2 take the value either 0 or 1. The sum $\mathsf{C}(r,\theta)g$ requires three function evaluations for each nonzero m_1, m_2, or m_3 node and six evaluations for each nonzero m_4 and m_5 node. Hence, let

$$n_1 = m_1 + m_2 + m_3 \quad \text{and} \quad n_2 = m_4 + m_5,$$

so that n_1 and n_2 denote the number of nodes with orbits of sizes 3 and 6, respectively. Then, an invariant cubature rule on $\triangle$ is of the form

$$\mathsf{C}(f) = m_0 \lambda_0 g(0,0) + \sum_{i=1}^{n_1} \lambda_i \mathsf{C}(r_i, 0) g + \sum_{i=n_1+1}^{n_1+n_2} \lambda_i \mathsf{C}(r_i, \theta_i) g.$$

The number of nodes of this cubature rule is

$$N = m_0 + 3n_1 + 6n_2 = m_0 + 3(m_1 + m_2 + m_3) + 6(m_4 + m_5).$$

By the Sobolev theorem, we only need to consider invariant polynomials in $\Pi_n^\triangle$. Working with the basis in (11.1), it is easy to verify that the cubature rule $\mathsf{C}(f)$ is of degree M if its nodes and weights satisfy the following system of equations,

$$\begin{aligned} m_0\lambda_0 + \sum_{i=1}^{n_1} \lambda_i + \sum_{i=n_1+1}^{n_1+n_2} \lambda_i &= v_{0,0}, \\ \sum_{i=1}^{n_1} \lambda_i r_i^j + \sum_{i=n_1+1}^{n_1+n_2} \lambda_i r_i^j \cos 3k_0\theta_i &= v_{j,3k_0}, \quad 2 \le j \le M, \\ \sum_{i=n_1+1}^{n_1+n_2} \lambda_i r_i^j (\cos 3k_0\theta_i - \cos 3k\theta_i) &= v_{j,3k_0} - v_{j,3k}, \quad (j,k) \in \Sigma_M, \end{aligned} \tag{11.2}$$

where $k_0 = 0$ if j is even and $k_0 = 1$ if j is odd,

$$\Sigma_M = \{(j,k) : j = 6 \text{ or } 8 \le j \le M,\ 6 \le 3k \le j,\ j+k \text{even}\},$$

and the numbers $v_{j,3k}$ are defined by

$$\begin{aligned} v_{j,3k} &= c_\triangle \int_\triangle r^{j+1} \cos 3k\theta\, W_\triangle(r\cos\theta, r\sin\theta) dr d\theta \\ &= 6c_\triangle \int_0^{\frac{\pi}{3}} \int_0^{\frac{1}{2\cos\theta}} r^{j+1} \cos 3k\theta\, W_\triangle(r\cos\theta, r\sin\theta) dr d\theta, \end{aligned}$$

where $W_\triangle$ is the weight function defined on $\triangle$ and $c_\triangle$ is the normalization constant so that $v_{0,0} = 1$ and, in particular, $c_\triangle = 4/3(\sqrt{3})$ if $W_\triangle(\mathbf{x}) = 1$.

For each M, the system contains $\dim \Pi_M^\triangle$ equations. The variables are r_i, θ_i, and λ_i, and there are two variables for each node of m_4 type and three variables for each node of m_5 type. Thus, we choose m_i so that the number of equations agrees with the number of variables. For the cubature rule of type $[m_0; m_1, m_2, m_3; m_4, m_5]$, this means

$$m_0 + m_1 + m_2 + 2m_3 + 2m_4 + 3m_5 = \dim \Pi_M^\triangle = \left\lfloor \tfrac{1}{12}(M^2 + 6M + 12) \right\rfloor. \tag{11.3}$$

For each M, there can be several integer solutions to the equation, which lead to several types of potential cubature rules of the same degree. However, since the system of equations is nonlinear, it may not have a solution for some or many types. To have enough parameters to solve the equations, one can deduce several consistency conditions that help reduce the choices. These conditions warrant that there are more variables than equations. The first one is $m_0 \le 1$ and the other two are

$$\begin{aligned} m_1 + m_2 + 2(m_3 + m_4) + 3m_5 &\ge \dim \Pi_M^\triangle - 1, \\ 2m_4 + 3m_5 &\ge \dim \Pi_{M-6}^\triangle. \end{aligned} \tag{11.4}$$

The first inequality follows from $m_0 \le 1$ and the second inequality is deduced by considering polynomials vanish on the boundary of the triangle. Moreover, there is also the necessary condition

$$m_5 > \begin{cases} (M-9)/4, & \text{if } m_4 \ne 0 \text{ and } M \ge 9, \\ (M-6)/4, & \text{if } m_4 = 0 \text{ and } M \ge 6, \\ (M-3)/4, & \text{if } m_3 = 0 \text{ and } M \ge 3, \end{cases} \tag{11.5}$$

that follows from considering polynomials that contain linear factors vanishing on the boundary, as well as medians of the triangle.

For fixed M, it is possible to identify all possible integer solutions of (11.3) that also satisfy (11.4) and (11.5), although the number of the solutions is large even for moderate M. For each type that satisfies (11.3), one can solve the nonlinear system of equations (11.2) numerically. If the system is solvable, there is an invariant cubature rule of degree M on $\triangle$.

Since the third group of the equations in (11.2) does not contain r_i and λ_i for $i = 1, 2, \ldots, n_1$, the system can be split into subsystems with independent variables. The smaller size of the subsystem makes it easier to solve. The third group consists of $\dim \Pi_{M-6}^\triangle$ equations, and it can be split whenever m_4 and m_5 satisfy

$$2m_4 + 3m_5 = \dim \Pi_{M-6}^\triangle.$$

This equation has integer solutions if $M \geq 7$, except $M = 10$. In particular, one subclass that admits the split consists of the type

$$[1;1,0,3m;m,m(m-1)] \quad \text{and} \quad [1;1,1,3m+1;m,m^2], \tag{11.6}$$

which correspond to cubature rules of degree $6m+2$ and $6m+5$, respectively, by (11.3). For small m, the nonlinear system of equations (11.2) for the two types (11.6) has been solved for the constant weight and the Chebyshev weight.

Of particular interest are cubature rules for the Chebyshev weight W_0 on the triangle since they correspond to cubature rules on the unit sphere as shown in Theorem 2.5.5. The solution of (11.6) for W_0 leads to cubature rules of degree $6m+2$ or $6m+5$ on the triangle, as well as cubature rules on the sphere $\mathbb{S}^2$ of degree $12m+5$ and $12m+11$. For $m = 1,2,3,4$, the solutions are known to exist and yield positive cubature rules. This leads to the following conjecture.

Conjecture 11.1.2 (V. Lebedev). *The invariant cubature rules of the types in* (11.6) *on the triangle and the unit sphere* $\mathbb{S}^2$ *exist for all* $m \geq 1$.

The conjecture illustrates well the virtue and limitation of finding cubature rules of lower degrees numerically. On the one hand, extensive numerical works help to identify the two types in (11.3) and the conjecture. On the other hand, numerical calculations offer little help in proving the conjecture.

We end this section with the following observation concerning the conjecture. Since we only need to consider invariant polynomials, it is essential to understand the structure of $\Pi_n^{\triangle}$. Let

$$t = r^2 = x_1^2 + x_2^2 \quad \text{and} \quad s = r^3 \cos 3\theta = x_1^3 - 3x_1^2 x_2.$$

Then, the space of invariant polynomials becomes

$$\Pi_n^{\triangle} = \operatorname{span}\left\{t^k s^j : 2k+3j \leq n\right\}.$$

Since the cubature rule in the conjecture has the number of parameters equal to the number of equations, taking a cue from the Gauss cubature rules, we may consider invariant polynomials of degree $\lfloor \frac{n}{2} \rfloor$. For the conjecture, let us consider, for example, the space $\Pi_{3m-1}^{\triangle}$.

Proposition 11.1.3. *Let* $\Pi_0^* = \operatorname{span}\{1\}$ *and* $\Pi_m^* = \Pi_{3m-1}^{\triangle}$ *for* $m = 1,2,3,\ldots$. *Then* $s\Pi_m^* \subset \Pi_{m+1}^*$ *and* $t\Pi_m^* \subset \Pi_{m+1}^*$.

This observation suggests that the space Π_m^* has a structure similar to the usual polynomial space Π_m^2. It offers hope that the relation between the cubature rule and the structure of Π_n^2, developed in the earlier chapters, could be useful.

11.2 Lobatto Cubature Rules on the Triangle

Triangulations are often used in the FEM or the hp-FEM, and cubature rules on triangles are needed for discretization on each triangle patch. Cubature nodes on the common boundary of two neighboring triangles are shared. Hence, it is desirable to have nodes on the boundary. As an analog of Lobatto quadrature rules in one variable, we call a cubature rule Lobatto if it has the maximal number of nodes on the boundary.

For discussing the boundary, it is more convenient to consider the triangle Δ^2. Since the vertex of a triangle is on two sides of the boundary, and it is shared by all triangles using the same vertex, we assume that all three vertexes are nodes of our cubature rule. Thus, we consider cubature rules of degree m that are of the form

$$\begin{aligned}\int_{\Delta^2} f(x,y)W(x,y)\mathrm{d}x\mathrm{d}y &= \sum_{k=1}^{N_0} \lambda_{k,0} f(x_{k,0},y_{k,0}) + \sum_{k=1}^{N_1} \lambda_{k,1} f(x_{k,1},0) \\ &\quad + \sum_{k=1}^{N_2} \lambda_{k,2} f(0,y_{k,2}) + \sum_{k=1}^{N_3} \lambda_{k,3} f(x_{k,3},1-x_{k,3}) \\ &\quad + \mu_0 f(0,0) + \mu_1 f(1,0) + \mu_2 f(0,1),\end{aligned} \tag{11.7}$$

where $(x_{k,0},y_{k,0})$ are distinct points in the interior of Δ^2; $(x_{k,1},0)$, $(0,y_{k,2})$, and $(x_{k,3},1-x_{k,3})$ are distinct points on the side $y=0$, $x=0$, and $x+y=1$ (but not on the corners) of Δ^2, respectively; and $\lambda_{k,j}$ and μ_i are cubature weights. Such a cubature has

$$N := N_0 + N_1 + N_2 + N_3 + 3$$

nodes. The most interesting weight functions are $W(x,y) = 1$ and the Chebyshev weight $W = W_0$. More generally, we consider the Jacobi weight $W = W_{\alpha,\beta,\gamma}$ defined in (3.27), for which the number of nodes N satisfy

$$N \ge \begin{cases} \frac{n(n+1)}{2} + \lfloor \frac{n}{2} \rfloor & \text{if } m = 2n-1, \\ \frac{n(n+1)}{2} & \text{if } m = 2n-2, \end{cases} \tag{11.8}$$

by (2.2) and (5.9). We can further deduce lower bounds for N_0 and $N_0 + N_i$.

Theorem 11.2.1. *If a cubature rule of the form* (11.7) *for* $W = W_{\alpha,\beta,\gamma}$ *exists with precision* m*, then*

$$N_0 \ge \begin{cases} \frac{n(n-1)}{2} & \text{if } m = 2n-1, \\ \frac{n(n-1)}{2} + \lfloor \frac{n-1}{2} \rfloor & \text{if } m = 2n. \end{cases} \tag{11.9}$$

Proof The cubature rule (11.7) integrates degree m polynomials of the form

$f(x,y) = xy(1-x-y)g(x,y)$ if

$$\int_{\Delta^2} g(x,y) W_{\alpha+1,\beta+1,\gamma+1}(x,y)\mathrm{d}x\mathrm{d}y = \sum_{k=1}^{N_0} \lambda_{k,0}^* g(x_{k,0}, y_{k,0}), \quad \forall g \in \Pi_{m-3}^2, \tag{11.10}$$

where $\lambda_{k,0}^* = \lambda_{k,0} x_{k,0} y_{k,0}(1 - x_{k,0} - y_{k,0})$, which is a cubature rule of precision $m-3$ for the weight function $W_{\alpha+1,\beta+1,\gamma+1}$ so that, by (11.8), N_0 satisfies the stated lower bound. □

Since the lower bound of N_0 is Möller's lower bound for the cubature rule for $W_{\alpha+1,\beta+1,\gamma+1}$ of degree $m-3$, it is likely sharp only for small m. More interestingly, a likely lower bound for N_i is

$$N_i \ge n-1, \qquad i = 1,2,3, \quad m = 2n-1 \text{ or } m = 2n-2. \tag{11.11}$$

Indeed, as shown in the above proof, a Lobatto cubature rule's interior nodes and weights are determined by (11.10), which is the restriction of (11.7) on the class of bubble functions $xyz\Pi_{m-3}^2$, where $z = 1-x-y$, that vanish on the boundary. A moment of reflection shows that Π_m^2 can be decomposed into a direct sum,

$$\Pi_m^2 = xyz\Pi_{m-3}^2 \oplus xz\Pi_{m-2}[x] \oplus yz\Pi_{m-2}[y] \oplus xy\Pi_{m-2}[x] \oplus \Pi_1^2, \tag{11.12}$$

where $\Pi_{m-2}[x]$ and $\Pi_{m-2}[y]$ denote the space of polynomials of one variable in x-variable and y-variable, respectively. Let us define

$$\mathcal{I}_1 g := \int_{\Delta^2} g(x) W_{\alpha+1,\beta,\gamma+1}(x,y)\mathrm{d}x\mathrm{d}y - \sum_{k=1}^{N_0} \lambda_{k,0} x_{k,0} z_{k,0} g(x_{k,0}),$$

$$\mathcal{I}_2 g := \int_{\Delta^2} g(y) W_{\alpha,\beta+1,\gamma+1}(x,y)\mathrm{d}x\mathrm{d}y - \sum_{k=1}^{N_0} \lambda_{k,0} y_{k,0} z_{k,0} g(y_{k,0}),$$

$$\mathcal{I}_3 g := \int_{\Delta^2} g(z) W_{\alpha+1,\beta+1,\gamma}(x,y)\mathrm{d}x\mathrm{d}y - \sum_{k=1}^{N_0} \lambda_{k,0} x_{k,0} y_{k,0} g(z_{k,0}).$$

where $z_{k,0} = 1 - x_{k,0} - y_{k,0}$. With $(x_{k,0}, y_{k,0})$ and $\lambda_{k,0}$ determined, these are linear functionals defined in the space of polynomials of one variable. In particular, if $f \in yz\Pi_{m-2}[x]$, then (11.7) becomes

$$\mathsf{C}_1 g = \sum_{k=1}^{N_1} \lambda_{k,1} y_{k,0} z_{k,0} g(x_{k,1}, 0), \qquad g \in \Pi_{m-2}[x], \tag{11.13}$$

which is a quadrature rule of degree $s-2$ for the linear functional $\mathcal{I}_1$. The heuristic argument of counting variables and the number of conditions indicates $N_1 \ge n-1$ for such a quadrature. Moreover, if $\mathsf{C}_1 g$ is a positive linear functional, then the lower bound $N_1 \ge n-1$ holds with equality held when the quadrature is Gaussian for C_1. The same argument works for N_2 and N_3. However, since we do not know if C_i is positive, the argument is not a proof of (11.11).

The above argument and the proof of (11.8) lead to an algorithm that can be used to construct Lobatto cubature rules on the triangle.

Algorithm 11.2.2. For $W_{\alpha,\beta,\gamma}$ on the triangle, a Lobatto cubature rule of degree m of the form (11.7) can be constructed as follows:

Step 1. Construct a cubature rule of degree $m-3$ for $W_{\alpha+1,\beta+1,\gamma+1}$ in the form of (11.10), with all nodes in the interior of Δ^2, and define

$$\lambda_{k,0} = \frac{\lambda^*_{k,0}}{x_{k,0}y_{k,0}(1-x_{k,0}-y_{k,0})}. \tag{11.14}$$

Step 2. Construct Gaussian quadrature rules C_1 in (11.13) and

$$\mathsf{C}_2 g = \sum_{k=1}^{N_2} \lambda^*_{k,2} g(y_{k,2}) \quad \text{and} \quad \mathsf{C}_3 g = \sum_{k=1}^{N_3} \lambda^*_{k,3} g(x_{k,3}), \qquad \forall g \in \Pi_{m-2},$$

with respect to the linear functional $\mathcal{I}_1$, $\mathcal{I}_2$, and $\mathcal{I}_3$, which gives the nodes of the cubature rule (11.7) on the boundary of $\Delta 62$, and define

$$\lambda_{k,1} = \frac{\lambda^*_{k,1}}{x_{k,1}(1-x_{k,1})}, \quad \lambda_{k,2} = \frac{\lambda^*_{k,2}}{y_{k,2}(1-y_{k,2})}, \quad \lambda_{k,3} = \frac{\lambda^*_{k,3}}{x_{k,3}(1-x_{k,3})}. \tag{11.15}$$

Step 3. Finally, determine the weights μ_0, μ_1, and μ_2 by setting $f(x,y) = 1,x,y$ in (11.7) and solve the resulting linear system of equations.

We make comments on each step of the algorithm for the Lobatto cubature rules of degree $s = 2n-1$. First, as long as a cubature rule of degree $2n-4$ with all nodes in the interior of Δ^2 is constructed in Step 1, regardless if it is a minimal one, Step 2 and Step 3 could be carried out. Second, cubature rules in Step 1 can be constructed by solving moment equations, and they can be, in particular, symmetric cubature rules as in the prior section; however, for the constant weight $W(x,y) = 1$, the cubature rule (11.10) is for integral against $W(x,y) = xy(1-x-y)$, which did not appear often in the literature but could be constructed using the existing method. Third, if $\mathcal{I}_1$, $\mathcal{I}_2$, and $\mathcal{I}_3$ are positive linear functionals, then Step 2 can be solved by the Gauss quadratures of degree $2n-3$ for the three linear functionals so that $N_1 = N_2 = N_3 = n-1$. Third, there is no guarantee that the weights μ_1, μ_2, and μ_3 are all positive in Step 3.

We illustrate the algorithm by two symmetric Lobatto cubature rules for the constant weight, which are of the form

$$\int_{\Delta^2} f(x,y)\mathrm{d}x\mathrm{d}y = \sum_{k=1}^{M_0} \lambda_{k,0}\left[f(x_{k,0},y_{k,0}) + f(y_{k,0},z_{k,0}) + f(z_{k,0},x_{k,0})\right] \tag{11.16}$$
$$+\sum_{k=1}^{M_1} \lambda_{k,1}\left[f(x_{k,1},0) + f(0,1-x_{k,1}) + f(1-x_{k,1},x_k)\right]$$
$$+\mu\left[f(0,0)+f(1,0)+f(0,1)\right],$$

where $z_{k,0} = 1 - x_{k,0} - y_{k,0}$. The two examples are the rules of degrees 5 and 7. For Step 1 of the algorithm, we need to find the symmetric cubature rules of degrees 2 and 4 of the form

$$\int_{\Delta^2} f(x,y)xy(1-x-y)\mathrm{d}x\mathrm{d}y = \sum_{k=1}^{M_0} \lambda_{k,0}^{*}\left[f(x_{k,0},y_{k,0}) + f(y_{k,0},z_{k,0}) + f(z_{k,0},x_{k,0})\right].$$

By Sobolev Theorem 2.5.2, we only need to consider f belonging to the space of symmetric polynomials of degrees up to 4, and the two rules can be established by solving the moment equations symbolically in terms of the symmetric basis $\{1, xy+yz+zx, xyz, (xy+yz+zx)^2\}$. In both cases, the linear functionals $\mathcal{L}_1, \mathcal{L}_2$, and $\mathcal{L}_3$ are positive definite, so that Step 2 can be solved using the Gaussian quadrature. The results are the following.

Example 11.2.3. Symmetric Gauss–Lobatto cubature rules of degree 5 with 12 nodes. This formula is in the form of (11.16) with $M_0 = 1$ and $M_1 = 2$. The nodes and weights are given below:

$$x_{1,0} = y_{1,0} = \tfrac{1}{21}(7-\sqrt{7}), \quad \lambda_{1,0} = \tfrac{7}{720}(14-\sqrt{7}),$$
$$x_{1,1} = \tfrac{1}{42}\left(21-\sqrt{21(4\sqrt{7}-7)}\right), \quad x_{1,2} = \tfrac{1}{42}\left(21+\sqrt{21(4\sqrt{7}-7)}\right),$$
$$\lambda_{1,1} = \lambda_{2,1} = \tfrac{1}{720}(7+4\sqrt{7}), \quad \mu = \tfrac{1}{720}(8-\sqrt{7}).$$

Example 11.2.4. Symmetric Gauss–Lobatto cubature rules of degree 7 with 18 nodes. This formula is in the form of (11.16) with $M_0 = 2$ and $M_1 = 3$. The nodes and weights are given below:

$$x_{1,0} = y_{1,0} = \tfrac{1}{18}(5-\sqrt{7}), \qquad x_{2,0} = y_{2,0} = \tfrac{1}{18}(5+\sqrt{7}),$$
$$\lambda_{1,0} = \tfrac{1}{17640}(1141-94\sqrt{7}), \quad \lambda_{2,0} = \tfrac{1}{17640}(1141+94\sqrt{7}),$$
$$x_{1,1} = \tfrac{1}{6}(3-\sqrt{3}), \quad x_{2,1} = \tfrac{1}{2}, \quad x_{3,1} = \tfrac{1}{6}(3+\sqrt{3}),$$
$$\lambda_{1,1} = \tfrac{3}{280}, \quad \lambda_{2,1} = \tfrac{4}{315}, \quad \lambda_{3,1} = \tfrac{3}{280}, \quad \mu = \tfrac{1}{315}.$$

The nodes of these two cubature rules are depicted in Figure 11.2. Because of the restriction of nodes on the boundary and vertices, Lobatto cubature rules

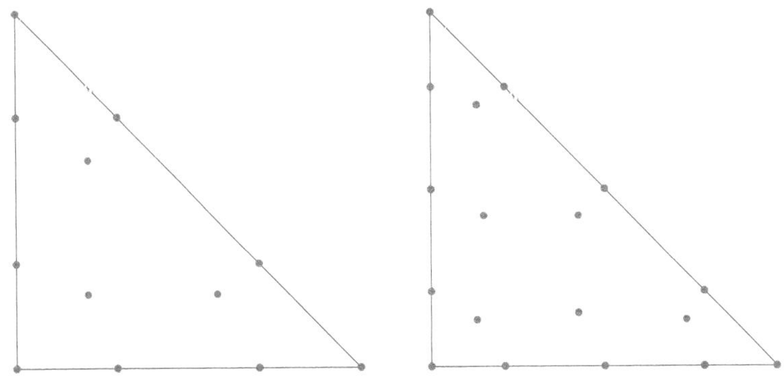

Figure 11.2 Nodes of symmetric Lobatto cubature rules of degrees 5 and 7.

are not minimal in general. The known minimal rule of degree 5 has 7 nodes and that of degree 7 has 13 nodes.

11.3 Invariant Cubature Rules on the Simplex

Compared to the triangle in $\mathbb{R}^2$, cubature rules on the simplex in $\mathbb{R}^d$ are more difficult to construct, and there are fewer studies. The most known results are cubature rules of lower degrees, and they are mainly established numerically by solving moment equations. To reduce the complexity of solving systems of nonlinear equations, it is common to consider cubature rules invariant under the symmetric group of the simplex.

11.3.1 Structure of Invariant Cubature Rules

As shown in Section 2.5, we consider homogeneous coordinates on the simplex and, instead of Δ^d at (2.9), we work with $\hat{\Delta}^{d+1}$ defined by

$$\hat{\Delta}^{d+1} = \{\mathbf{y} = (y_1, \ldots, y_{d+1}) : y_1 \ge 0, \ldots, y_{d+1} \ge 0, |\mathbf{y}| = 1\},$$

with $|\mathbf{y}| = y_1 + \cdots + y_{d+1}$, of which Δ^d is the special parametrization with $y_{d+1} = 1 - y_1 - \cdots - y_d$. The Jacobi weight function $W_{\boldsymbol{\kappa}}$ with $\boldsymbol{\kappa} = (\kappa_1, \ldots, \kappa_{d+1}) \in \mathbb{R}^{d+1}$, defined in (2.10), becomes $\mathbf{y}^{\boldsymbol{\kappa}} = y_1^{\kappa_1} \cdots y_{d+1}^{\kappa_{d+1}}$ on the simplex $\hat{\Delta}^{d+1}$ in the homogeneous coordinates. Let $\mathcal{P}_m^{d+1}$ be the space of homogeneous polynomials of degree m in $d+1$ variables.

Proposition 11.3.1. *The cubature rule of degree m on Δ^d,*

$$\int_{\Delta^d} f(\mathbf{x}) W_{\boldsymbol{\kappa}}(\mathbf{x}) d\mathbf{x} = \sum_{k=1}^{N} \lambda_k f(\mathbf{x}_k), \qquad f \in \Pi_m^d, \tag{11.17}$$

with all $\mathbf{x}_k \in \Delta^d$ is equivalent to the cubature rule for the integral on $\hat{\Delta}^{d+1}$,

$$\int_{\hat{\Delta}^{d+1}} f(\mathbf{y}) \mathbf{y}^{\boldsymbol{\kappa}} d\mathbf{y} = \sum_{k=1}^{N} \lambda_k f(\mathbf{y}_k), \qquad \forall f \in \mathcal{P}_m^{d+1}, \tag{11.18}$$

where $\mathbf{y}_k$ and $\mathbf{x}_k$ are related through $\mathbf{y}_k = (\mathbf{x}_k, 1 - |\mathbf{x}_k|) \in \hat{\Delta}^{d+1}$.

Proof Every homogeneous polynomial $F \in \mathcal{P}_m^{d+1}$ can be written as

$$F(x_1, \ldots, x_{d+1}) = \sum_{j=0}^{n} f_j(x_1, \ldots, x_d) x_{d+1}^{n-j} \quad \text{with} \quad f_j \in \Pi_j^d,$$

which implies, in particular, that $f(\mathbf{x}) := F(x_1, \ldots, x_d, 1 - |\mathbf{x}|_1) \in \Pi_n^d$. Indeed, on the one hand, applying (11.17) to such f and using

$$\int_{\hat{\Delta}^{d+1}} f(\mathbf{y}) \mathbf{y}^{\boldsymbol{\kappa}} d\mathbf{y} = \int_{\Delta^d} f(\mathbf{x}, 1 - |\mathbf{x}|) W_{\boldsymbol{\kappa}}(\mathbf{x}) d\mathbf{x},$$

gives the cubature rule (11.18). On the other hand, for each $f \in \Pi_k^d$ with $0 \le k \le m$, applying (11.18) to the homogeneous polynomials $f(x_1, \ldots, x_d)|\mathbf{x}|^{m-k} \in \mathcal{P}_n^{d+1}$ gives (11.17). □

The invariant cubature rule of degree m on the simplex is given in (2.22), which we rewrite as, by the above proposition,

$$c_\kappa \int_{\hat{\Delta}^{d+1}} f(\mathbf{y}) \mathbf{y}^{\kappa} d\mathbf{y} = \sum_{k=1}^{M} \frac{\lambda_k}{|\mathbf{y}_k \mathcal{S}_{d+1}|} \sum_{\mathbf{y} \in \mathbf{y}_k \mathcal{S}_{d+1}} f(\mathbf{y}), \qquad f \in \mathcal{P}_m^{d+1}, \tag{11.19}$$

where $\mathcal{S}_{d+1}$ is the symmetric group of permutations acting on $\mathbf{y} \in \mathbb{R}^{d+1}$, and $\mathbf{y}\mathcal{S}_{d+1}$ and $|\mathbf{y}\mathcal{S}_{d+1}|$ denote the orbit of $\mathbf{y}$ and its cardinality, and

$$c_\kappa = \frac{\Gamma((d+1)(\kappa+1))}{[\Gamma(\kappa+1)]^{d+1}},$$

is the normalization constant so that $\mathcal{I}_\kappa 1 = 1$. As in the case of the triangle, the most interesting case is the constant weight for $\kappa = 0$, and the Chebyshev weight for $\kappa = -\frac{1}{2}$, the latter because of its relation to cubature rules on the unit sphere.

By Sobolev Theorem 2.5.2, we only need to verify the cubature rule (11.19) for polynomials that are invariant under the group $\mathcal{S}_{d+1}$. These are symmetric polynomials, which are indexed by partitions as discussed in part B of Section 3.4.5.

Again, we only need partitions $\boldsymbol{\ell} \in \mathbb{N}_0^{d+1}$ of at most $d+1$ part, which satisfies $\boldsymbol{\ell} = (\ell_1, \ldots, \ell_{d+1})$ with $\ell_1 \ge \ell_2 \ge \cdots \ge \ell_{d+1}$. If $|\boldsymbol{\ell}| = m$, then $\boldsymbol{\ell}$ is a partition of the integer m. It is customary to drop the trail of zeros and write $\boldsymbol{\ell} = (\ell_1, \ldots, \ell_k)$, where k is the number of positive elements in $\boldsymbol{\ell}$.

Example 11.3.2. There are seven partitions for the integer 5, given by

$$(5), \quad (4,1), \quad (3,2), \quad (3,1,1), \quad (2,2,1), \quad (2,1,1,1), \quad (1,1,1,1,1),$$

of which the first five are partitions of at most three parts.

For $m \in \mathbb{N}$, we denote the set of partitions of m of at most part $d+1$ as

$$\Lambda_m^{d+1} := \{\boldsymbol{\ell} = (\ell_1, \ldots, \ell_k) : |\boldsymbol{\ell}| = m, \ \ell_1 \ge \ell_2 \ge \cdots \ge \ell_k > 0, \ k \le d+1\}.$$

Let $\mathcal{P}_m^{\mathcal{S}_{d+1}}$ be the space of homogeneous symmetric polynomials of degree m. A basis of this space consists of polynomials defined by, for $\boldsymbol{\ell} \in \Lambda_m^{d+1}$,

$$S_{\boldsymbol{\ell}}(\mathbf{y}) = \frac{1}{|\boldsymbol{\ell}\mathcal{S}_{d+1}|} \sum_{\sigma \in \mathcal{S}_{d+1}} \mathbf{y}^{\boldsymbol{\ell}\sigma} = \frac{1}{|\boldsymbol{\ell}\mathcal{S}_{d+1}|} \sum_{\sigma \in \mathcal{S}_{d+1}} y_{\sigma(1)}^{\ell_1} \cdots y_{\sigma(d+1)}^{\ell_{d+1}}, \tag{11.20}$$

which is clearly homogeneous with degree $|\boldsymbol{\ell}| = m$. Thus,

$$\mathcal{P}_m^{\mathcal{S}_{d+1}} = \operatorname{span}\left\{S_{\boldsymbol{\ell}} : \boldsymbol{\ell} \in \Lambda_m^{d+1}\right\} \quad \text{and} \quad \dim \mathcal{P}_m^{\mathcal{S}_{d+1}} = \left|\Lambda_m^{d+1}\right|,$$

where $\left|\Lambda_m^{d+1}\right|$ denotes the cardinality of Λ_m^{d+1}. By Proposition 11.3.1, the invariant cubature rule 11.19 is of degree m if it holds for all polynomials in $\mathcal{P}_m^{\mathcal{S}_{d+1}}$. The dimension of this space is the number of integer partitions of m in at most $d+1$ parts. This number can be complicated to compute for large d. For $d = 2$ and $d = 3$, they are given by explicit formulas

$$\dim \mathcal{P}_m^{\mathcal{S}_3} = \left\lfloor \frac{1}{12}(m^2 + 6m + 12) \right\rfloor,$$

$$\dim \mathcal{P}_m^{\mathcal{S}_4} = \frac{1}{144} m(m-1)(m+7) + \frac{1}{4}\left\lfloor \frac{m}{4} \right\rfloor + \frac{1}{3}\left\lfloor \frac{m}{3} \right\rfloor + \frac{m+5}{8}\left\lfloor \frac{m}{2} \right\rfloor + \frac{1}{4}\left\lfloor \frac{m+1}{4} \right\rfloor + 1.$$

For m large, they are in the order of $\frac{1}{(d+1)!}\binom{m+d}{m} = \frac{1}{(d+1)!} \dim \Pi_m^d$, in agreement with the order $|\mathcal{S}_{d+1}| = (d+1)!$ of the symmetric group.

The nodes of the invariant cubature rules can be more specific according to the symmetry since if $\mathbf{y} \in \hat{\Delta}^{d+1}$ is a node of (11.19), then what matters is the orbit of $\mathbf{y}$. Thus, we can choose $\mathbf{y}$ more specifically. The set of partitions of a positive integer k is Λ_k^k. We further define

$$\Lambda_{d+1} = \bigcup_{k=1}^{d+1} \Lambda_k^k,$$

which contains all partitions of positive integers up to $d+1$.

Definition 11.3.3. For $1 \le k \le d+1$, let $\mathbf{n} = (n_1, n_2, \ldots, n_k) \in \Lambda_{d+1}$. We call $\mathbf{y} \in \hat{\Delta}^{d+1}$ type $[\mathbf{n}] = [n_1, \ldots, n_k]$ if

$$\mathbf{y} = \mathbf{y}[\mathbf{n}] = \big(\overbrace{y_1, \ldots, y_1}^{n_1}, \overbrace{y_2, \ldots, y_2}^{n_2}, \ldots, \overbrace{y_k, \ldots, y_k}^{n_k}, 0, \ldots 0\big) \in \mathbb{R}^{d+1},$$

where $y_i \ne y_j$, $y_j > 0$ and $|\mathbf{y}| = 1$.

If $\mathbf{y}$ is a type $[n_1, \ldots, n_k]$ point in $\hat{\Delta}^{d+1}$, then the orbit of $\mathbf{y}$ has the order

$$|\mathbf{y}\mathcal{S}_{d+1}| = \frac{(d+1)!}{n_1!n_2!\cdots n_k!} = \frac{(d+1)!}{\mathbf{n}!}.$$

For each $\mathbf{n} \in \Lambda_{d+1}$, an invariant cubature rule could contain several nodes of type $\mathbf{n}$. Thus, the invariant cubature rule (11.21) can be written as

$$c_\kappa \int_{\hat{\Delta}^{d+1}} f(\mathbf{y})\mathbf{y}^\kappa d\mathbf{y} = \sum_{\mathbf{n}\in\Lambda_{d+1}} \sum_{j=1}^{k(\mathbf{n})} \lambda_{j,\mathbf{n}} \frac{\mathbf{n}!}{(d+1)!} \sum_{\mathbf{y}\in\mathbf{y}_j[\mathbf{n}]\mathcal{S}_{d+1}} f(\mathbf{y}), \tag{11.21}$$

where $k(\mathbf{n})$ denotes the number of type $[\mathbf{n}]$ nodes. To construct such a cubature rule, we only need to work with invariant polynomials in $\mathcal{P}_m^{\mathcal{S}_{d+1}}$. Since

$$\frac{1}{|\mathbf{y}\mathcal{S}_{d+1}|} \sum_{\mathbf{y}\in\mathbf{y}\mathcal{S}_{d+1}} f(\mathbf{y}) = f(\mathbf{y}), \qquad f \in \mathcal{P}_m^{\mathcal{S}_{d+1}}, \tag{11.22}$$

the cubature rule 11.21 for invariant polynomials becomes

$$c_\kappa \int_{\hat{\Delta}^{d+1}} f(\mathbf{y})\mathbf{y}^\kappa d\mathbf{y} = \sum_{\mathbf{n}\in\Lambda_{d+1}} \sum_{j=1}^{k(\mathbf{n})} \lambda_{j,\mathbf{n}} f(\mathbf{y}_j[\mathbf{n}]), \qquad f \in \mathcal{P}_m^{\mathcal{S}_{d+1}}. \tag{11.23}$$

With this setup, constructing an invariant cubature rule (11.21) of degree m comes down to solving the nonlinear system of equations

$$\sum_{\mathbf{n}\in\Lambda_{d+1}} \sum_{j=1}^{k(\mathbf{n})} \lambda_{j,\mathbf{n}} S_\ell(\mathbf{y}_j[\mathbf{n}]) = c_\kappa \int_{\hat{\Delta}^{d+1}} S_\ell(\mathbf{y})\mathbf{y}^\kappa d\mathbf{y}, \qquad \ell \in \Lambda_m^{d+1},$$

which is equivalent to, by (11.20) and (11.22), the system of moment equations

$$\sum_{\mathbf{n}\in\Lambda_{d+1}} \sum_{j=1}^{k(\mathbf{n})} \lambda_{j,\mathbf{n}} (\mathbf{y}_j[\mathbf{n}])^\ell = m_\ell, \qquad \ell \in \Lambda_m^{d+1}, \tag{11.24}$$

where m_ℓ is the integral of $\mathbf{y}^\ell$, since by (11.20) and the beta integral,

$$m_\ell^\kappa = c_\kappa \int_{\hat{\Delta}^{d+1}} S_\ell(\mathbf{y})\mathbf{y}^\kappa d\mathbf{y} = c_\kappa \int_{\hat{\Delta}^{d+1}} \mathbf{y}^\ell \mathbf{y}^\kappa d\mathbf{y} = \frac{(\kappa+1)_{\ell_1}\cdots(\kappa+1)_{\ell_{d+1}}}{((\kappa+1)(d+1))_{|\ell|}},$$

where $(a)_m := a(a+1)\dots(a+m-1)$ is the Pochhammer symbol. In particular, for the constant weight function, $\kappa = 0$, the moments are

$$m_{\boldsymbol{\ell}}^0 = \frac{\ell_1!\dots\ell_{d+1}!}{(d+1)_{|\boldsymbol{\ell}|}} = d!\frac{\ell_1!\dots\ell_{d+1}!}{(|\boldsymbol{\ell}|+d)!}.$$

The system (11.24) consists of $|\Lambda_m| = \dim \mathcal{P}_m^{\mathcal{S}_{d+1}}$ equations and has

$$\sum_{\mathbf{n}\in\Lambda_{d+1}} k(\mathbf{n})\ell(\mathbf{n})$$

variables, where $\ell(\mathbf{n})$ is the number of nonzero elements in $\mathbf{n}$, and it is equal to the number of distinct nonzero elements in $y_j[\mathbf{n}]$, so that each summand in (11.24) corresponds to $\ell(\mathbf{n})$ variables, one for $\lambda_{j,\mathbf{n}}$ and $\ell(\mathbf{n})-1$ for $\mathbf{y}_j[\mathbf{n}]$ because of $|\mathbf{y}_j[\mathbf{n}]| = 1$.

To solve the system of equations 11.24 numerically, one often assumes the system has more variables than equations. That is,

$$\sum_{\mathbf{n}\in\Lambda_{d+1}} k(\mathbf{n})\gamma(\mathbf{n}) \ge \left|\Lambda_m^{d+1}\right| = \dim \mathcal{P}_m^{\mathcal{S}_{d+1}}.$$

There are other such conditions by considering various subsystems. These are called *consistency conditions* in the literature. They are linear inequalities that warrant that each subsystem will have no fewer unknowns than the number of equations. However, for a nonlinear system of equations, they are neither necessary nor sufficient for the existence of a solution.

11.3.2 Symmetric Cubature Rules on Tetrahedra

We specialize in the tetrahedra Δ^3 in $\mathbb{R}^3$ or $\hat{\Delta}^4$ in homogeneous coordinates. For $d = 3$, the set $\Lambda_{d+1} = \Lambda_4$ contains 11 partitions of at most 4 parts,

$$\begin{array}{lllll} [4], & [3,1], & [2,2], & [2,1,1], & [1,1,1,1] \\ [3], & [2,1], & [1,1,1] & & \\ [2], & [1,1], & & & \\ [1]. & & & & \end{array}$$

To simplify the presentation, we introduce a notation

$$f_{\mathcal{S}}(\mathbf{y}) = \sum_{\sigma\in\mathcal{S}_4} f(\mathbf{y}\sigma).$$

Then, an invariant cubature rule (11.21) on the tetrahedra is of the form,

$$c_\kappa \int_{\hat{\Delta}^4} f(\mathbf{y})\mathbf{y}^\kappa d\mathbf{y} = n_0 f\left(\tfrac{1}{4}\mathbf{1}\right) + \sum_{j=1}^{n_1} \lambda_{j,1} \frac{1}{4} f_S(\boldsymbol{\alpha}_j) + \sum_{j=1}^{n_2} \lambda_{j,2} \frac{1}{6} f_S(\boldsymbol{\beta}_j)$$
$$+ \sum_{j=1}^{n_3} \lambda_{j,3} \frac{1}{12} f_S(\boldsymbol{\gamma}_j) + \sum_{j=1}^{n_4} \lambda_{j,4} \frac{1}{24} f_S(\boldsymbol{\delta}_j), \qquad (11.25)$$

where the notes are given by $\frac{1}{4}\mathbf{1} = \left(\frac{1}{4}, \frac{1}{4}, \frac{1}{4}, \frac{1}{4}\right) \in \mathbb{R}^4$, which is type [4],

$$\begin{aligned} \boldsymbol{\alpha}_j &= (\alpha_{j,1}, \alpha_{j,1}, \alpha_{j,1}, \alpha_{j,2}), && \text{type } [3,1] \\ \boldsymbol{\beta}_j &= (\beta_{j,1}, \beta_{j,1}, \beta_{j,2}, \beta_{j,2}), && \text{type } [2,2] \\ \boldsymbol{\gamma}_j &= (\gamma_{j,1}, \gamma_{j,1}, \gamma_{j,2}, \gamma_{j,3}), && \text{type } [2,1,1] \\ \boldsymbol{\delta}_j &= (\delta_{j,1}, \delta_{j,2}, \delta_{j,3}, \delta_{j,4}), && \text{type } [1,1,1,1], \end{aligned}$$

and the elements in each node are distinct if their subindexes are different, and each node satisfies $|\mathbf{y}| = 1$. The last requirement implies, for example, $\alpha_{j,2} = 1 - 3\alpha_{j,1}$, and it further implies $\alpha_{j,1} \neq \frac{1}{4}$ by $\alpha_{j,1} \neq \alpha_{j,2}$. We do not assume, however, that the components of nodes are all different from zero, which is what allows us to restrict to only five types of nodes, namely $[4], [3,1], [2,2], [2,1,1]$, and $[1,1,1,1]$, out of 11 types. To see that this agrees with the definition of (11.21), we observe that the sum over type $[3,1]$ nodes on the right-hand side of (11.25) can be written as

$$\sum_{j=1}^{n_1} \lambda_{j,1} f_S(\boldsymbol{\alpha}_j) = m_{1,1} f_S(1,0,0,0) + m_{1,2} f_S\left(\tfrac{1}{3}, \tfrac{1}{3}, \tfrac{1}{3}, 0\right) + \sum_{j=1}^{m_{1,3}} \lambda_{j,1} f_S(\boldsymbol{\alpha}_j^*),$$

where $\boldsymbol{\alpha}_j^* = (\alpha_{j,1}^*, \alpha_{j,1}^*, \alpha_{j,1}^*, \alpha_{j,2}^*)$ with $\alpha_{j,1}^*$ and $\alpha_{j,2}^*$ nonzero, which further implies by $\alpha_{j,2}^* = 1 - 3a_{j,1}^* > 0$ that $0 < \alpha_{j,1}^* < 1/3$. The first and the second terms on the right-hand side of the displayed expression correspond to the type $[1]$ and type $[3]$ nodes. Without the restriction of $y_j > 0$ in Definition 11.3.3, type $[1]$ and type $[3]$ nodes can be regarded as type $[3,1]$. In short, we call $[1]$ and $[3]$ subordinate to $[3,1]$. Similarly, $[2]$ is subordinate to $[2,2]$ so that

$$\sum_{j=1}^{n_2} \lambda_{j,1} f_S(\boldsymbol{\beta}_j) = m_{2,1} f_S\left(\tfrac{1}{2}, \tfrac{1}{2}, 0, 0\right) + \sum_{j=1}^{m_{2,2}} \lambda_{j,1} f_S(\boldsymbol{\beta}_j^*),$$

with $\boldsymbol{\beta}_i^* \neq 0$, and $[1,1]$ and $[2,1]$ are subordinate to $[2,1,1]$ with

$$\sum_{j=1}^{n_3} \lambda_{j,3} \frac{1}{12} f(\boldsymbol{\gamma}_j) = \sum_{j=1}^{m_{3,1}} \lambda_{j,3} \frac{1}{12} f(\boldsymbol{\gamma}_j^1) + \sum_{j=1}^{m_{3,2}} \lambda_{j,3} \frac{1}{12} f(\boldsymbol{\gamma}_j^2) + \sum_{j=1}^{m_{3,3}} \lambda_{j,3} \frac{1}{12} f(\boldsymbol{\gamma}_j^*),$$

where $\boldsymbol{\gamma}_j^1 = (\gamma_1^1, \gamma_1^1, \gamma_2^1, 0)$ and $\boldsymbol{\gamma}_j^2 = (0, 0, \gamma_3^2, \gamma_4^2)$.

The number of nodes of the cubature rule (11.25) is

$$N = n_0 + 4n_1 + 6n_2 + 12n_3 + 24n_4,$$

which becomes, if we use all 11 types,

$$N = n_0 + 4(m_{1,1} + m_{1,2} + m_{1,3}) + 6(m_{2,1} + m_{2,2}) + 12(m_{3,1} + m_{3,2} + m_{3,3}) + 24n_4.$$

If this cubature rule is of degree m, then it exists if the system of moment equations (11.24) is solvable. The latter is of the form

$$n_0\left(\frac{1}{4}\right)^{\ell} + \sum_{j=1}^{n_1} \lambda_{j,1}(\boldsymbol{\alpha}_j)^{\ell} + \sum_{j=1}^{n_2} \lambda_{j,2}(\boldsymbol{\beta}_j)^{\ell} + \sum_{j=1}^{n_3} \lambda_{j,3}(\boldsymbol{\gamma}_j)^{\ell} + \sum_{j=1}^{n_4} \lambda_{j,4}(\boldsymbol{\delta}_j)^{\ell} = m_{\ell}, \quad \ell \in \Lambda_m^4. \tag{11.26}$$

The system (11.26) has

$$n_0 + 2n_1 + 2n_2 + 3n_3 + 4n_4$$

many variables if we assume that none of $\alpha_{j,i}$, $\beta_{j,i}$, $\gamma_{j,i}$, and $\delta_{j,i}$ are preassigned as in, for example, type $[1]$ or $[3]$ nodes, and it has $|\Lambda_4|$ many equations. For the consistency conditions that guarantee the system of moment equations having more variables than equations, this gives

$$n_0 + 2n_1 + 2n_2 + 3n_3 + 4n_4 \ge \left|\Lambda_m^4\right| = \dim \mathcal{P}_m^{\mathcal{P}_4},$$

and another condition is $0 \le n_0 \le 1$. There are many more such conditions. Let us give another one by setting $V(\mathbf{x}) = [(x_1 - x_2)(x_2 - x_3)(x_3 - x_4)(x_4 - x_1)]^2$, an invariant homogeneous polynomial of degree 8, and consider the space of polynomials

$$V\mathcal{P}_{m-8}^{\mathcal{S}_4} = \left\{ VP_{m-8} : \quad P_{m-8} \in \mathcal{P}_{m-8}^{\mathcal{S}_4} \right\},$$

which is a subset of $\mathcal{P}_m^{\mathcal{S}_4}$. If we apply the cubature rule on VS_{ℓ}, with $\ell \in \Lambda_{m-8}^4$, then we obtain the system of equations

$$\sum_{j=1}^{n_4} \lambda_{j,4}^* V(\boldsymbol{\gamma}_j)(\boldsymbol{\gamma}_j)^{\ell} = c_{\mathbf{k}} \int_{\hat{\Delta}^4} V(\mathbf{y}) S_{\ell}(\mathbf{y}) \mathbf{y}^{\kappa} \mathrm{d}\mathbf{y}, \quad \ell \in \Lambda_{m-8}^{d+1},$$

since V vanishes on types $[4]$, $[3,1]$, $[2,2]$, and $[2,1,1]$ nodes. This is a system that has $4n_4$ variables and $|\Lambda_{m-8}^4|$ equations, which leads to another consistency condition

$$4n_4 \ge \left|\Lambda_{m-8}^4\right| = \dim \mathcal{P}_{m-8}^{\mathcal{S}_4}, \qquad \text{if } m \ge 8.$$

Writing the invariant cubature rule in terms of all 11 types of nodes, we could write several other consistency conditions by considering subspaces of invariant polynomials vanishing on some of the nodes.

However, to reiterate once more, consistency conditions ensure more variables than equations for a system and some of its subsystems of nonlinear equations, but they are neither sufficient nor necessary for solving the system. The nonlinear system of equations may not have a solution or could have complex solutions for nodes and/or have negative cubature weights.

11.4 Notes and Further Results

The setup of symmetric cubature rules on the triangle in Section 11.1 follows from that of [97], which contains a list of symmetric cubature rules of degree up to 11 for the integral against constant weight function. Further search was carried out in [46] for cubature rules of degrees up to 20 and in [157] for degrees up to 30. Other works for computing cubature rules on the triangle, including asymmetric ones, can be found in [123, 151, 152, 159], for example. A more recent extensive search for symmetric cubature rules on the triangle and the tetrahedra is carried out in [184]. For integral against the Chebyshev weight, a list of symmetric cubature rules is calculated in [63], motivated by the connection to cubature rules on the sphere.

Conjecture 11.1.2 is based on a sequence of positive cubature rules, invariant under the octahedral group, on the unit sphere constructed by V. Lebedev and his collaborators, starting from degrees 25 and 29 in [86] until degree 131 in [88], see [87, 88] and references therein. The version over the triangle is formed accordingly by Corollary 2.5.6.

The Lobatto cubature rules in Section 11.2 follow [177], which was motivated by [61] showing that there exist no cubature rules of degree $2n-1$ with $(n-1)(n-2)/2$ nodes in the interior, $n-1$ nodes on the boundary, and three on the vertices, a result that follows from (11.9). They are not minimal. For example, the cubature rules of degree 5 with 7 nodes and degree 7 with 13 nodes can be found in [97].

Apart from the product cubature rules given in Subsection 2.4.2, there are fewer cubature rules of high degrees for the simplex Δ^d for $d \geq 3$. One such family is given in explicit formula in [57] for the constant weight and in [62] for the Jacobi weight, but the cubature rules are not positive. All others are lower degrees. Some of the known rules of lower degree are nonpositive. It was shown in [135] that if an invariant cubature rule of degree $n \geq 4$ for the

Jacobi weight on the simplex uses only nodes

$$\left(\tfrac{1}{d+1},\dots,\tfrac{1}{d+1}\right) \quad \text{and} \quad (r_k\mathbf{v}_k, 1-|r_k\mathbf{v}_k|)\,\mathcal{S}_{d+1} \in \Delta^d, \quad 1 \le k \le d+1,$$

where $r_k > 0$ and $\mathbf{v}_k = \mathbf{e}_1 + \cdots + \mathbf{e}_k = (1,\dots,1,0,\dots,0)$, then it cannot be positive.

Symmetric polynomials have been extensively studied and so are integer partitions; see, for example, [99]. For the cardinality of integer partitions of at most three or four parts, see [122]. Cubature rules of lower-to-moderate degrees on the tetrahedra Δ^3 or $\hat{\Delta}^4$ have been constructed numerically; we refer to [74, 75, 159, 184] and references therein. Many authors determined the consistency conditions on simplexes; see [75, 97, 100] and, most recently, [158], and their references.

Finally, we mention two internet resources for cubature rules on the triangle and simplex. The first one is the large repository [136] of cubature rules in Python that include the case of the triangle and simplex. The second one [145] contains cubature rules on the triangle in MATLAB files.

References

[1] Abramowitz M., and Stegun, I. (1970). *Handbook of Mathematical Functions*. New York: Dover.

[2] Agahanov, S. A. (1965). A method of constructing orthogonal polynomials of two variables for a certain class of weight functions. *Vestnik Leningrad Univ.* **20**, 5–10.

[3] Alexandersson, P., and Shapiro, B. (2014). Around multivariate Schmidt-Spitzer theorem. *Linear Alg. Appl.*, **446**, 356–368.

[4] Andrews, G. E., Askey, A. and Roy, R. (1999). *Special Functions*, Encyclopedia of Mathematics and its Applications **71**. Cambridge: Cambridge University Press.

[5] Bannai E., and Bannai, E. (2009). A survey on spherical designs and algebraic combinatorics on spheres. *European J. Comb.* **30**, 1392–1425.

[6] Baran, M. (1995). Complex equilibrium measure and Bernstein type theorems for compact sets in $\mathbb{R}^n$. *Proc. Amer. Math. Soc.* **123**, 485–494.

[7] Bacry, H. (1986). Zeros of polynomials and generalized Chebyshev polynomials. In *Group-Theoretic Methods in Physics*, Vol. **2** (Russian) (Jurmala, 1985), 239–249, Moscow: "Nauka".

[8] Bayer C., and Teichmann, J. (2006). The proof of Tchakaloff's theorem. *Proc. Amer. Math. Soc.* **134**, 3035–3040.

[9] Bedford E., and Taylor, B. A. (1986). The complex equilibrium measure of a symmetric convex set in $\mathbb{R}^n$. *Trans. Amer. Math. Soc.* **294**, 705–717.

[10] Beerends, R. J. (1991). Chebyshev polynomials in several variables and the radial part of the Laplace-Beltrami operator. *Trans. Amer. Math. Soc.* **328**, 779–814.

[11] Berens, H., and Schmid, H. J. (1992). On the number of nodes of odd degree cubature formulae for integrals with Jacobi weights on a simplex. In *Numerical Integration*, ed., T. O. Espelid and A. Genz, 37–44. Dordrecht: Kluwer Academic Publishers.

[12] Berens, H., Schmid, H. J. and Xu, Y. (1995a). Multivariate Gaussian cubature formula. *Arch. Math.* **64**, 26–32.

[13] Berens, H., Schmid, H. J. and Xu, Y. (1995b). On two-dimensional definite orthogonal systems and a lower bound for the number of nodes of associated cubature formulae. *SIAM J. Math. Anal.* **26**, 468–487.

[14] Bojanov, B., and Petrova, G. (1997). On minimal cubature formulae for product weight function. *J. Comput. Appl. Math.* **85**, 113–121.

[15] Bondarenko, A., Radchenko, D. and Viazovska, M. (2013). Optimal asymptotic bounds for spherical designs. *Ann. Math.* **178**, 443–452.

[16] Bos, L., De Marchi, S. and Vianello, M. (2006). On the Lebesgue constant for the Xu interpolation formula. *J. Approx. Theory* **141**, 134–141.

[17] Bos, L., De Marchi, S. and Vianello, M. (2017). Trivariate polynomial approximation on Lissajous curves. *IMA J. Numer. Anal.* **37**, 519–541.

[18] Bos, L., De Marchi, S., Vianello, M. and Xu, Y. (2006). Bivariate Lagrange interpolation at the Padua points: the generating curve approach. *J. Approx. Theory* **143**, 15–25.

[19] Bos, L., De Marchi, S., Vianello, M. and Xu, Y. (2007). Bivariate Lagrange interpolation at the Padua points: the ideal theory approach. *Numer. Math.* **108**, 43–57.

[20] Brass H., and Petras, K. (2011). *Quadrature Theory: The Theory of Numerical Integration on a Compact Interval. Mathematical Surveys and Monographs*, Vol. **178**. Providence: American Mathematics Society.

[21] Caflisch, R. E. (1998). Monte Carlo and quasi-Monte Carlo methods. In *Acta Numerica*, Vol. **7**, 1–49. Cambridge: Cambridge University Press.

[22] Caliari, M., De Marchi, S. and Vianello, M. (2005). Bivariate polynomial interpolation on the square at new nodal sets. *Appl. Math. Comput.* **165**, 261–274.

[23] Cao, L., Ghimire, S. and Wang, X. S. (2022). Bivariate Lagrange interpolation at the checkerboard nodes. *Proc. Amer. Math. Soc.* **150**, 2153–2163.

[24] Chihara, T. S. (1978). *An Introduction to Orthogonal Polynomials. Mathematics and Its Applications*, Vol. **13**. London: Gordon & Breach Science Publishers.

[25] Conway, J. H., and Sloane, N. J. A. (1998). *Sphere Packings, Lattices and Groups*, 3rd ed. New York: Springer.

[26] Cools, R. (1997). Constructing cubature formulae: the science behind the art. *Acta Numerica* **6**, 1–54.

[27] Cools, R. (1999). Monomial cubature rules since "Stroud": a compilation – part 2. *J. Comput. Appl. Math.* **112**, 21–27.

[28] Cools, R. (2003). An encyclopedia of cubature formulas. *J. Complexity* **19**, 445–453. nines.cs.kuleuven.be/ecf.

[29] Cools, R., and Poppe, K. (2011). Chebyshev lattices, a unifying framework for cubature with Chebyshev weight function. *BIT Numer. Math.* **51**, 275–288.

[30] Cools R., and Rabinowitz, P. (1993). Monomial cubature rules since "Stroud": a compilation. *J. Comput. Appl. Math.* **48**, 309–326.

[31] Cools, R., and Schmid, H. (1989). Minimal cubature formulae of degree $2k-1$ for two classical functionals. *Computing* **43**, 141–157.

[32] Cools, R., and Schmid, H. J. (1993). A new lower bound for the number of nodes in cubature formulae of degree $4n+1$ for some circularly symmetric integrals. In *Numerical Integration, IV (Oberwolfach, 1992)*, eds., H. Brass and G. Hämmerlin, Vol. **12** of *Internat. Ser. Numer. Math.*, 57–66. Basel: Birkhäuser.

[33] Cox, D., Little, J. and O'Shea, D. (1997). *Ideals, Varieties, and Algorithms*, 2nd ed., Berlin: Springer.

[34] Curto, R. E., and Fialkow, L. A. (2002). A duality proof of Tchakoloff's theorem. *J. Math. Anal. Appl.* **269**, 519–532.

[35] Dai, F., and Xu, Y. (2013). *Approximation Theory and Harmonic Analysis on Spheres and Balls. Springer Monographs in Mathematics*. New York: Springer.

[36] Davis, P. J., and Rabinowitz, P. (2007) *Methods of Numerical Integration*. Corrected reprint of the second (1984) edition. New York: Dover.

[37] Della Vecchia, B., Mastroianni, G. and Vértesi, P. (2009). Exact order of the Lebesgue constants for bivariate Lagrange interpolation at certain node-systems. *Studia Sci. Math. Hungar.* **46**, 97–102.

[38] Delsarte, P., Goethals, J. M. and Seidel, J. (1977). Spherical codes and designs. *Geom. Dedicata* **6**, 363–388.

[39] De Marchi, S., Vianello, M. and Xu, Y. (2009). New cubature formulae and hyperinterpolation in three variables. *BIT Numer. Math.* **49** (2009), 55–73.

[40] Dencker, P., and Erb, W. (2017). Multivariate polynomial interpolation on Lissajous-Chebyshev nodes. *J. Approx. Theory* **219**, 15–45.

[41] Dick, J., Kuo, F. Y. and Sloan, I. H. (2013). High-dimensional integration: the quasi-Monte Carlo way. *Acta Numerica* **22**, 133–288.

[42] Dick, J., and Pillichshammer, F. (2010). *Digital Nets and Sequences. Discrepancy Theory and Quasi-Monte Carlo Integration.* Cambridge: Cambridge University Press.

[43] van Diejen, J. F., and Emsiz, E. (2019). Exact cubature rules for symmetric functions. *Math. Comp.* **88**, 1229–1249.

[44] van Diejen, J. F., and Emsiz, E. (2021). Cubature rules from Hall-Littlewood polynomials. *IMA J. Numer. Anal.* **41**, 998–1030.

[45] Dudgeon, D., and Mersereau, R. M. (1984). *Multidimensional Digital Signal Processing.* Englewood Cliffs: Prentice Hall.

[46] Dunavant, D. A. (1985). High degree efficient symmetrical Gaussian quadrature rules for the triangle. *Int. J. Numer. Methods Eng.* **21**, 1129–1148.

[47] Dunkl, C. F., and Xu, Y. (2014) *Orthogonal Polynomials of Several Variables. Encyclopedia of Mathematics and Its Applications*, Vol. **155**, Cambridge: Cambridge University Press.

[48] Dunn, K. B., and Lidl, R. (1980). Multi-dimensional generalizations of the Chebyshev polynomials. I, II. *Proc. Japan Acad.* **56**, 154–165.

[49] Eier, R., and Lidl, R. (1982). A class of orthogonal polynomials in k variables. *Math. Ann.* **260**, 93–99.

[50] Engels, H. (1980). *Numerical Quadrature and Cubature.* London-New York: Academic Press.

[51] Erdélyi, A., Magnus, W., Oberhettinger, F. and Tricomi, G. (1981). *Higher Transcendental Functions. Vol. II. Based on notes left by Harry Bateman.* Reprint of the 1953 original. Melbourne: Robert E. Krieger Publishing Co.

[52] Fedorov, E. S. (1953). Elements of the study of figures (Russian). *Zap. Mineral. Imper. S. Peters-burgskogo Obsc.* **21** (1885), 1–279. Izdat. Akad. Nauk SSSR.

[53] Förster, K., and Petras, K. (1990). On estimates for the weights in Gaussian quadrature in the ultraspherical case. *Math. Comp.* **55**, 243–264.

[54] Fuglede, B. (1974). Commuting self-adjoint partial differential operators and a group theoretic problem. *J. Funct. Anal.* **16**, 101–121.

[55] Gautschi, W. (2004). *Orthogonal Polynomials: Computation and Approximation.* Numerical Mathematics and Scientific Computation. New York: Oxford University Press.

[56] Gautschi, W. (2021). *A Software Repository for Gaussian Quadratures and Christoffel Functions.* Philadelphia: Society for Industrial and Applied Mathematics (SIAM).

[57] Grundmann, A., and Möller, H. M. (1978). Invariant integration formulas for the n-simplex by combinatorial methods. *SIAM. J. Numer. Anal.*, **15**, 282–290.

[58] Hardin, R. H., and Sloane, N. J. A. (1996). McLaren's improved snub cube and other new spherical designs in three dimensions. *Discrete and Comp. Geometry.* **15**, 429–441.

[59] Harris, L. (2015). Lagrange polynomials, reproducing kernels and cubature in two dimensions. *J. Approx. Theory* **195**, 43–56.

[60] Harris, L., and Simanek, B. (2021). Interpolation and cubature for rectangular sets of nodes. *Proc. Amer. Math. Soc.* **149**, 3485–3497.

[61] Helenbrook, B. T. (2009). On the existence of explicit hp-finite element method using Gauss-Lobatto integration on the triangle. *SIAM J. Numer. Anal.* **47**, 1304–1318.

[62] Heo, S., and Xu, Y. (2000). Invariant cubature formulae for spheres and balls by combinatorial methods. *SIAM J. Numer. Anal.*, **38** (2000), 626–638.

[63] Heo, S., and Xu, Y. (2001). Constructing fully symmetric cubature formulae for the sphere. *Math. Comp.* **70**, 269–279.

[64] Higgins, J. R. (1996). *Sampling Theory in Fourier and Signal Analysis, Foundations.* New York: Oxford Science Publications.

[65] Hinrichs, A., and Novak, E. (2007). Cubature formulas for symmetric measures in higher dimensions with few points. *Math. Comp.* **76**, 1357–1372.

[66] Hubert, E., Metzlaff, T., and Riener, C. (2024) Orbit spaces of Weyl groups acting on compact tori: a unified and explicit polynomial description. *SIAM J. Appl. Algebra & Geometry* **8**, 612–649.

[67] Masatake, H., Nozaki, H., Sawa, M., and Vesselin, V. (2012). A new approach for the existence problem of minimal cubature formulas based on the Larman-Rogers-Seidel theorem. *SIAM J. Numer. Anal.* **50**, 2716–2728.

[68] Hirao, M., and Sawa, M. (2009). On minimal cubature formulae of small degree for spherically symmetric integrals. *SIAM J. Numer. Anal.* **47**, 3195–3211.

[69] Horn, R. A., and Johnson, C. R. (1985). *Matrix Analysis.* Cambridge: Cambridge University Press.

[70] Hrivnák, J., and Motlochová, L. (2014). Discrete transforms and orthogonal polynomials of (anti)symmetric multivariate cosine functions. *SIAM J. Numer. Anal.* **52**, 3021–3055.

[71] Hrivnák, J., Motlochová, L. and Patera, J. (2016). Cubature formulas of multivariate polynomials arising from symmetric orbit functions. *Symmetry* **8**, no. 7, Art. 63., 22 pp.

[72] Iosevich, A., Katz, N., and Tao, T. (2003). The Fuglede spectral conjecture holds for convex planar domains. *Math. Res. Lett.* **10**, 559–569.

[73] Ismail, M. (2005). *Classical and Quantum Orthogonal Polynomials in One Variable.* With two chapters by Walter Van Assche. With a foreword by Richard A. Askey. Encyclopedia of Mathematics and its Applications, **98**. Cambridge: Cambridge University Press.

[74] Jaśkowiec, J., and Sukumar, N. (2020). High-order cubature rules for tetrahedra. *Int. J. Numer. Methods Eng.* **121**, 2418–2436.

[75] Keast, P. (1986). Moderate degree tetrahedral quadrature formulas, *Comput. Methods Appl. Mech. Engrg.* **55**, 339–348.

[76] Klimyk, A., and Patera, J. (2006). Orbit functions, *Symmetry Integrability Geom. Methods Appl.* **2**, Paper 006, 60pp.

[77] Klimyk, A., and Patera, J. (2007) Antisymmetric orbit functions, *SIGMA Symmetry Integrability Geom. Methods Appl.* **3**, Paper 023, 83 pp.

[78] Kolountzakis, M. N. (2004). The study of translational tiling with Fourier analysis. In *Fourier Analysis and Convexity*, Appl. Numer. Harmon. Anal., pp. 131–187. Boston: Birkhäuser.

[79] Koornwinder, T. (1974a). Orthogonal polynomials in two variables which are eigenfunctions of two algebraically independent partial differential operators. I, II. *Indag. Math.* **36**, 48–58, 59–66.

[80] Koornwinder, T. (1974b). Orthogonal polynomials in two variables which are eigenfunctions of two algebraically independent partial differential operators. III, IV. *Indag. Math.* **36**, 357–369, 370–381.

[81] Koornwinder, T. (1975). Two-variable analogues of the classical orthogonal polynomials, In *Theory and Applications of Special Functions*. ed. R.A. Askey. New York-London: Academic Press.

[82] Krasikov, I. (2006). On extreme zeros of classical orthogonal polynomials. *J. Comput. Appl. Math.* **193**, 168–182.

[83] Korevaar, J., and Meyers, J. L. H. (1994). Chebyshev-type quadrature on multidimensional domains. *J. Approx. Theory* **79**, 144–164.

[84] Lasserre, J. B. (2012). The existence of Gaussian cubature formulas. *J. Approx. Theory* **164**, 572–585.

[85] Larcher, H. (1959). Notes on orthogonal polynomials in two variables. *Proc. Amer. Math. Soc.* **10**, 417–423.

[86] Lebedev, V. I. (1977). Spherical quadrature formulas exact to orders 25-29. *Siberian Math. J.* **18**, 99–107.

[87] Lebedev, V. I., and Lauikov, D. N. (1999). A quadrature formula for a sphere of the 131st algebraic order of accuracy. (Russian) *Dokl. Akad. Nauk.* **366**, 741–745.

[88] Lebedev, V. I., and Skorokhodov, A. L. (1992). Quadrature formulas for a sphere of orders 41,47 and 53. (Russian) *Dokl. Akad. Nauk* **324**, 519–524; translation in *Russian Acad. Sci. Dokl. Math.* **45**, 587–592 (1993).

[89] Lev, N., and Matolcsi, M., (2022). The Fuglede conjecture for convex domains is true in all dimensions. *Acta Math.* **228**, 385–420.

[90] Levenberg, N. (2006). Approximation in $\mathbb{C}^N$. *Surveys in Approx. Theory* **2**, 92–140.

[91] Levenshtein, V. I. (1983). Bounds for packings of metric spaces and some of their applications. (Russian) *Problemy Kibernet.* **40**, 43–110.

[92] Li, H., and Xu, Y. (2009). Discrete Fourier analysis on a dodecahedron and a tetrahedron. *Math. Comp.* **78**, 999–1029.

[93] Li, H., and Xu, Y. (2010). Discrete Fourier analysis on fundamental domain of A_d lattice and on simplex in d-variables. *J. Fourier Anal. Appl.* **16**, 383–433.

[94] Li, H., Sun, J. and Xu, Y. (2008). Discrete Fourier analysis, cubature and interpolation on a hexagon and a triangle, *SIAM J. Numer. Anal.* **46**, 1653–1681.

[95] Li, H., Sun, J. and Xu, Y. (2009). Cubature formula and interpolation on the cubic domain. *Numer. Math. Theory, Methods, Appl.* **2**, 119–152.

[96] Li, H., Sun, J. and Xu, Y. (2010). Discrete Fourier analysis with lattices on planar domains. *Numer. Algorithm* **55**, 279–300.

[97] Lyness, J. N., and Jespersen, D. (1975). Moderate degree symmetric quadrature rules for the triangle. *J. Inst. Math. App.* **15**, 19–32.

[98] Macdonald, I. G. (1992). Schur functions: theme and variations. *Séminaire Lotharingien de Combinatoire (Saint-Nabor, 1992)*, 5–39. Publ. Inst. Rech. Math. Av., **498**, Strasbourg, Univ. Louis Pasteur.

[99] Macdonald, I. G. (1995). *Symmetric Functions and Hall Polynomials*, 2nd ed. With contributions by A. Zelevinsky. Oxford Mathematical Monographs. Oxford Science Publications. New York: Oxford University Press.

[100] Maeztu, J., and Sainz de la Maza, E. (1995). Consistent structures of invariant quadrature rules for the n-simplex. *Math. Comp.* **64**, 1171–1192.

[101] Macaulay, F. S. (1916). *The Algebraic Theory of Modular Systems*. Cambridge Tracts in Mathematics and Mathematical Physics, Vol. **19**. Cambridge: Cambridge University Press.

[102] McMullen, P. (1980). Convex bodies which tile space by translation. *Mathematika* **27**, 113–121.

[103] Mhaskar, H. N., Narcowich, F. J. and Ward, J. D. (2021). Spherical Marcinkiewicz-Zygmund inequalities and positive quadrature. *Math. Comp.* **70**, 1113–1130. (Corrigendum: *Math. Comp.* **71**, 453–454.)

[104] Moody, R. V., and Patera, J. (2011). Cubature formulae for orthogonal polynomials in terms of elements of finite order of compact simple Lie groups. *Adv. Applied Math.* **47**, 509–535.

[105] Moody, R. V., Motlochová, L. and Patera, J. (2014). Gaussian cubature arising from hybrid characters of simple Lie groups. *J. Fourier Anal. Appl.* **20**, 1257–1290.

[106] Möller, H. M. (1973). Polynomideale und Kubaturformeln. University of Dortmund. D.Phil. Thesis.

[107] Möller, H. M. (1976). Kubaturformeln mit minimaler Knotenzahl. *Numer. Math.* **25**, 185–200.

[108] Möller, H. M. (1979). Lower bounds for the number of nodes in cubature formulae. In *Numerical Integration*, ISNM Vol. **45**, ed., G. Hämmerlin. Basel: Birkhäuser.

[109] Möller, H. M., and Sauer, T. (2000). H-bases for polynomial interpolation and system solving. *Adv. Comput. Math.* **12**, 335–362.

[110] Morrow, C. R., and Patterson, T. N. L. (1978). Construction of algebraic cubature rules using polynomial ideal theory. *SIAM J. Numer. Anal.* **15**, 953–976.

[111] Munthe-Kaas, H. (2006). On group Fourier analysis and symmetry preserving discretizations of PDEs. *J. Phys. A,* **39**, 5563–5584.

[112] Munthe-Kaas, H., Nome, M. and Ryland, B. (2013). Through the kaleidoscope: symmetries, groups and Chebyshev-approximations from a computational point of view. In *Foundations of Computational Mathematics, Budapest 2011*, London Math. Soc. Lecture Note Ser. **403**, 188–229. Cambridge: Cambridge University Press.

[113] Mysovskikh, I. P. (1975). On Chakalov's theorem. *USSR Comput. Math. Math. Phys.* **15**, 221–227. Translation of *Zh. Vychisl. Mat. i Mat. Fiz.* **15**, 1589–1593.

[114] Mysovskikh, I. P. (1976). Numerical characteristics of orthogonal polynomials in two variables. *Vestnik Leningrad Univ. Math.* **3**, 323–332.

[115] Mysovskikh, I. P. (1981). *Interpolatory Cubature Formulas*, Moscow: "Nauka".

[116] Nikolov, G. (2019). New bounds for the extreme zeros of Jacobi polynomials. *Proc. Amer. Math. Soc.* **147**, 1541–1550.

[117] Nikolov, G. (2023). On the extreme zeros of Jacobi polynomials. *Numerical Methods and Applications*, 246–257. Lecture Notes in Comput. Sci., **13858**. Cham: Springer.

[118] *NIST Digital Library of Mathematical Functions.* (2023). ed. F. W. J. Olver, A. B. Olde Daalhuis, D. W. Lozier, B. I. Schneider, R. F. Boisvert, C. W. Clark, B. R. Miller, B. V. Saunders, H. S. Cohl, and M. A. McClain. https://dlmf.nist.gov/ Release 1.1.12 of 2023-12-15.

[119] Novak, E., and Ritter, K. (1999). Simple cubature formulas with high polynomial exactness. *Constr. Approx.* **15**, 499–522.

[120] Noskov, M. V., and Schmid, H. J. (2004). On the number of nodes in n-dimensional cubature formulae of degree 5 for integrals over the ball. *J. Comp. Applied Math.* **169**, 247–254.

[121] Nozaki, H., and Sawa, M. (2012). Note on cubature formulae and designs obtained from group orbits, *Canad. J. Math.* **64**, 1359–1377.

[122] OEIS (2022). The online encyclopedia of integer sequences. Entry A001399 (https://oeis.org/A001399) and Entry A001400 (https://oeis.org/A001400).

[123] Papanicolopulos, S. (2016). New fully symmetric and rotationally symmetric cubature rules on the triangle using minimal orthonormal bases. *J. Comp. Applied Math.* **294**, 39–48.

[124] Peherstoefer, F. (1981). Characterization of positive quadrature formulas. *SIAM J. Math. Anal.* **12**, 935–942.

[125] Peherstoefer, F. (1984). Characterization of quadrature formula II. *SIAM J. Math. Anal.* **5**, 1021–1030.

[126] Peherstoefer, F. (2008). Positive quadrature formulas. III. Asymptotics of weights. *Math. Comp.* **77**, 2241–2259.

[127] Petrushev, P., and Xu, Y. (2008). Localized polynomial kernels and frames on the ball. *Constr. Approx.* **27**, 121–148.

[128] Pólya, G., and Szegő, G. (1976). *Problems and Theorems in Analysis*, Vol. II. Berlin: Springer-Verlag.

[129] Putinar, M. (1997). A Note on Tchakaloff's Theorem. *Proc. Amer. Math. Soc.* **125**, 2409–2414.

[130] Radon, J. (1948). Zur mechanischen Kubatur. *Monatsh. Math.* **52**, 286–300.

[131] Saff, E., and Totik, V. (2010). *Logarithmic Potentials with External Fields.* Berlin: Springer.

[132] Sauer, T. (2000). Gröbner bases, H-bases and interpolation. *Trans. Amer. Math. Soc.* **353**, 2293–2308.

[133] Sauer, T. (2021). *Continued Fractions and Signal Processing*. Berlin: Springer.

[134] Sauer, T., and Xu, Y. (2023). Continued fractions and orthogonal polynomials in several variables. arXiv:2303.16030.

[135] Sawa, M., and Xu, Y. (2014). On positive cubature rules on the simplex and isometric embeddings. *Math. Comp.* **83**, 1251–1277.

[136] Schlomer, N. Numerical integration (quadrature, cubature) in Python. github.com/nschloe/quadpy#triangle.

[137] Schmid, H. J. (1983). Interpolatorische Kubaturformeln. *Diss. Math.* **CCXX**, 1–122.

[138] Schmid, H. J. (1989). A note on positive quadrature rules. *Rocky Mountain J. Math.* **19**, 395–404.

[139] Schmid, H. J. (1995). Two-dimensional minimal cubature formulas and matrix equations. *SIAM J. Matrix Anal. Appl.* **16**, 898–921.

[140] Schmid, H. J., and Xu, Y. (1994). On bivariate Gaussian cubature formula. *Proc. Amer. Math. Soc.* **122**, 833–841.

[141] Shapiro, B., and Shapiro, M. (2009). On eigenvalues of rectangular matrices. *Proc. Steklov Math. Inst.* **267**, 248–255.

[142] Sobolev, S. L. (1962). Cubature formulas on the sphere which are invariant under transformations of finite rotation groups. *Dokl. Akad. Nauk SSSR* **146**, 310–313.

[143] Sobolev, S. L. (1992). *Cubature Formulas and Modern Analysis. An Introduction.* Translated from the 1988 Russian edition. Montreux: Gordon and Breach Science Publishers.

[144] Sobolev, S. L., and Vaskevich, V. L. (1997). *The Theory of Cubature Formulas.* Translated from the 1996 Russian original and with a foreword by S. S. Kutateladze. Revised by Vaskevich. Mathematics and its Applications, **415**. Dordrecht: Kluwer Academic Publishers Group.

[145] Sommariva, A. Cubature rules on the triangle (simplex). www.math.unipd.it/ alvise/SETS_CUBATURE_TRIANGLE/rules_triangle.html

[146] Stroud, A. H. (1971). *Approximate Calculation of Multiple Integrals.* Englewood Cliffs: Prentice-Hall.

[147] Stroud, A. H., and Secrest, D. (1966). *Gaussian Quadrature Formulas.* Englewood Cliffs: Prentice-Hall.

[148] Szegő, G. (1975). *Orthogonal Polynomials*, 4th ed., American Mathematical Society Colloquium Publication **23**. Providence: American Mathematical Society.

[149] Szili, L., and Vértesi, P. (2009). On multivariate projection operators. *J. Approx. Theory* **159**, 154–164.

[150] Tao, T. (2004). Fuglede's conjecture is false in 5 and higher dimensions. *Math. Res. Lett.* **11**, 251–258.

[151] Taylor, M. A. (2008). Asymmetric cubature formulas for polynomial integration in the triangle and square. *J. Comp. Appl. Math.* **218**, 184–191.

[152] Taylor, M. A., Wingate, B. and Bos, L. (2007). A cardinal function algorithm for computing multivariate quadrature points. *SIAM J. Numer. Anal.* **45**, 193–205.

[153] Tchakaloff, V. (1957). Formules de cubatures mécaniques à coefficients non négatifs. (French) *Bull. Sci. Math.* (2) **81**, 123–134.

[154] Venkov, B. (1954). On a class of Euclidean polyhedra (Russian). *Vestnik Leningrad. Univ. Ser. Mat. Fiz. Him.* **9**, 11–31.

[155] Verlinden, P., and Cools, R. (1992). On cubature formulae of degree $4k+1$ attaining Möller's lower bound for integrals with circular symmetry. *Numer. Math.* **61**, 395–407.

[156] Victoir, N. (2004). Asymmetric cubature formulae with few points in high dimension for symmetric measures. *SIAM J. Numer. Anal.* **42**, 209–227.

[157] Wandzura, S., and Xiao, H. (2003). Symmetric quadrature rules on a triangle. *Comput. Math. Appl.* **45**, 1829–1840.

[158] Wang, W., and Papanicolopulos, S. (2023). Explicit consistency conditions for fully symmetric cubature on the tetrahedron. *Eng. Comput.* **39**, 4013–4024.

[159] Xiao, H., and Gimbutas, Z. (2010). A numerical algorithm for the construction of efficient quadrature rules in two and higher dimensions. *Comput. Math. Appl.* **59**, 663–676.

[160] Xu, Y. (1992). Gaussian cubature and bivariable polynomial interpolation. *Math. Comp.* **59**, 547–555.

[161] Xu, Y. (1993). On multivariable orthogonal polynomials. *SIAM J. Math. Anal.* **24**, 783–794.

[162] Xu, Y. (1994a). Quasi orthogonal polynomial, quadrature, and interpolation, *J. Math. Anal. Appl.* **182**, 779–799.

[163] Xu, Y. (1994b). Characterization of positive quadrature formulae. *Math. Comp.* **62**, 703–718.

[164] Xu, Y. (1994c). Block Jacobi matrices and zeros of multivariate orthogonal polynomials. *Trans. Amer. Math. Soc.* **342**, 855–866.

[165] Xu, Y. (1994d). On zeros of multivariate quasi-orthogonal polynomials and Gaussian cubature formulae. *SIAM J. Math. Anal.* **25**, 991–1001.

[166] Xu, Y. (1994e). A class of bivariate orthogonal polynomials and cubature formulas. *Numer. Math.* **69**, 233–241.

[167] Xu, Y. (1994f). *Common Zeros of Polynomials in Several Variables and Higher Dimensional Quadrature*, Pitman Research Notes in Mathematics Series. Essex: Longman.

[168] Xu, Y. (1995). Christoffel functions and Fourier series for multivariate orthogonal polynomials. *J. Approx. Theory*, **82**, 205–239.

[169] Xu, Y. (1996). Lagrange interpolation on Chebyshev points of two variables. *J. Approx. Theory*, **87**, 220–238.

[170] Xu, Y. (1997). On orthogonal polynomials in several variables. In *Special Functions, q-Series, and Related Topics*. Fields Institute Communications, Amer. Math. Soc., Providence, RI, Vol. **14**, 247–270.

[171] Xu, Y. (1998a). Orthogonal polynomials and cubature formulae on spheres and on balls. *SIAM J. Math. Anal.* **29**, 779–793.

[172] Xu, Y. (1998b). Orthogonal polynomials and cubature formulae on spheres and on simplices, *Methods Anal. Appl.* **5**, 169–184.

[173] Xu, Y. (1998c). Minimal Cubature formulae for a family of radial weight functions. *Advances in Comp. Math.* **8**, 367–380.

[174] Xu, Y. (1999). Cubature formula and polynomial ideals. *Advances in Appl. Math.*, **23**, 211–233.

[175] Xu, Y. (2001). Orthogonal polynomials and cubature formulae on the balls, simplices, and spheres. *J. Comp. Appl. Math.* **127**, 349–368.

[176] Xu, Y. (2003). Lower bound for the number of nodes of cubature formulae on the unit ball. *J. Complexity*, **19**, 392-402.

[177] Xu, Y. (2011). On Gauss-Lobatto integration on the triangle. *SIAM J. Numer. Anal.* **49**, 541–548.

[178] Xu, Y. (2012a). Orthogonal polynomials and expansions for a family of weight functions in two variables. *Constr. Approx.* **36**, 161–190.

[179] Xu, Y. (2012b). Minimal Cubature rules and polynomial interpolation in two variables. *J. Approx. Theory* **164**, 6–30.

[180] Xu, Y. (2015). Generalized characteristic polynomials and Gaussian cubature rules. *SIAM. J. Matrix Anal.* **36**, 1129–1142.

[181] Xu, Y. (2017)., Minimal Cubature rules and polynomial interpolation in two variables, II. *J. Approx. Theory* **214**, 49–68.

[182] Xu, Y. (2021). Approximation and localized polynomial frame on conic domains. *J. Functional Anal.* **281**, no. **12**, Paper no. 109257, 94 pp.

[183] Yudin, V. A. (1997). Lower bounds for spherical designs. (Russian. Russian summary) *Izv. Ross. Akad. Nauk Ser. Mat.* **61**, no. 3, 213–223. Translation in *Izv. Math.* **61** (1997), no. 3, 673–683.

[184] Zhang, L., Cui, T. and Liu, H. (2009). A set of symmetric quadrature rules on triangles and tetrahedra. *J. Comp. Math.* **27**, 89–96.

Index

For EU product safety concerns, contact us at Calle de José Abascal, 56–1°,
28003 Madrid, Spain or eugpsr@cambridge.org.

www.ingramcontent.com/pod-product-compliance
Lightning Source LLC
LaVergne TN
LVHW011523200926
848484LV00001B/127

* 9 7 8 1 0 0 9 6 6 3 9 2 2 *